Black '41

Black '41

The West Point Class of 1941
and the American Triumph in World War II

Bill Yenne

JOHN WILEY & SONS, INC.

New York • Chichester • Brisbane • Toronto • Singapore

Copyright © 1991 by Bill Yenne

Published by John Wiley & Sons, Inc.

All rights reserved. Published simultaneously in Canada.

Library of Congress Cataloging in Publication Data

Yenne, Bill, 1949–
   Black '41 : The West Point class of 1941 and the American Triumph in World War II / Bill Yenne.
     p. cm.
   Includes bibliographical references and index. ISBN 0-471-54197-4
  1. United States Military Academy. Class of 1941--History. 2. United States. Army--History--World War, 1939–1945. 3. United States. Army--Officers--History--20th century. 4. United States. Army--History--20th century. 5. United States--History, Military--20th century.
U410.N1 1941b
355'.0071'173--dc20   91–16640

Printed in the United States of America

10 9 8 7 6 5 4 3 2 1

Foreword

In late May 1991, the members of the United States Military Academy Class of 1941 gathered at West Point to celebrate the 50th anniversary of their graduation. Class members, wives, widows, other family members, and friends joined together for another occasion to reflect on our ties to one another and our experiences together stretching back more than 50 years. On May 28 we assembled at the Kelleher-Jobes Memorial Arch (built by the class in memory of two classmates who died as cadets) to honor the memories of all of those classmates who have died since our June 11, 1941, graduation. And, in other less solemn ways, we retold old stories, relived our times together, and renewed our dedication to one another as classmates and to the institution, the Academy, that brought us together and molded our lives.

In this book, *Black '41*, Bill Yenne tells the story of America's coming-of-age as a world power through the eyes and experiences of our class that graduated on the eve of America's entry into World War II. He takes many of the stories we have told one another, our families, and friends over the years and weaves them together into a uniquely vivid tapestry of an American experience.

We came together in that faraway summer of 1937 from all over these United States. From the first day, through the ensuing four years of West Point's demanding academic and military routine, we were bound together as we were prepared for the service to come. At our graduation, Secretary of War Henry L. Stimson told us that we were ". . . going into the new army of the United States—a great army which is now in process of enlistment and formation and training." We rushed into the army and the Army Air Corps as brand new lieutenants, and within weeks were caught up in the exhilaration of the task. Within six months some were in combat, and within seven months we had suffered the first "killed in action"—Sandy Nininger, who subsequently was awarded the first Medal of Honor of World War II. By the end of the war we were majors, lieutenant colonels, and a few colonels, and 40 of the class had been killed in action, with 11 more killed in Air Corps training accidents. In the 1950s, as we went into the Korean action, another four were killed. In a sense it seemed that the Class of 1941 was indeed "the class upon which the world fell." We did not give it much thought then because we were too caught up in the demands of the times and our assignments to devote attention to speculation on history.

The story of the USMA Class of 1941 comes alive here for the first time in the public realm. We have told it for years to whomever would listen. Bill Yenne now brings it to all—from the first day to the present. We are proud to be '41ers and proud of the story. We hope this comes through in Bill Yenne's telling, and, in the words of our alma mater, "When our work is done, our course on earth is run, it may be said, 'Well done' . . ."

Michael J. L. Greene
Brigadier General, US Army, Retired
President, USMA Class of 1941

Preface

They graduated into the eerie twilight of a nation at peace in the midst of a world at war. They graduated from the school that had trained America's military leaders since 1802, and they began their service careers in a US Army that had yet to prepare itself for what they all knew was coming.

It has been said that the Class of 1915, which included men such as Dwight Eisenhower, Omar Bradley, George Stratemeyer, and James Van Fleet, was the "class upon which the stars fell." If that is so, then the Class of 1941 is the class upon which the entire world fell.

They were known at the time as the "Black Class," or simply "Black '41." Nobody knew why. One theory now holds that they were a class that had an excessive number of black marks against it for unusual enthusiasm in playing the usual undergraduate pranks. This was not necessarily true. The Class of 1941 was probably no more mischievous than its predecessors. But somewhere in the course of their Plebe Year, former classmates believe, a now-anonymous upperclassman tarred the men of this class with the epithet that made them forever West Point's *classe noire*. It was but one of a number of distinctions that separated Black '41 from other classes.

In his commencement address to the graduates on a sunny June day in 1941, Secretary of War Henry L. Stimson remarked, "Usually commencement is a time of rejoicing and congratulations as we elders give good wishes to the young men who are beginning life's journey. But that is hardly the atmosphere which surrounds our country today. And I have the feeling that I should be false to the responsibility which is laid upon me by the invitation to meet here if I did not try to help you to understand the nature of the crisis which confronts us all today, and to give you encouragement in meeting it. The work of meeting it may fall, in large measure, upon your shoulders."

Indeed it was time. It was *their* time.

Black '41 is the story of the experiences of this remarkable class at West Point and in the four long years of war that followed. It is adapted from what *Black '41's* classmates wrote about themselves in 1941 and in all the years leading to the present, when I was privileged to correspond with, and speak at length with, the surviving members of the class. It is an unusual story, the story of how some of our best and brightest went off to war to lead our troops in battle against the German armies in Europe and the Japanese troops in the Pacific. Some of these men died, and many, serving valiantly, learned from their experiences and became, in a way they might never have otherwise, men.

Through this process, I have come to know this class better and more intimately than I know my own college classmates; I am proud and humbled to have talked with the men upon whose shoulders Henry Stimson laid the burden of a nation at war.

Bill Yenne
San Francisco
August 1991

Acknowledgments

The author wishes to thank all the members of the Class of 1941 with whom he worked during the preparation of this book, especially Mike Greene, Joe Gurfein, Lynn Lee, George Pickett, Paul Skowronek, and Ben Spiller, who devoted a great deal of their own time to the project and made items from their personal collections available for my use. I'd also like to extend a special thanks to Gail Rolka, without whom I could never have completed this book. She not only typed the manuscript, but also transcribed many days' worth of tapes and devoted long hours to helping organize and collate the various elements of this project. Finally, I'd like to thank my agent, Jane Dystel, who made this book possible, and my editor, Roger Scholl, who helped to hone the final work to a polished finish.

Contents

Prologue:
Rites of Passage

June 11, 1941, was a very pleasant spring day. June is still springtime in the Hudson River Valley, but just barely. The trees had leafed out and the hillsides were green, but the nights were still cool, the days still just warm rather than sticky hot.

June 1941 was still the springtime of our century, but just barely. The springtime of our century had ended for Europe in 1940, and it would end for the United States inside of six months.

It had been a year since France had fallen, and the war was no longer in every headline. Russia would remain at peace for another 11 days, but even when Hitler's panzers invaded on June 22, little concern was aroused in the United States.

Within the gray walls of the United States Military Academy at West Point, where another corps of First Class cadets was about to become second lieutenants in the United States Army, there was a certain ambivalence about the world situation. The war seemed so far away. It was, after all, still spring at West Point—as it had been for every graduating class for two decades.

There wasn't much folderol this graduation day. Everything about it was simple and direct—much more so than a West Point graduation today. All of the cadets had been sworn in as second lieutenants early that morning, just after their final breakfast in Washington Hall. Technically, even though they still wore the gray uniforms of West Point cadets, these 424 young men were already second lieutenants when they filed into the Field House to receive their diplomas.

That would be their last act in West Point careers that dated back to a humid July first in 1937. When they received their diplomas they would be *gone*, and they were itching to get going. "Our primary interest was to get the hell out of there because they had told us we had to be gone before dark," recalled George Pickett of Cadet Company H. "Although we were the graduating class—and second lieutenants—we had to have our tails off the post before dark!"

For the cadets there was a sense of this day being the culmination of a great many things, but as a practical matter, June 11, 1941, began more like a typical duty day than like the dawn of a brave new world. It was an exceedingly hectic day; there wasn't time to stand around or become nostalgic. The men had crates and boxes to move and administrative activities to complete. There would be no prolonged farewells, except between roommates who would see one another once more when they returned to their barracks to gather their belongings. There was no more time.

The graduation exercise would be the last time that these men, who had bonded like brothers, would ever assemble as a unit in one place. Within a few hours they would all be gone. There would be reunions 10, 20, and 50 years down the road, but the coming war would claim 53 lives, and the class could never again be complete. Indeed, even among the living members of the class, no such gathering as today's breakfast in Washington Hall, once commonplace, would ever be repeated.

At 9:30 A.M., the cadets filed into the cavernous Field House. Of the graduates on that day, there were 62 men whose fathers or grandfathers had graduated from West Point. Many of the graduates would one day send their sons into the valley of the Hudson.

They came from every one of the 48 states in the Union, as well as from the Territories of Hawaii and Puerto Rico. The largest number, 43, were from New York. There were 14 from New York City alone. There were 33 men from California, 8 of those from San Francisco. Among the other states, 28 cadets had been appointed from Pennsylvania, 23 from Georgia, and 21 from Texas. Ten of the men had been appointed from the District of Columbia. There was one cadet, Atanacio Torres "Tony" Chavez, who came from the Philippines, the Pacific commonwealth of the United States, a land that was to see the first blood of this class spilled in wartime within seven months.

Nearly every year since before World War I, it had been customary for the secretary of war—the army's man in the presidential cabinet—to give the commencement address. Indeed, it seemed as though Newton Baker was *always* there when he was secretary of war, and even after he retired. There were exceptions, of course. General John J. "Blackjack" Pershing (Class of 1886), the army's greatest hero since George Washington and the man who had led it to victory in World War I, had done the honors in 1920, 1923, 1924, and 1936. Another exception was Franklin D. Roosevelt, who had come in 1935 and 1939, and whose presence was especially unusual because he was a navy man.

On June 11, 1941, Henry L. Stimson came north to deliver the graduation address. Secretary of war in the Roosevelt cabinet, the 74-year-old Stimson was a gentleman's gentleman of the patrician Ivy League aristocracy that Franklin D. Roosevelt had gathered about him. A former cabinet officer under two presidents—secretary of war under Taft and secretary of state under Hoover—he was one of Roosevelt's closest advisors, preparing for a conflagration that would make the nation all but forget Blackjack Pershing's war.

As he looked out at the Class of 1941, Stimson knew that these men were the ones who would fight and die in the coming war. More than that, he knew that they would also be the ones who would *lead* the countless thousands who would fight and die in the coming war.

Stimson must have felt a sense of sadness when he looked out at that sea of faces over which he presided, but it must have been a sense of sadness tempered by a commitment to purpose. Stimson was a crusader, or, more properly, the leader of a crusade. As Hoover's secretary of state when the Japanese went into Manchuria in 1931, he had authored the Stimson Doctrine under which the United States would not recognize land taken by military conquest. So much more territory had come under fascist boots since 1931. It was Stimson's job—his mission in life—to take back every square inch of land that the fascists had seized.

Stimson came to address the West Point graduates as more than just a figurehead in a ritual. He was one of the key architects of the policy that would shape the destinies of those who were gathered before him. Indeed, Stimson recognized that in this season—this final spring—the *entire* nation was about to undergo a rite of passage.

Stimson stood ramrod straight before the audience as he began his speech. His dark blue suit was in stark contrast to the shock of white hair that framed his stern and determined face.

He began by commenting that his words that morning would be of an "unusual and perhaps unconventional character. Usually commencement is a time of rejoicing and congratulations as we elders give good wishes to the young men who are beginning life's journey. But that is hardly the atmosphere which surrounds our country today."

Stimson went on to frame the events of that day against the backdrop of a world in torment, noting that beyond the borders of a geographically sheltered nation, there was a world "where justice and law have been overthrown, where mutual tolerance has been replaced by cultivated hatred, and where the doctrines of humanity and religion have been trampled under by ruthless barbarity and by the organized slavery of fellow men. In all that world today, only the British Commonwealth of Nations is still fighting for

the old standards of freedom. All other nations are either cowed or conquered."

Stimson was obsessed with the righteousness of his crusade, *the* crusade.
It was an effort that went straight to the heart of the doctrine of *Pax Amer-
icana*, of America's future role as superpower.

At the grassroots level, however, most Americans, while not "capital-
letter Isolationists," were instinctively "small-letter isolationists." There were
probably very few parents or other guests at West Point on June 11, 1941,
who had any desire to take part in the conflict.

"If Japan had not attacked Pearl Harbor," Merritt Hewitt of Cadet Company M would later observe, "I doubt that Congress would have ever declared
war. In those days, Europe was still very far away. I believe that without the
United States in the war, the British and the Germans would have negotiated
a peace. Neither was willing to go through 1914–1918 again."

The secretary of war saw it differently. He went on to emphasize that
anyone who believed that the United States could come to terms with the
Nazi system and peacefully live in the same world with theirs "simply has
not thought this matter through. The world today is divided between two
camps," Stimson went on tersely, "and the issue between those camps is
irreconcilable. It cannot be appeased. It cannot be placated. Humanity cannot
permanently make terms with injustice, with wrong, and with cruelty."

It was an era of upheaval and a time when old traditions and ideologies
were moved by the winds of change. This change was also washing over the
US Army as it awoke to the challenge before it. The graduates, like those
before them, were the American military elite, but the army's manpower
needs were suddenly so great that in the restructuring that was taking place,
the elite of West Point would be the exception rather than the rule in the
ranks of officers. The Class of 1941 would not be going into the old "regular
army," but into the new army of the United States—a great army that was
then in the process of enlistment, formation, and training.

But among the 424 graduates in the audience, sitting with their parents,
fiancées, and girlfriends, thoughts were turned not so much to the battlefields
of Europe as to the extended leave that was about to begin.

George Pickett's thoughts were on Montgomery, Alabama. "I don't think
any of us paid any attention to the graduation address—really remembered
it," he said later. "It was just something that sounded good. None of us,
throughout our military careers, was ever inspired by a *speech*. . . . he might
as well have stood up and said, 'Fellas, you did good. We need you. Your
country needs you. Get out there and fight!' "

For Jesse Thompson, remembered by his classmates as a "master of the
art of indifference," Stimson's speech was something to be endured before
the men could conclude graduation ceremonies. Thompson was impatiently
waiting to collect his diploma so he could run back to the barracks to get his

bags, get into his car, and head south to Haileyville, Oklahoma, for a few weeks of celebration.

While many of the men believed in the inevitability of the war, few were truly inspired by Stimson's call to the crusade. It would take an attack on a Hawaiian naval base, already in preparation half a world away, to rouse the class to Stimson's call.

It was still springtime in the Hudson River Valley when Henry L. Stimson folded his notes and returned to Washington and the gathering storm. It was a day that saw 424 brand-new second lieutenants climb into roadsters and railroad cars, racing off to six weeks of leave and the world beyond.

It was still springtime in America, but the winds of war were already blowing.

1

You're in the Army Now

It was a wonder that the Greene boys found each other in the surging crowds in the cavernous terminal. In 1937, everybody traveled by train, and New York's Grand Central Station, in its sprawling, ostentatious splendor, was America's crossroads.

They came together at dawn on the first day of July with that assured, backslapping familiarity that is part of the special bond between brothers.

Michael Joseph Lenihan Greene and Lawrence Vivans Greene did not *look* like brothers. Much to the contrary. To identify them as such would be to invite incredulity. Larry was 6'2" and weighed 200 pounds, while Mike was 5'7" and weighed just 110. Larry was an imposing figure with a strong jaw, sharply chiseled features, and a high forehead. Mike's round face, centered beneath a shock of sandy hair, was the face of a kid on his first big adventure in life. Mike looked a lot younger than Larry, and few would have guessed how close they were in age. Many people who knew both of them didn't realize they were related. If you didn't *know* it was true, you'd think they were pulling your leg. It was one of their favorite inside jokes.

The two brothers were on their way to West Point to enter the United States Army Military Academy. For Larry, it had been a long haul. At the age of 19, two years of out high school, he had been doing everything possible—including joining the National Guard and calling on members of Congress—to snare an appointment to West Point. He had known he had wanted to go to West Point since he was a child. Mike, two years younger, was much less certain about the Academy. He had finished high school in 1936, when he was only 16, and had joined his brother at the Millard Prep School in Washington, D.C., arguably the foremost military prep school in the country. Indeed, at Millard the Greenes would meet no fewer than three dozen men who would later become their classmates at West Point. After six weeks, however, Mike withdrew to attend the Drexel Institute of Technology in Philadelphia, where his father, Colonel Douglass Greene, was a professor of Military Science.

Colonel Greene, himself a graduate of the Military Academy, had always hoped that Larry and Mike would join the corps of cadets. But Mike didn't necessarily agree with his father's plans for him.

It was during the summer of 1936, while Mike was attending the Millard Prep School, that an Illinois senator offered Mr. Millard a second alternate appointment to the Military Academy. "Beanie" Millard asked Mike's father if he wanted to put Larry's name in for the appointment. Colonel Greene replied, "Thanks, but that's not necessary. Larry is already competing for a Pennsylvania congressional appointment, a Pennsylvania National Guard appointment, and a presidential appointment." But he quickly caught himself. "Oh, what the hell. If it's only second alternate, you might as well put Mike's name in."

On May 30 of the following year, Mike received a telegram at his home near Philadelphia, informing him that the principal as well as the first alternate had failed, and therefore, if Mike passed the physical exam and the validating exams, he could have the West Point appointment from Illinois.

Mike told his family he didn't think he wanted to go to West Point. "The next thing I knew," Mike related facetiously, "I was on the living room floor and my dad was kicking me, my grandfather was kicking me, my grandmother was kicking me, and my mother was kicking me. So I said, 'Okay, I give up. I'll go to West Point!' "

Mike went to Walter Reed Army Hospital for the physical exam, which he passed without difficulty. When he was finished he went on to Portsmouth, New Hampshire, to visit his girlfriend, Ruth Ann Greenlee, whose father was a navy captain stationed there. On June 21, Mike arrived at West Point to take the validating exams. After his year at Drexel, he was already validated in most subjects, so he took only the history and English exams.

Nine days later, he got a telephone call from his father. "You've been appointed to West Point," his father told him, "and you have to report tomorrow."

"I really didn't know what the hell I was getting into," Mike said later. "I was just seventeen years old, and this would be the first time I had ever lived away from home."

Mike told his father, "I *still* don't really think I want to go to West Point."

Colonel Greene, who had six children and little money to educate them, said, "What do you really want to do?"

Mike said, "I think I want to go into the foreign service, and I want to go to Georgetown University."

His father said, "You know as well as I do, Mike, that we can't afford that. I don't have that kind of money."

"Dad," Mike replied, "other people go to college and don't have any money. Maybe I could get a job, or maybe I could find somebody—grandparents or somebody—who could help me."

"You have this appointment, Mike," Colonel Greene said to his son. "It's a sure thing. I understand that you think it's not what you want to do. All I'm asking you is if you will try it, and if after the end of six weeks you *really* don't want to go on, I will see what I can do to get you into Georgetown."

It didn't take six weeks for Mike to decide that West Point was for him. "I guess I knew within the first day," he recalls, "that this was what I wanted to do."

In the meantime, Larry had won a Pennsylvania National Guard appointment to West Point.

Mike and Larry would be the third generation of their family to attend West Point. Their father had graduated in 1913. Their paternal grandfather, Lewis Greene of Illinois, graduated in 1878 and was immediately assigned to the frontier, where he served as a lieutenant for 19 years, fighting the Sioux and the Ute in Colorado and Montana. For two years, he was an aide to General George Crook, the legendary Indian fighter. The Greenes' maternal grandfather for whom Mike was named, Michael Lenihan of Massachusetts, graduated in 1887. He served in Montana, Nebraska, and Texas and in 1901 was sent to the Philippines to help quell the insurrection. He came home to serve on the Army General Staff and during World War I was made brigadier general, serving as brigade commander in France with the 42d ("Rainbow") Division. He stayed on active duty until his 64th birthday, in 1929.

Mike and Larry Greene left Manhattan on the New York Central and headed north into the broad valley of the Hudson River. Nestled in the Catskills—verdant, rolling hills formed a billion years before by slow-moving glaciers—the rich woodlands along the Hudson were once home to the Mohawk. Henry Hudson, the first white man to behold the river's craggy cliffs, anchored the *Half Moon* here in 1609 while searching for the Northwest Passage.

Summer stock theater, a phenomenon scarcely known a decade before, was in full swing in the Hudson River Valley as Mike and Larry Greene traveled to West Point. It was a hot and muggy day, overcast and threatening rain, with temperatures in the high 70s. Before long the train passed Dobbs Ferry, Tappan Zee, and the villages made famous by Washington Irving's tales of sleepy hollows and sleeping rascals. Mike Greene mentioned that the Yankees were playing Philadelphia that afternoon. The Yankees would win 12–7, thanks to a Joe DiMaggio homer, and would continue to lead the American League, five games ahead of the Chicago White Sox.

People on the train were already twittering about a third term for Franklin D. Roosevelt, and he'd only just been inaugurated for his second term in

March. In Germany, Adolf Hitler had just declared that the German government would take over 966 Catholic schools in Bavaria and would close the rest. Simultaneously in Spain, the Spanish Civil War—which in reality involved almost as many Germans and Italians fighting Russians as it did Spaniards fighting Spaniards—was emerging from the spring doldrums as Nationalist forces under Francisco Franco shelled leftist positions in Madrid and captured Bilbao. King George VI, who had ascended the British throne the previous winter upon the abdication of his brother, announced that he had changed his birthday from December 14 to June 9. George (whose name had been Albert until he was crowned king) believed it wasn't sporting for his subjects to have to celebrate his birthday in the foggy chill of December and that such a celebration in the "midst of the Christmas shopping rush is poor business." Meanwhile, American society pages were abuzz with the event of the summer as Franklin D. Roosevelt's son and namesake exchanged wedding vows with Ethel Du Pont—daughter of millionaire Eugene Du Pont of the Delaware Du Ponts—at tiny Christ Church in Christiana Hundred, Delaware.

Yet for the Greene brothers, the news of the day held little sway. For them, a great adventure was about to begin. For the next four years, the grounds at West Point would be their home, and there they would forge the friendships and the professional alliances that would form the framework of the rest of their lives. Once through the United States Military Academy's gray stone gates, the outside world would suddenly become a distant memory. The silliness of the British monarchy, the comfortable conviviality of summer stock, and even the uneasiness of the war in Spain and the posturing of dictators was no nearer, nor more real, than ghosts dancing on the moon.

Gibraltar on the Hudson

Well over a century before the men of the Class of 1941 were born, George Washington had come to be enamored of the strategic position of the fortress at West Point. "The importance of this post is so great," General Washington commented in 1783, "as to justly have been considered the key to America."

West Point had, in fact, been the keystone in the defense of the Hudson during the Revolutionary War, when its presence prevented the British from splitting the colonial forces into two easily digestible portions. It was, indeed, an American Gibraltar, but the accuracy of this appellation held for only a brief moment in time. When the war was over and the colonies united, the site was relegated to a secondary role with no real defensive function. A company of engineers was stationed there, but little else. By the end of the eighteenth century, West Point was already a relic of the past.

Washington himself recommended to Congress in 1793 that a cadet training program be instituted, with West Point as its venue. Nevertheless, it was not until March 16, 1802, that the United States Military Academy (USMA) was officially founded. It was placed under the control of the Corps of Engineers (also formally created in 1802), with Corps of Engineers Major (later Lieutenant Colonel) Jonathan Williams as its first superintendent. Williams was to be the only nongraduate to ever head the USMA, but it would not be until after the Civil War that a nonengineer would hold the post.

The exemplary performance of West Point–trained officers in the generally disastrous War of 1812 clearly demonstrated that the new military academy fulfilled a vital function within a US Army composed mainly of untrained, hastily conscripted citizen soldiers. Among those officers who proved themselves in the War of 1812 was the USMA's 33rd graduate, Major Sylvanus Thayer of the Class of 1908, who became West Point's superintendent on July 28, 1817.

Major (later General) Thayer is considered to be the father of the United States Military Academy and is so identified on the pedestal of his statue, located on the northwest corner of the parade ground. How a man could be the father of an institution from which he himself graduated is explained by the fact that Thayer created many of the institutions, programs, affectations, and traditions that have come to define the USMA for well over a century. When Thayer stepped down in 1833, the USMA had a clearly articulated gray uniform, a rigid code of personal and group integrity and identity, and a strict honor code summarized by the motto "Duty, Honor, Country." In the century following Thayer's tenure, there was little change in the basic structure of the Academy.

Since its inception, West Point has produced a spectacular lineage of both heroes and rascals in whose footsteps the men of the Class of 1941 would walk. Among them were Douglas MacArthur, who graduated at the head of the Class of 1903, and George Armstrong Custer, who graduated at the very bottom of the Class of 1861. The Class of 1913—the "class the stars fell on"—ultimately produced two five-star generals, Dwight Eisenhower and Omar Bradley. Among their classmates, 40 percent would wear stars as major generals (two stars) or above. This is in contrast to most other classes, where only about ten percent would ultimately wear the single star of a brigadier general. As the Class of 1941 was entering the challenging days of Plebe Year, the men of 1913 were already percolating to the apogee of the US Army command structure.

The Class of 1941 lived in what was called the "MacArthur" West Point. General Douglas MacArthur, fresh from his success during the First World War as Commander of the 42d ("Rainbow") Division, had taken over as superintendent for three years beginning in June 1919. Except for the legendary O. O. Howard (1881–1882), MacArthur was the only general to head the USMA. He strengthened West Point in the aftermath of World War I and ended the controversial practice of early graduation brought about by war hysteria. Under this policy the cadets of the Class of 1917, who had entered West Point in 1913, were graduated two months early, on April 20, 1917, just after the United States' declaration of war. The cadets of the Class of 1918, who had entered in 1914, graduated on August 30, 1917. The Class of 1919 graduated on June 12, 1918. The Class of 1920, which entered in June 1916, completed a "War Emergency Course" and graduated on November 1, 1918, less than two weeks before the Armistice. Cadets who entered in 1917 and who normally would have been the Class of 1921, also graduated on November 1, 1918. However, when the war ended, these men were recalled as "student officers" and regraduated in June 1919.

The idea of a two-year course was put to rest by Douglas MacArthur. Most of the cadets who entered the War Emergency Course in 1918 were graduated in June 1920, but 17 men were not graduated until June 1921, making the Class of 1921 the smallest in over a century. Another 102 men who had entered in 1918 were held until June 13, 1922, and the following day, West Point graduated yet another 20 men, most of whom had entered in 1919, but who were allowed—under an act of Congress on March 30, 1920— to graduate after three years. West Point officially became a four-year school again with the graduation on June 12, 1923, of the lion's share of the class that had entered in 1919.

MacArthur brought back discipline, system, and method that would temper and mold the fledgling Plebes of 1937 into the men of Black '41 by defining the structure of their lives for four years.

Although physically the West Point "reservation" had grown to 16,000 acres—including nearby Camp Buckner, the site of most cadet field training— the heart of the United States Military Academy was, and still is, a relatively compact, half-mile-square campus located on a bend in the Hudson River. The centerpiece of the campus is a fortresslike crescent of imposing gray stone buildings, dating from the mid–nineteenth century, that surround "The Plain," the cadet parade grounds. Visible from the Hudson, and indeed from most places on the campus, the landmark building at the Academy is the neo-Gothic cadet chapel, which is located on a hill above the granite academic buildings. Other facilities and faculty housing are located on the streets that radiate from this center, but the academic buildings and the chapel are the heart of West Point.

Mike and Larry Greene disembarked from their train at the New York Central's West Point station, located near river level beneath the granite cliffs on which the Academy stands. As they stood on the platform, they looked about them and saw for the first time a gaggle of unfamiliar faces soon to become their classmates. There were a few handshakes and introductions, but for the most part all young minds were fixed upon the task at hand: lugging their baggage 200 yards up the hill to Thayer Hall, beyond whose enormous doors lay their futures.

Beast Barracks

Into the maw of the beast walked Mike and Larry Greene. Into the maw of the beast walked 50 dozen others. Beast Barracks. Hell on earth, courtesy of the West Point classes of 1938, 1939, and 1940—especially the Yearlings (sophomores) of 1940, who had themselves been relieved of their Plebe (freshman) status only a few weeks before.

Beast Barracks is a grueling six-and-a-half-week period of indoctrination, training, and hazing for incoming cadets. The culture shock and physical challenge of it can prove too intense for some. The purpose of Beast Barracks was, and is, multifold. First and foremost, it is an opportunity for the Academy, by way of the upperclassmen of its corps of cadets, to teach the neophytes in as graphic a way as possible the meaning of the phrase "rank has its privileges," and that rank *itself* is a privilege earned through sacrifice and endurance of adversity.

"Mister, you're in the ARMY now!"

Among the upperclassmen who would "supervise" the Class of 1941 throughout the next 60 days of hell were Henry Harley Arnold, Jr. (Class of 1940), son of the Army Air Corps chief; Andrew Jackson Goodpaster (Class of 1939), who would attain four-star rank and command all Allied forces in Europe from 1969 to 1974, then become a superintendent at West Point from 1977 to 1981; and Leon Robert Vance, Jr. (Class of 1939), who would win the Congressional Medal of Honor in 1944 as a Bomb Group Commander in Europe, only to die in a plane crash on the way home.

"Mister, you're in the ARMY now!" shouted John Dale Ryan, Class of 1938, to George Scratchley Brown, the man who would succeed him as the US Air Force Chief of Staff in 1973, but who now stood before him as a would-be Plebe. In less than five minutes, future members of the Class of 1941 would learn that they were not young men destined to be officers and gentlemen, they were swine, cockroaches, and maggots or, in some cases, the excrement of such fauna.

"Mister, you're in the ARMY now!"

Shortly after 9:00 A.M. on July 1, 1937, all of the new cadets lined up to report at a desk set up at the east sally port of Central Barracks, where each man checked in and then was turned over to an upperclass cadet, who escorted him out into a sea of other upperclassmen indiscriminately shouting conflicting orders.

Those who do survive the nightmarish weeks of Beast Barracks—and many would not—will no longer be the runny-nosed mama's boys that the upperclassmen have reminded them they were.

"Mister, you're in the ARMY now!"

"You will be referred to as men, not boys!" insisted the upperclassmen between insults, as though to call them "men" was to suggest that they were mere parodies of men. Indeed, "men" was not strictly correct—most of them *were* really boys, and they *knew* it. It was the mission of the Academy to make them men.

Richard Kline had served three years in the Oklahoma National Guard in addition to a year in the regular army at Fort Sill before arriving at West Point. Kline already knew close order drill, how to roll a field pack, and how to care for himself under field conditions. "I was stunned," he recalled, to be faced with upperclassmen who were younger and less experienced than he was and to "find the puerile children of the upper classes being presented as seasoned role models. Yet, the West Point reputation was clear, so I held my peace."

There were 324 men who walked through Thayer Gate on July 1 to become part of the Class of 1941. Counting late arrivals and "turnbacks" from the Class of 1940, their number would grow to 579, but of those only 424 would make it all the way to graduation.

After being shaved and shorn, the cadets were issued piles of equipment, from bedding to rifles, packs, field gear, and uniforms for every possible occasion, including full dress grays with "feather duster" hats and the full dress coat that had "44 brass buttons" and a "cast iron collar."

By five o'clock that afternoon, when they had been completely equipped, the new men put on their uniforms and marched in formation to Battle Monument on Trophy Point overlooking the Hudson River to be sworn in. They marched there in perfect step only eight hours after they had arrived.

"They ran our asses off," George Pickett remembers. "We didn't *walk* all day until they marched us over to the swearing-in ceremony and back. The rest of the time we were going at a dead run. They ran us up to the cadet store with lists in our hands, where they handed us each a mattress, two pillows, a down comforter, sheets, and pillowcases in bags. Then they told us to carry it all back. It was impossible for anything but a mule to carry all of that stuff, but we did. And we had to run with it."

For Jesse Thompson, the first day at West Point was very nearly what he'd had been told to expect. There was so much going on that he could hardly take everything in. Luckily, he was in very good physical condition, because the new men were ordered about "on the double" nearly everywhere they went.

"We were learning to do what one is told to do quickly, without question," Thompson recalls. "I can still visualize dozens of new cadets running across the open area lugging newly issued bedding. I remember seeing one new cadet walk into the confusion and, without hesitation, turn around and walk back out."

Inspection accomplished, with sighs of relief they turned off the lights and were soon fast asleep.

In many respects, the first 24 hours were the most important of each cadet's life. They taught him that no matter how tough things were, or how seemingly impossible, he could do what was ordered. If anyone had told the cadets that they would carry a pack mule's worth of equipment from the cadet store to the barracks—up and down and up several flights of stairs—at a dead run, and do that three times during the day and then drill, they would not have believed it. Nothing that ever happened after that first day ever measured up to the abuse and culture shock that they experienced during the initial 24 hours they were at West Point.

The first *full* day of Beast Barracks began at 5:50 A.M. with the enlisted buglers and drummers, known affectionately as "Hell Cats," blowing reveille as they marched through the barracks area. Some would-be cadets failed to fall out, and they were out the big gates by noon, never to be seen again.

"Mister, you're in the ARMY now!"

As reveille sounded on the morning of July 2, the new cadets endured 90 minutes of "physical training"—calisthenics—and were then "formed" in ranks and told to report in their underpants with a bar of soap and a towel to march down to the showers. West Point has worked out the number of minutes it takes a human being to shower; cadets had exactly that many minutes to shower and get out. There was an upperclass detail yelling, shouting, and hazing the cadets the whole time. They were then marched back to their barracks to get dressed in their uniforms.

Breakfast would come each day at 8:00 A.M., followed by the first training and/or class session at exactly 8:45. Lunch was at 1:00 P.M., also for 45 minutes, organized sports at 4:00 for 90 minutes, and a 45-minute dinner break at 6:00 P.M. Aside from the diversion of meals and sports, the cadets were in class or in training continually from 8:45 A.M. until 9:00 P.M., leaving them an hour of "personal time" before taps.

The cadets were "sized"—assigned to companies according to height and stature—which made the formations at parade appear to be identical. Cadets in the middle of the field could look to the right and look to the left with the illusion that every cadet was the same height. The Plebes were divided into three battalions of four companies each, for a total of twelve companies, with three dozen men in each company. As the saying went, "The second batt is short and fat." The first and third battalions, which had the tallest cadets—including those over 6'1"—were called the flankers because they were on the flanks of the marching order.

Mike Greene, at 5'7", ended up in Company G, one of the short companies. His brother Larry ended up in Company A, one of the flanker companies. Because everything at West Point was done by companies, Mike and Larry would have little association almost the entire four years that they were cadets.

Larry Greene eventually became a center on the Army football team and played lacrosse, baseball, and tennis as well. Mike had no particular interest in athletics and played intramural sports only—if a cadet wasn't on the varsity, he *had* to play intramural.

During Plebe Year, when Larry came to see his brother Mike, the upperclassmen in Company G—who were, of course, as short as Mike was— stood on chairs to make themselves as tall as Larry. When Mike went to visit Larry, the upperclassmen in Company A put *him* on a chair so he was as tall as they were. Larry didn't think this was very funny, and after a while he stopped going to see his brother.

Mike—Michael Joseph Lenihan Greene—soon became known as the "little man with the big name." This amused the upperclassmen. Every time he was asked who he was, the upperclassmen expected him to give his full name. Before long, everybody knew it.

Not everyone in his class, however, knew Mike Greene's full name. Joe Gurfein of Cadet Company B recalled that he used to mistake Mike for James Oscar "Jog" Green—one "e" and no relation—of Company F. Mike, who knew Gurfein, would respond to Gurfein's "Hello, Jog" with a "Hi, Bob" or "Hello, Ed," until Joe finally realized that he'd goofed. They later became great friends.

Gurfein had grown up in the Flatbush area of Brooklyn, around the corner from Ebbetts Field, a vehement Dodger fan. Before coming to West Point, he had gone to Public School 92 on Parkside Avenue and for a full four years to Erasmus Hall High School. He took four years of math, three years of German, and three years of Latin. In 1936, he won a four-year, full-tuition scholarship to Cooper Union. He attended for only one year because of his appointment to West Point, but it was here that he met the woman who ultimately would become his wife. For him, the first days of Beast Barracks were a shock. Having been brought up in a quiet, white-collar Brooklyn home where personal feelings were respected by others, the demeaning treatment was traumatic. "My whole dream of high-minded, gentlemanly cadets collapsed into a pile of dirt," he recalled.

One of the first guys Larry Greene met in Company A was Lynn Cyrus Lee, the son of a naval officer who had run away from home at 17 to enlist in Teddy Roosevelt's Great White Fleet. Shy, good-natured, but memorably bowlegged, Lee arrived at Company A with a stack of Hawaiian records under one arm. "It's a bad guess whether he's a seaman trying out his land legs or a bronc buster from way out West," someone whispered. But his mathematical brilliance and general likeableness quickly made "the Sloppy Mr. Lee," as he was known, one of the most popular men in his company.

As they sat around the barracks during their one hour of free time before taps, the men often talked about why—and how—they had come to attend West Point. The story that Lynn Lee told was fairly typical and clearly illustrative. For him, getting to West Point was a case of his father, a navy officer based at Pearl Harbor, having five offspring to put through college on a lieutenant's pay. When they started graduating from high school, he had told them, "It is essential that some of you go to 'Uncle Sam's school for boys' on one river—the Severn in Maryland—or the other—the Hudson in New York."

In Hawaii, where Lynn Lee was living, the entrance exam to Annapolis, the "school for boys on the Severn River," was conducted by the United States Civil Service Board at the post office in Honolulu in February. Lee arrived to take the test and quickly determined that the monitor knew nothing. He had answered the one or two questions he was asked by reading from the same printed instructions that everyone had. Lee found out later that the monitor had been instructed to do this: No other words were to be spoken by him. This was to insure that people all over the world who took the exam had exactly the same instructions, and that no one had one extra word of help.

After six months of practice exams every two weeks, Lee was not the least bit nervous. The morning segments consisted of math and algebra, with a section on geometry in the afternoon. At 8:06 A.M., Lee started working the four-hour exam. He had finished all his practice tests in less than 90 minutes, and when he wrote the final answer to the final question of this exam, it was 9:10. He looked around at the other 15 or 16 applicants taking the test. They were all still working furiously. He checked his work for obvious errors only, knowing from experience that he tended to make mistakes when he reworked problems.

The monitor's eyes caught his several times, and it was obvious that Lynn Lee was doing nothing. No one else was about to leave, but he knew if he stayed much longer he would start reworking the exam, so he folded it, cleared his desk, walked up to the monitor, and handed him the envelope. Having watched Lee do nothing for 30 minutes, the monitor was convinced that the envelope contained a blank paper. "Better luck next year," he said.

The geometry and physics exams were no more of a problem, nor were the history and English. Lee felt comfortable that he had easily passed. He then went to the navy hospital to take his physical. When the doctor dilated his eyes, he discovered that Lee was farsighted. Cadets with eyes that were not normal had to receive a waiver from the surgeon (as all military doctors were then called) at the US Naval Academy when, and if, they were ordered to report there.

Soon afterward, the army's Letterman Hospital in San Francisco found the same abnormality, but the army doctor immediately wired the War Department a recommendation that Lee be given a waiver. Lynn Lee would have preferred to go to Annapolis, but he realized that he would not know if he had passed the navy physical until the day he arrived at the Naval Academy, hoping to enter. On the other hand, he had already been approved to enter West Point. For Lee it was Army 1, Navy 0.

Paul Skowronek, with his thick blond hair, ready smile, and penetrating eyes, symbolized the easy, self-assured confidence that characterized many of the men. They were, after all, the cream of the crop back home. The USMA didn't admit losers.

As a freshman at the University of Pittsburgh, Skowronek had begun to study chemical engineering and enrolled in ROTC. When his newly elected congressman offered to reward Paul's mother—who had helped with his campaign—by offering her son an appointment to West Point, Paul readily accepted without giving much thought to the military career that would result. "It was not a decision based on years of anticipation and planning," Skowronek admitted, "but it was the most rewarding decision of my life."

Joseph Tuck Brown arrived at West Point from Plymouth, Illinois, "his chin out just a little farther than anyone else's and with a telltale glint of red hair creeping from under his cap." He had been impressed and awed upon his arrival. He found his room changed three times in three hours that first day, and each time he had to carry his mattress and his gear with him. A tall man with a strong chin and dark, piercing eyes, Tuck had the kind of movie-star good looks that, to his great dismay, attracted the wrath and insults of every upperclassman in Beast Barracks.

Ira Cheaney was born in Alabama but later moved to El Monte, California. Cheaney was a study in contrasts, a man who loved the sun and warmth of southern California, but who left it for the gray, cold gates on the Hudson that had become the austere goal of his ambitions. He loved music, Chippendale furniture, Kipling, and polemics as much as he loathed the early morning, snow, and math; yet he had pursued a Congressional appointment to West Point with a peculiar determination that his classmates never really understood. Although naturally introspective and more comfortable with matters of the intellect, he deliberately pushed himself into football—which he basically hated—as though he were testing himself with strange and unfriendly fires.

Plebes underwent an intensive course of self-defense in the course of their training, culminating in a furious, final boxing bout in which they could prove their mettle against a randomly chosen classmate. Thomas Cleary was the classmate who was to be Cheaney's sparring partner. Having boxed in collegiate competition before entering the Academy, Cleary squared off against the obviously awkward and inexperienced Cheaney with confidence.

At the sound of the bell, Cleary eased toward Cheaney in a way that would impress boxing coach Billy Cavanaugh. To Cleary's astonishment, he encountered a veritable windmill—Cheaney was all over him. He beat a tattoo on Cleary's skull and punished his middle while Cleary retreated in disarray.

For three rounds Cheaney treated Cleary to a fighting lesson while Cleary tried to recover his boxing reputation. At the final bell, to Cleary's surprise and chagrin, Cavanaugh remarked, "Mild guys are often wildcats in disguise." After the fight, Cleary and Cheaney formed a friendship that would endure throughout their years at West Point.

Some came to West Point after careful consideration. Hugh Foster of New York City had been attending Cooper Union's engineering school at night, working in a bank during the day, and doing most of his studying on subway trains. He concluded there had to be a better way of life. His brother had talked about West Point, and Hugh had begun to listen. In 1935, the life of an army officer looked interesting and glamorous, and West Point had an impressive image. Beyond this, the Foster family was still experiencing the effects of the depression, and the opportunity of getting one of the best educations in the country while being paid to do so was very attractive.

Hugh eventually applied and was presented with an enviable choice. He could have a Congressional appointment to the US Naval Academy at Annapolis in 1936 or a Congressional appointment to West Point in 1937. When he was asked by a classmate one night early in Plebe Year why he had chosen West Point over the Naval Academy, Foster explained that "under normal conditions, I could reasonably expect to have a 30-year career and retire as a colonel. However, if, during a 30-year career there would be more years of war than of peace, I wanted to be in the navy with dry beds, warm food, medical care close at hand, no barbed wire and poison gas. On the other hand, if during a 30-year career there were to be more years of peace than of war, I wanted to be in the army and raise a family. A review of American history suggested involvement in a major war about every 20 years, which meant one—at most two—wars in my career."

Duward "Pete" Crow, with his innocent blue eyes, Southern drawl, and an abundance of self-confidence, had been born on a large, general-purpose farm near Fort Payne, Alabama, that had been a gathering stockade in 1838 for the Cherokees who were sent on the Trail of Tears. The Crow family raised cotton, fruits, and vegetables. Their livestock consisted of mules, horses, pigs, and cattle, with flocks of chickens, guinea hens, and turkeys, and never less than half a dozen foxhounds. In the summer, fresh vegetables and poultry dominated their diet, and in winter they feasted on canned fruit and vegetables, smokehouse slabs of bacon and ham, and an occasional beef. Pete grew up loving farm life and the outdoors, carrying water to the fields for his father and older brothers, bringing in wood.

When Pete was six, his father traded the big farm for a general merchandise store in Fort Payne. Life in town was easier, but within a year and a half Pete's father had his family back on a farm, this time a smaller one nearer town, although he kept the store as well. As a result, the senior Crow had "the best of both worlds," and his daughter and five sons worked with him. "It was cotton, corn, and 'taters all over again, rounded out by a commercial dairy called Hillside," Crow laughed. "Whoever heard of child labor laws?"

Pete was in "city" school until grade five, when he was sent to a little one-room schoolhouse a quarter mile from his home. While in class, one could listen to the instruction for all six grades. By Christmas, Pete was taking sixth grade work. Throughout his school years he made good grades; he was on a high school debating team that won second place in statewide competition at the University of Alabama, played basketball, and graduated as valedictorian and president of his class. Just before high school graduation, Pete was called to the principal's office, where he found his father and Congressman Joe Starnes.

"Ah'm embarrassed not one of muh five prior appointments to West Point or Annapolis has made it past Plebe Year," Starnes drawled. "Pete, y'all are to be next."

Before Crow attended West Point, however, it was decided that a year at the University of Alabama would be useful. The home of the Crimson Tide was a big step socially for a country boy like Pete Crow, but the classroom work came easily. By July 1937, Pete was ready for the next step in Joe Starnes' plan. Through a mix-up, however, Crow's appointment was not in order, and it was July 13 before he joined the Class of 1941. He brought a laugh to Beast Barracks when he missed his first formation, saying "Suh, I am not familiar with the routine." But fellow Company A cadets William Kromer and David Woods took him in hand for a little straightening out.

"I needed 'straightening out' for a long time," Crow recalled. "Once, while in full dress formation with the rest of Company A Plebes out early for a parade, I smiled at the OAO [one and only] of the company commander's roommate. She smiled back and told him what a 'cute' Plebe she had seen. I had KP calls for months."

One of the youngest men in the class was Horace Brown, who was assigned to Company C barracks. Brown had always wanted to attend the USMA. He came from a patriotic family, and his mother was a leading member of the Daughters of the American Revolution and the United Daughters of the Confederacy. Following his sophomore year in high school, Brown obtained a working scholarship to Riverside Military Academy, from which he graduated at the age of 16. His father wanted him to go to The Citadel, which he attended for two years. During his sophomore year, fueled by his financial situation, Brown decided to apply to West Point. He received an appointment in March 1937.

Alfred "Al" Judson Force Moody, the son of a Yale University professor, entered the army at Fort H. G. Wright before attending the army prep school at Fort Peeble, Maine. Moody won an appointment to West Point in the competitive exams, so he had already taken many of the courses he would encounter during Plebe Year. Nevertheless, he had what many classmates would call "a miraculous mind," and it seemed to require very little study to keep him far ahead of all his classmates. As roommate George Pittman observed, Moody spent more time coaching classmates than he did on his own studies.

James Fowler was the only black man among the gaggle of Plebes. He was to be, in fact, the only black man in the entire cadet corps for the next two years. It would not be easy. There had been only one black graduate of the Academy previously in the twentieth century—Benjamin O. Davis (Class of 1936), who later retired as a three-star Air Force general, and no one had made any serious effort to recruit black cadets.

At Dunbar High School in Washington, D.C., Fowler had been an honor student, president of the Honor Society, battalion commander of the ROTC, and manager of the swimming team, and he had been voted the most dependable member of the senior class. He had enrolled at Howard University as a pre-med student with the intention of becoming a neurosurgeon. However, in 1937, during his last year at Howard, Jim was persuaded by a congressional representative from Illinois to give up his career in medicine in order to pioneer an attempt to crack open what many people saw as a segregated US Army. He easily passed the entrance exam for West Point and graduated from Howard as the outstanding cadet in the ROTC program. He received an A.B. degree, magna cum laude, in June 1937.

Fowler had agreed to enter West Point at a time when blacks were expected to not ride in the elevators in public buildings in Washington, D.C. Within the army, the black enlisted men of the 10th Cavalry—the "Buffalo Soldiers"—were the only non-whites. Jim Fowler's entrance into West Point was met with what was later described by a classmate as "all the fervor reserved for a holy war. He may have slipped in, but he would surely not stay."

At every formation, five or six upperclass cadets would descend on Fowler, accusing him of a dirty cap visor, stubble on chin, lint on braid, cuffs not pinned properly, dirt in his fingernails, spots on trousers, and unshined shoes. Half a page of demerits filled the delinquency report for "Fowler, J. D." almost every day. He was lucky if he received only five demerits at a time.

Through Plebe summer camp and into the fall, Fowler endured the cruel torment. As one of his classmates recalled afterward: "His classmates at this time, of course, suffered the usual indignities of new cadets, but even we could distinguish between our treatment and that reserved for 'Mister Fowler'."

As the autumn of Plebe Year went on, Fowler's demerits went up and his grades went down. The common expectation was that his stay at West Point would not be long. However, in January 1938, after the first turnout of those poor souls who had failed to make the grade, Jim Fowler was still there.

Throughout his Plebe Year, Jim was awakened during the night and kept awake so that he would fall asleep in class the next day. His classmates were advised to refrain from speaking to him and were threatened with ostracism themselves if they tried to give him any support or encouragement. Fowler would room alone during Plebe Year and for the next three years as well.

George Pickett had wanted to go to West Point from the time he was able to look at Civil War pictures and read world history. His grandmother used to sit on the porch and talk about Confederate General James Longstreet, and he grew up amidst the Southern tradition of military service as the province of gentlemen.

"I was sort of brainwashed into it before I was able to walk," he would say. His grandmother decided that he was going to military school, and since she owned all the land and thereby controlled the family, she was able to make it stick with George's father. Pickett went to Tennessee Military Institute in 1932, graduating in 1936. He had the alternate appointment to West Point in 1936 but didn't get in. In 1937, he got a Congressional appointment— not an easy appointment to win. This was during the Great Depression, and in Alabama times were hard. Appointments to West Point were very competitive because everyone in the state saw the service academies as an opportunity for a career in the army and a free education.

There were 300 men from Alabama who took the competitive examination in 1937, and only two of them were selected. One went to Annapolis, and George Pickett went to West Point.

Pickett was a little guy: Some would say that he was the quintessential little guy. Mike Greene, who was also quite small, underscored the kind of Plebe Pickett was with an amusing and illustrative tale: "As a cadet, I used to think that even if I wasn't good in gymnastics, I knew that George Pickett was poorer at it than I was. At least in the rope climb I could pull myself up off the floor, but George Pickett couldn't. The joke in our riding class was that George Pickett always made it over his jump, but his horse *never* did."

Ever the rascal of Company K, Joseph Scranton Tate, Jr., was born in Marfa, Texas, while his father, Joseph Tate, Sr. who had just graduated with the Class of April 1917 was busy chasing Pancho Villa across the Mexico-New Mexico border with General John J. "Blackjack" Pershing's expeditionary force. Growing up an army brat, Joe Tate loved everything about army life—the outdoors, the horses, camping, scouting, hunting, and, most of all, athletics. He grew up on army posts such as Fort Sam Houston, Fort Sill, Fort Myer, and even West Point when his father was transferred to the USMA in the late 1920s. Joe loved every minute there. He was able to indulge his penchant for collecting army memorabilia, and his room resembled a miniature museum, with insignia, bits of uniforms, broken hockey sticks, old polo balls, and medals from many wars and many countries. His brother Dan swears that Joe knew more people on the post than any other kid at West Point.

Ruth Alexander—who later married Tate's Company D classmate Bradish Smith—grew up at West Point while Joe Tate was there. Her father was a professor of "drawing," as it was called in the 1920s, but he later became the professor of military topography and graphics and ultimately the first dean at the USMA. Like the Tates, the Alexanders lived on "professors' row," not far from the Catholic chapel, and next door to the chapel were the quarters where Danny and Joe Tate lived.

Joe was older than Ruth, and she remembered him as a "large, hefty young man who scared all the girls to death." They loved to play football, and Ruth and her friends organized a girls' football team and played against the Tate boys and their pals, albeit, as Ruth recalls, "rather disastrously."

The Tates had a wonderful cook who made taffy, "the best taffy you ever tasted." The kids would get a telephone call, or a message by the grapevine, that the cook was making taffy, and children from all around the neighborhood would suddenly appear for a lovely time pulling taffy and eating it.

Joe's high school years were spent at the Kent School in Connecticut, where he was an above-average student and a competitive athlete, playing tackle on Kent's undefeated football team. Following his graduation in 1936, Joe spent a year at the Stanton Preparatory School in Cornwall, New York, preparing for the West Point entrance exams. He passed them and was able to get an appointment from Texas.

George Brown came to Cadet Company L from Montclair, New Jersey. His father, Thoburn Brown, graduated from West Point in 1913 and had been a career cavalry officer who rode with Joe Tate's father in Blackjack Pershing's expeditionary force against Pancho Villa in 1917. He was an instructor at the USMA until 1922 and an assistant professor between 1928 and 1932. Like Joe Tate, George Brown grew up on the campus where he now found himself a cadet.

One of Black 41's oldest cadets, Ernest Durr, was born in Newport, Rhode Island, the son of a naval officer. He grew up against a backdrop of tragedy. His only sister died of spinal meningitis at the age of three; a maternal uncle, Fran Sidney Long (Class of 1917) was killed in action in the Meuse-Argonne in 1918; and Ernie's mother died in the great flu epidemic during the same year. Ernie graduated from Washington High School in Los Angeles at the age of 15 and, at the age of 16, left home. With his grandfather acting as his sponsor and guardian-in-fact, he attended Urban Military Academy in West Los Angeles (1933-1934) and the University of California at Los Angeles (1935-1937) before being accepted to West Point.

When William Gillis arrived at West Point, his roommate described him as a "tall, handsome, confused product of Texas." Despite all the honors of his past, Bill was not so sure of his future. Beast Barracks was rugged and strict, and few moments were available for the carefree life to which he had been accustomed.

At Cameron High School, Gillis had captained the football team and was the mainstay of the track team. At Schreiner Institute in Kerrsville, Texas, he was again captain of the football team and starred in track. Bill Gillis was a natural leader. He was constantly among the honor students of his high school class and was a leader of student government. He was also more than a little mischievous, setting off firecrackers under the house of a prominent local family during an elegant dinner party and inviting girls to dances with the mock-Shakespearean, "Prithee, canst thou to with me to the shindig this eventide?"

Alexander "Sandy" Nininger, Jr., was born in Atlanta, Georgia, the son of Mr. and Mrs. A. R. Nininger. As a small boy, he spent some time at his mother's old family home in Central Valley, New York, just across the hills from West Point. One day, while Sandy was still a preschooler, a group of cadets happened to pass the house during an exercise. When he waved to them, they waved back. He cheered until they disappeared from sight, then quickly ran to his mother, wanting to know all he could about those "boys in gray." It was then that he made up his mind that one day he, too, would go to West Point. His family moved to Florida in 1926 and finally established a home in Fort Lauderdale, where he went through grammar school and high school, graduating in June 1937. He received his appointment from Congressman Mark Wilcox.

Sandy Nininger had grown up a quiet boy of gentle disposition, deeply interested in his studies, with a strong, natural sense of duty. When an older friend said to him just before he left for the USMA, "Sandy, why do you want to be a soldier? You couldn't hate and *kill* people, could you?," Nininger replied, "I wouldn't kill out of hate but I *would* kill out of love for my country."

The longest six and a half weeks in the Plebes' lives ended on a sweltering Friday the 13th in August 1937, and then it was time to embark on the first academic year at West Point. For some, it would be their last. The attrition rate in Beast Barracks had been high. Dozens of men had packed up and left—some of them in the first week—and throughout their Plebe Year others would also fall by the wayside.

Most of the men of Black '41 hadn't come to West Point because of the free education. They were there because they were dedicated to the service. Consequently, the things that meant the most to them were not what West Point taught about Chaucer or what they learned about engineering. For most of the men, their attitude was that they were at West Point not just to learn history and math, but to become officers in the United States Army. It was the first step in an army career, and everything that happened in Beast Barracks and over the coming four years was put within that context.

2

Duty, Honor, West Point

The cadets in the Class of 1941 were divided into two halves, and these groups would define their social life for the next four years. Companies A through F made up one half of the class, and Companies G through M made up the other. When the first half of the class went to math, the second half went to physical education, and vice versa. Because of this system, the two halves rarely mixed, even for social or athletic events.

Sometimes a cadet got to know classmates from the other half because he was dating the friend of another classmate's girlfriend, or through some other circuitous route, but in effect the system created two classes, and while cadets knew each other by name and could recognize fellow classmates, it was difficult to feel the same kind of bond for cadets in the other half that one felt for those in his own half.

Mike Greene, for example, at the end of his second year, took a list of the Class of 1941 and checked off the names of the cadets he knew well enough to picture. He was determined to get to know each of his classmates personally in the next two years. It was a feat that served him well over the coming years, including a half century later when he became president of Black '41's alumni. But Mike Greene was more the exception than the rule.

One incident involving George Pickett illustrates the long-term rewards in war of being able to recognize fellow classmates. "I remember once during World War II I was told to report to Colonel Murray, the G-3 of the 87th Division. I went in and reported, and right away, I recognized Jack Murray. Instead of being Colonel and Major, he told me where he wanted me to take the tanks, and I got them there."

The cadets at West Point in the years before World War II had to complete a standard curriculum, even though many of them had been in college previously and already had completed some of the same courses. The USMA curriculum included English, foreign language, and history—with a focus on military history—and was heavily skewed to math and engineering. As George Pittman later described it, this system of study created a "Jack of all trades, master of none." Pittman, among others, came to believe that concentration on a specialty would have been a better method for retention. "Fast force feeding was too short."

The cadets paid no tuition. Instead, they pledged up to 30 years of their lives in service to their country. In an era when the average family of four lived on $3,000 a year, the government annually paid each cadet $780—the average price of a new car—to attend the USMA. Cadets were also given a daily ration. With this money, which was retained by the treasurer until a cadet went on leave, everyone purchased additional rations, uniforms, books, and supplies.

Cadet life, whether for Fourth Classmen (Plebes) or First Classmen (Seniors), was a minefield of rules and regulations, each of which carried the threat—indeed the promise—of demerits if transgressions occurred. Uniform violations were common, as were demerits for failing to pass a white-glove room inspection. Carrying a red blanket on the arm or sitting in a parked car also brought demerits, although these could be worked off by punishment tours—one hour per demerit—of marching under arms.

For example, George Pickett recalled that Bill Hoge seemed to have trouble with athlete's foot. Part of the treatment for this malady at the West Point hospital included a prescription to wear plain white socks instead of the regulation black, which reacted with the ointment the hospital also prescribed.

Pickett was in class with Hoge on several occasions when the P (instructor) would ask, "Mr. Hoge, do you have permission to wear white socks?"

On many of these occasions Bill's reply would be, "No sir."

The P would then skin him (report him delinquent) for not being in proper uniform.

Certain things were simply never done. No cadet could possess a horse, a car or a mustache, and radios were frowned on until 1939. To have any of the prohibited accouterments was grounds for expulsion, although by 1941 rules had changed to permit First Classmen to have cars.

The most serious transgression was, and still is, violation of the Cadet Honor Code. A violation of the Code meant expulsion. No amount of punishment touring could save the military career of a cadet who lied, cheated, stole, or tolerated anyone who did.

Occasionally the Honor Code created difficult and divisive moments for the men. Charlie Murrah's worst moment at West Point was having to "silence" his best friend, Charles Reed, for an Honor Code violation. Reed was a "brilliant, but maverick, cadet" who was found guilty of lying and refused to resign when asked to by the Honor Committee. Classmates in Company

I were split on whether or not to silence Reed, to essentially condemn him to an institutional "cold shoulder" in which no cadet would speak to him. Ultimately, it was Murrah who did the deed, and turned Reed in.

Even though he was living one room away, Murrah didn't even notice when Charles Reed was whisked away by the MPs in the dead of night. "I have never seen Charlie again . . . I still revere his memory. We had fun as cadet buddies, mostly shooting craps and playing poker."

Plebe Summer Camp

Immediately after Beast Barracks, the cadets were marched across the West Point Parade Ground to the Plebe Summer Camp at Fort Clinton. It was a memorable, if excruciating, experience, characterized by marches in hot, humid Hudson River Valley weather, biting bugs, and frequent physical and verbal harassment by the Yearlings, who were newly in charge and feeling their oats.

The regular army tactical officers during that summer were well remembered, perhaps none more so than Major Bowes, who seemed to relish with perverse delight his power to deride and degrade the new Plebes. At one point, he skinned George Pittman for a spot on his crossed belt webbing. Pittman volunteered that the laundry had put it there, whereupon Bowes proceeded to gig him (give him demerits) for about a dozen more infractions he would not have mentioned had Pittman not spoken.

Captain Joe Costello, another of the tactical officers, was frequently in charge on the pistol range. Once he gave Hal Tidmarsh, who kept missing the target completely, a small handful of rocks, kicked him lightly, and told him to throw the rocks at the target. Hal Tidmarsh did not miss the target with the rocks.

"I thought that the treatment of Plebes was petty and degrading, as was the army in general," Bill Hoge said. But he added, "I must admit that I was about the rawest of raw material and I needed to form a lot of new habits, such as keeping my mouth shut and making efficient use of my time."

At that West Point was particularly good. Cadets learned to do things on the double, with a minimum of lost time.

During guard duty at Camp Clinton, the tac officers often tricked the Plebes who were "guarding" the post. Fortunately for the tac officers, the Plebes were not issued live ammo. Even so, one of the Plebe guards once tackled an officer who was trying to evade him. At other times, however, less vigilant Plebes left their posts to talk to the young women who visited West Point from time to time. Plenty of walking tours and other discipline tended to solve that problem.

For many, the most torturous element of Plebe Summer Camp was the hike to Crow's Nest peak, a forced march for all Plebes under full pack and rifle. The leaders were dressed in lightweight clothes, in which they could move with relative ease. The Plebes, under the weight of full field gear, had difficulty pulling themselves up many of the steeper inclines. By the end of the march, some of the Plebes had difficulty undressing to shower.

"Each time I lifted my leg to remove my boots, I would cramp up," complained George Pittman.

It was not until later that the Plebes were told that they had gained the distinction of being the first in many years to have successfully endured the Crow's Nest hike.

There were high points during that Plebe summer. George Pittman met his future wife on a blind date. Plebe summer was also a time of song and dance, when the cadet choir and glee club members were selected. Each cadet had to sing "Nearer My God to Thee," and many were eliminated after the first word. The training of officers also included dance lessons under the watchful eye of Mrs. Roberts. Class members who had never danced before were a source of endless delight for Mrs. Roberts, who enthusiastically yelled instructions. Mr. Roberts wore a hearing aid and often turned it down to avoid hearing the lessons. However, at the Saturday night dances over the next three years, Mrs. Roberts' instructions saved many a Black '41 classmate from embarrassment—and many a young lady's foot.

There were a few disagreements and fights, though they often had the effect of bringing classmates closer together. One such fight involved Dick von Schriltz, who was the biggest man in the class, weighing in at well over 200 pounds. He had aspirations as a boxer and was the only man in the class to qualify as a heavyweight. "He wasn't too good, but he had a terrific punch," Bill Hoge later said.

Von Schriltz had been the butt of a lot of jokes in Company M, and that began to get on his nerves. He finally announced that the next person who teased him would end up fighting him. Ernie Durr immediately accepted the challenge.

Moments later, the two went to the gym to put on gloves. Von Schriltz outweighed Durr by 40 pounds, and the fight wasn't even close—Durr ended up with a split lip, a black eye, and minor cuts and abrasions. The fight didn't particularly dampen Durr's spirits, but it did have the desired effect. The other cadets quit teasing von Schriltz, and in turn he became a lot less touchy. And Ernie Durr came to be known as the kind of gutsy guy who would not hesitate to accept a challenge.

At one point during the summer, a Yearling burst into Mike Greene's tent and ordered him to put on full dress uniform and report in front of the tent. The Yearling proceeded to berate Greene, marching him to the middle of the parade grounds, where another Plebe was standing rigidly at attention in full dress uniform. He marched Mike up to face the other Plebe.

Mike looked at Larry and Larry back at Mike, neither of them betraying any hint of recognition.

"Mr. Greene, I want you to meet Mr. Greene. Shake hands," the Yearling demanded.

They did.

"About face," he then commanded. "Go back to your tents."

Mike went back to his tent, and about a half an hour later the upper-classman came in. "I appreciate that I've been pretty rough on you for the last month you've been here," he said, "but you've got to admit I did go out of my way to introduce you to another Plebe with the same last name as yours."

"Sir," Mike asked, "may I make a statement?"

"What's that?"

"Sir, that cadet was my brother."

The upperclassman went through the roof. He had intended to humiliate *them*, and the Greene brothers had humiliated *him*. Mike answered reveille calls for the next six weeks.

Plebe Year

Beast Barracks and Plebe Summer Camp had been like reaching up to touch bottom. When August came to a close, those who had persevered finally found themselves at the bottom. They were *official* Plebes, a name adapted from the Latin word denoting the lowest social class. They were Fourth Class-men amid four classes, the lowest rung of the cadet corps totem pole. Yet having endured, those who reached bottom felt a certain sense of superiority over the weak-kneed souls outside the Academy. The first seeds of pride in the corps had been sown.

The Plebes of 1941 may have been outranked by three classes of cadets, but they had a sense of their place in the world. The same upperclassmen who had so colorfully characterized them as scum and phlegm, and who would continue to verbally harass them, had also taught them that even Plebes outranked some of God's creatures. There were not many in this category, but at any moment an upperclassman could stop a Plebe, call him to full attention, and demand that he answer—in proper order—the question "What do Plebes outrank?"

"Sir!" the cadet would reply, for the foremost mistake was *not* to address an upperclassman by this title, "Plebes outrank the Superintendent's dog, the Commandant's cat, the waiters in the Mess Hall, the Hell Cats, and all the admirals in the whole blamed navy." (Actually, it was usually spoken as "damned navy," but in polite company it was the "blamed navy.") Presum-ably the omission of any of the listed entities would suggest that a Plebe

considered *that* specific creature to be his superior. Ten or 20 push-ups did
wonders for a Plebe's memory.

On the other hand, the Plebes discovered that cadet gray—regardless of
the number of stripes on the sleeve—commanded respect in the outside world.
A few weeks later, when they were bussed down to New York City for the
Army–Notre Dame football game, the cadets had a few free hours. Nobody
was supposed to drink, but they were bored after the game and didn't have
anyplace to go, so George Pickett and Riley King went to Jack Dempsey's
restaurant on Broadway. In those days, a cadet could not keep money in his
quarters, but he was issued $5 when he left the post to go to a football game.
If he didn't spend it all, he had to turn back what was left. Even in those
days, as Pickett recalls, "You couldn't buy a glass of water in a 'good' restaurant for $5."

When the cadets timidly went into Dempsey's, someone came up to them
and said that it had been suggested they order so and so. They did, being
careful not to overspend, but overspending nonetheless. When they got ready
to leave, Jack Dempsey came out and picked up the check.

That impressed Pickett, a kid from Alabama in the big city. "This famous
man, a former heavyweight champion, saw these young kids out there who
obviously didn't have any money and were looking around the big city. I
never forgot it."

A similar thing would happen in 1939 in Boston, where the class had
gone for the Harvard game. A group of cadets was in a restaurant in one of
the downtown hotels, and somebody picked up the check. They didn't even
know who did it, only that someone had taken care of it. Wherever they
went, people seemed to appreciate the cadets. "I guess they thought that we
were dedicating our lives in service to the country, and they wanted to be
nice to us," Pickett said later. "In today's world, that attitude doesn't make
any sense, but that was the attitude that people had toward the professional
military in those days. In 1937, any second lieutenant in the army could walk
in, produce his ID card, and every bank in town would cash his check. That
was the reputation of the service and how people looked at the service."

Back at West Point, however, Plebe Year was very tough. The cadets had
no off-campus vacation breaks all year and no dances, parties, or boodlers
(the cadet restaurant) except at Christmas. As Ben Spiller explained it, "You
have to understand that the cadets were treated pretty much as children in
those days at West Point. There were no privileges at all, only complete
supervision."

The autumn of Plebe Year was to be the worst experience most of the
men had ever had. The shock treatment of Beast Barracks would be a constant
reminder of potential disaster, but as Jack Murray said later, "The loss of
my first two roommates because of academic failure was hard to take, but

the greatest thrill was to realize, after a few tough and doubtful months, that I *could* survive and eventually graduate."

The hazing by the upperclassmen only seemed to intensify. It was as though they—especially the Yearlings—had an obsession with breaking as many Plebes as possible. "Although it was officially against the rules governing hazing, we did get our share from certain hateful upperclassmen who used to get personal," George Pittman recalled. There were clothing formations (quick changes to odd mixes); push-ups; balancing acts; "swimming to New-burgh"—half squatting while holding a bayonet by its point under your backside; and having an increasing number of books piled onto your outstretched arms.

Finally, however, the Plebes began to strike back—in guerrilla fashion, of course, because to confront an upperclassman was unheard of. There were special occasions, such as the last day of Mechanical Drawing class, when all drawings were torn to bits and spread throughout the upperclassmen's rooms, in shoes, pockets of clothes, rolled up window shades, and rifle barrels. Sometimes the Plebes even went so far as to mix all of the shoes of a class together, then randomly redeposit them in various rooms. Some men never retrieved their own shoes.

Though the Plebes were not permitted to go home on leave, Christmas provided a short escape from the incessant "Plebe regime." For many, this was their first Christmas away from home and family. For Paul Gray, it was an especially miserable time because he was obliged to undergo eight hours of "turnout" exams—four on Christmas morning, and four on December 27. By passing the exams, he escaped expulsion or, in cadet parlance, being "turned out."

Among the men who had been turned out of the Class of 1940 the previous year, and who joined the Class of 1941 at midterm, was Thomas Reagan. Reagan was born near Chicago, but being an army brat, he had attended various schools and graduated from high school in St. Louis, Missouri. Early in his youth, Tom had set his sights on West Point. He entered Columbia Preparatory School in Washington, D.C., to prepare himself for the entrance examination and the subsequent rigors of the West Point academic life. He passed the examination and entered the USMA on July 1, 1936. Unfortunately, he found himself in for a surprise at the end of his Plebe Year, and that surprise was the USMA Math Department. He "took a cut at an outside curve and became an ex-1940."

Undaunted, he immediately began to satisfy the requirements for readmission, and January 1, 1938, found him a member of the Class of 1941, assigned to Company I, where he became known as "Thos" because Company I already had another Tom. In Company I—described by one of his classmates as the "best of the best"—Thos Reagan found the relaxed excellence that exactly suited his personality. Having learned his lesson from the Math De-

partment, he stayed proficient without further difficulty, and even the Tactical
Department was unsuccessful in disturbing further his academic composure.

Several other cadets from the Class of 1940 were turned back to Black
'41. One was Hamilton Avery, Jr., who shared distant relatives—including
President James Polk and Lieutenant General Leonidas Polk, the colorful
"Fighting Bishop of Louisiana" who fought with Confederate General Robert
E. Lee—with Company F classmate Dick Polk.

Ham Avery had been born in Memphis, Tennessee, but his family moved
to New Orleans when he was six. Since Avery's mother spoke fluent French,
all of the Avery "young'uns" adapted well to the Creole/Cajun atmosphere
of the Crescent City.

In New Orleans, the Averys lived around the corner from Governor Huey
Long's mansion, and Ham attended school with Long's eldest son, Russell,
who later became a state senator. His academic achievements in high school
provided Avery with a needed scholarship to Tulane University, his father's
alma mater. Nearly everyone had suffered financial ruin after the stock market
crash in 1929, and Avery's father was no exception. To offset the effects of
sudden poverty, young Ham had a newspaper route, sold magazines, and
worked as a drugstore carhop.

Through his father's appeal to Senator Ransdell and the efforts of his
mother's family patron, Senator Ellender, Avery eventually achieved his goal
of being able to apply for an appointment to West Point. In the meantime,
he had joined the Louisiana National Guard, where he saw combat duty in
1934 when his machine gun company was federalized during the long, violent
New Orleans public transportation strike. Avery took the National Guard
competitive exam for West Point in 1936 and scored first among 22 competitors, qualifying for an appointment that year. "May my high-ranking
ancestors [in the Confederate Army] roll over in their graves," Avery said
when he received his appointment, "but I never hoped to achieve such an
honor."

Being turned out in 1936—the result of what Avery characterized as a
"head-on" collision with the Math Department—was a major blow to his
pride. But instead of going home to lick his wounds, he enrolled in the
Sullivan Prep School in Washington, D.C., which was ably run by Gerald
Sullivan (Class of 1923). Scott Peddie, Al Muzyk, Felix Gerace, Paul Day,
and Sam Barrow also attended the Sullivan School and accompanied Avery
on his return trip to the Hudson and to Black '41. Dozens of other cadets
who were turned out, however, were not so fortunate. The rigors of the West
Point classrooms quickly became notorious among new cadets.

Yearling Year

On June 14, 1938, a strange transformation came over the Plebes of Black
'41. As the Class of 1938 graduated, the tormented Fourth Classmen suddenly
became Yearlings. Human beings again at last!

Six weeks later, when the 326 new cadets of the Class of 1942 made their way through the gate at Thayer Hall, it was Black '41's turn to remind them: "You're in the ARMY now!" Beast Barracks was something to be inflicted upon a new batch of Plebes. Summer camp was a lark, at least until its last week, when the First Classmen of the Class of 1939 returned.

It was about this time—the 1938 summer camp—that the mysterious "Black '41" epithet began to make the rounds at West Point. Nobody knows why. It was just suddenly there. Nevertheless, the Yearling school year began happily enough. With the wrath of the upperclassmen directed at the new Plebes and a familiar routine to follow, it was a relatively relaxed time. Army beat Navy 14–7 and ended the football season with eight wins against only two losses. Three men from Black '41—Bill Kelleher, Joe Grygiel, and Bill Gillis—had made the varsity. After football season, the Yearlings could look forward to their first time out of uniform in 18 months. Plebes were not permitted a Christmas break, so December 1938 would be Black '41's first opportunity to go home since the class had arrived at West Point.

However, Sandy Nininger did not go home for Christmas. He told his parents that he had "certain duties" to attend to. It was not until much later that his parents learned Sandy had used his travel money to help a school friend pay tuition at a theological college. Quiet and shy, Nininger was always remembered as one of the most selfless of the men of Black '41. As he wrote to his father on his father's birthday, "Let my present then be a promise, that I will not disappoint you."

During their second year at West Point, all the cadets were enrolled in the basic engineering course that was standard at most engineering schools nationwide in the late 1930s. For Mike Greene, the academic part of West Point was almost exactly what he had already taken at Drexel. Because of this, when they finished the first year, Mike outranked his brother academically. Larry hadn't been to college. He had been in prep school getting ready to pass the West Point entrance exams. However, Larry went on to catch up with Mike, so that by graduation he ranked Mike by 11 places. Mike Greene would survive academically at USMA because, although he was at the bottom of the class in chemistry and physics and in the middle in math, he was always at the top of his class in French, history, and drawing. He would nearly fail his second year of physics, but he managed to pass the turnout exam after his brother's roommate, Ed Rowny, gave up his Christmas vacation to tutor him.

Every month or six weeks the cadets were divided according to academic average and arranged in order of merit, so that the 20 cadets with the best academic records were placed in the first section and the 20 with the lowest grades were put in the last section. Typically, however, some of the best combat leaders come from the middle of a class. They tended to be more realistic. This is true of most classes anywhere: The more well-rounded people tend to come from the middle rather than from the top of the class.

George Pickett remembers being placed in the first section of Military History, having gotten there by academic standing. As he walked in, the professor said, "Mr. Pickett, you absolutely *do not* belong up here in the first section."

Pickett replied, "Well, sir, my grades put me in the first section according to the rules."

"All right," the professor growled, "but you can rest assured that you *won't* be here next month."

He wasn't. "The idea of having a guy with my pitiful academic record in the first section of anything would defeat the system!" Pickett laughed.

Pickett's Company H roommate was Ernie Poff from Albany, Missouri. They were best friends at West Point. "You just don't live with a guy for four years and share all of your hopes and ambitions and *not* become best friends," Pickett said. When the cadets graduated, if someone had asked them who their best friend was, most would have said their roommates.

"That was the way the system worked," Pickett added. "If your roommate *wasn't* your best friend, you never graduated, because when you got ready to go to parade, someone had to dust your uniform off in back, somebody had to see that your cartridge box was two fingers below your waist belt, somebody had to see that there was no gray showing between your waist belt and your cartridge belt, and somebody had to make sure that your trousers were the proper distance off the ground in the back. When you got dressed to go out for any formation, your roommate had to check you before you left, and if your roommate wasn't the kind of guy who was going to cooperate with you, you were in trouble. In those days, they even checked to see if a cadet had a dress coat under his overcoat at reveille formation."

There was a white-glove inspection every day in the barracks. The officers ran their fingers along the rails. Cadets had to have their rooms orderly. One roommate swept, dusted, and cleaned for a week, then the other did it for the next week. Of course, each man made his own bed and kept his own bookcase, but the care for the communal parts of the room was rotated back and forth.

Each cadet had his name in a little holder on his side of the room, and when the room orderly—the cadet's slang for it was "room bitch"—came in to inspect and found something he didn't like, the cadet's name was up there for him to see. So if a cadet's roommate wanted to screw him up, he could throw a piece of paper on the floor or let something drop at the wrong time, and his roommate was in trouble. Cadets had to have a roommate that they got along with, or they just didn't make it.

As Yearlings, the men of Black '41 became more and more part of the USMA. No longer were they outsiders, mere Plebes. As Kelleher, Gillis, and Grygiel made the football team, Ben Spiller made the choir. There were 160 members in the cadet choir, 40 voices in each section—first and second tenors, first and second basses. "We were good," remembers Ben Spiller. "It would raise the hair on the back of your neck to hear us sing in the barracks after a choir trip in winter, with the air crisp and the sound carrying across the Hudson."

There were two choir trips to New York City each spring, one in February, when the choir sang at St. Thomas Cathedral, the huge Episcopal church on Fifth Avenue, and the second at Columbia University Chapel, which was smaller and more intimate. The Columbia trip occurred in April, usually in great weather. The cadets sometimes ducked into a studio at Radio City Music Hall to cut a cheap record or sat in the bar of the Picadilly Hotel, drank beer, and sang to entertain the customers.

The darkness of winter had given way to the spring of 1939, and members of the Class of 1941 were looking forward to adding the Second Classman's service stripe to their dress grays when a pair of tragedies struck.

Bill Kelleher, with his red crewcut and winning smile, was a true standout among the men of Black '41. Ever popular, he easily won election as class president. He was also an outstanding baseball player and was picked to be part of the USMA team that would play the annual exhibition game against the New York Giants on April 19.

In the meantime, however, Kelleher got sick. He had a mild cold and sore throat at first, nothing more. He didn't realize it was pneumonia and avoided sick call because his greatest fear was that he would be sidelined from the game.

Bill got his wish. He played outfield against the Giants—or rather he stood alternately sweating and shivering in the outfield. The next morning the pneumonia gripped his body like an iron fist, and he staggered to the hospital, where he was immediately placed in intensive care. By now the pneumonia had overwhelmed him, and, despite efforts to save him, he died two days later on April 22, 1939.

It was to be first time that the Class of 1941 heard the term "Long Gray Line" applied to one of their own, but it would not be the last. The Long Gray Line is the term West Point cadets and graduates use to denote those of their number who have died and form a long line of men in the gray uniform of the USMA.

Bill Kelleher's death was a tragic and sobering experience, but nobody in Black '41 expected that the second of their number to join the Long Gray Line would do so in only three short weeks. Among those who mourned Bill was Charles Jobes, late of Kansas City and now a popular member of Cadet Company L.

Like Bill, Chuck Jobes was an athlete. He had played intramural football, but his real passion was lacrosse. It was the afternoon of May 12, during a break in practice, when Chuck decided to try for a front circle on the goal. The goal was a good deal flimsier than it looked, and as Chuck executed his maneuver, the goal cage collapsed on him, killing him instantly.

"I knew Charley Jobes pretty well," recalled Bill Hoge. "We were in the same battalion and we kept bumping into each other at the gym. He was a hearty, boisterous, irreverent roughneck, always ready with some wisecrack or joke. Although he was a natural athlete and always good company, I don't know where he stood academically, but I don't think he was very high in the class. I don't think he gave a damn either. I can imagine Jobes being the idol of a company of young soldiers. They would have seen him as one of themselves."

Stunned by the uncanny experience of losing two of their friends in less than a month, the men of Black '41 decided to raise funds to erect a monument to them. The Kelleher-Jobes Memorial Arch, which stands on Flirtation Walk, was completed before the class graduated. At each reunion, members of Black '41 gather there for a ceremony to commemorate all classmates who have died since graduation. In recent years, an arched wall has been added, and the memorial was rededicated at the Class of 1941's fiftieth reunion in 1991.

The Polo Army

The US Army of the late 1930s bore little resemblance to that which would be fielded in the early 1940s. It was indeplorable shape to face the challenges that lay ahead. The army of 1938 thought of itself (not officially, of course) as a "garrison army." The USMA, too, was long on spit and polish and short on practicality.

"I was somewhat disappointed in the Academy," Bill Hoge said. "I think I had too many glorified ideas about the corps of cadets. I suppose my views were rather similar to those of General MacArthur when he was appointed superintendent. I have read that he used to ask his chief of staff, 'Chief, when are we going to quit preparing for the War of 1812?' "

The US Army ranked 18th in size among the world's armies in 1938. The Soviet Union had over a million soldiers and sailors, while Germany had nearly a million. With 120,000 troops, the United States had fewer soldiers than Greece or Belgium. Of those 120,000, over one-quarter were stationed overseas in the Philippines, Hawaii, and the Panama Canal Zone, another one-third were in administrative positions, and another one-fifth were assigned to guard the Mexican border. This left an effective strength of barely 30,000. The army had 20,000 horses—the cavalry was still the principal means of combat mobility—but ten percent of those horses were polo ponies.

During Richard Kline's one-year enlistment in the Field Artillery at Fort Sill prior to attending West Point, he had served as an orderly for officers during a field exercise. The officers chatted about the marvelous invention of the motorized split trail drawn artillery, but not a single piece was fired. The trucks broke down, and field telephone equipment was crude.

At West Point, the cadets' exposure to weaponry was equally meager. The largest weapon the cadets fired was a 37mm gun mounted on the tube of a 75mm cannon, because the army could not afford to provide anything better.

"The training given by the Tactical Department wasn't very imaginative," Bill Hoge remembered. "For example, the Class of 1941 did not once fire the M1 service rifle while at the Academy. We got all our marksmanship training with .22 caliber weapons. We did fire the .45 caliber pistol at a regular range, and we did fire the .30 caliber machine gun, but only at 1,000 inches."

As Hugh Foster of Cadet Company A described it, the army was a "poor man's army." The cadets trained with stovepipes, and the army itself was not in very much better condition, even in 1941.

Eventually, the acquisition of more field ranges and training sites would change this situation. But at the time, many in the country advocated a *smaller* army—or indeed no army at all—pointing out that the United States had two great oceans and friendly Canada as buffers and that the National Guard could raise 540,000 men in 12 months, if necessary. The problem with this argument, aside from the obvious question of what to do for the year it took to recruit and train an army, was the expense of maintaining facilities. A war-strength division of 20,000 troops could be cut to 7,000 in peacetime, but the equipment and infrastructure for 20,000 had to be maintained. In 1918 it had taken 5.6 million Allied troops to hold the Western Front against 3.9 million Germans, and it was only the pressure of the sizable American Expeditionary Force that finally broke the hellish stalemate.

Among the army of 120,000, there were only about 100 generals, although many analysts saw this as at least double the number required. With an average age of 61, the general officer corps was composed entirely of men on the verge of retirement. When General Douglas MacArthur was Chief of Staff in 1935, he was the youngest general in the army at age 55. It wasn't that the generals spent a long time *being* generals—the average age of the army's colonels was 58—it was simply that the process of promotion was a long, sluggish climb. The average age for first lieutenants was 34 (although there was one who was almost 60). This meant that West Point graduates, who graduated as second lieutenants, could spend over a decade in the army before they received a promotion.

There were fewer than 5,000 officers between the ranks of first lieutenant and lieutenant colonel in 1938, and only 14 percent—almost all of them West Point men—had been commissioned since World War I.

The National Defense Act of 1920, which MacArthur had hailed as the "Magna Carta of National Defense," had called for an army of 280,000, a modest enough force, but a goal that had yet to be reached in 1938, despite the fact that the world had become a pincushion of flashpoints, each with the potential to affect America's national interests.

The army was organized differently in the years before World War II than it is today. In 1938, the Department of War oversaw the army. Both the Department of War and the Department of the Navy had a secretary with full cabinet rank equal to that of the Secretary of State.

The military equivalent of Secretary of War was the Chief of Staff of the Army. The Chief of Staff was the direct inheritor of the powers that for a century had been exercised by the Commanding General of the Army. The chiefs of the various bureaus and branches *within* the War Department were analogous to lower-echelon staff officers reporting to a commander.

On the operational side, there were four numerically designated armies within the US Army, each of which corresponded to a particular region of the country. The First Army was headquartered in New York, the Second Army in Chicago, the Third Army in San Antonio, and the Fourth Army at the Presidio of San Francisco, the oldest military post in the continental United States. Each of these armies was assigned a varying complement of infantry, cavalry, artillery, and support units.

Also part of the "line of the army," but not necessarily subservient to any numbered army, were three corps: the Corps of Engineers, the Signal Corps, and the Air Corps. The Air Corps had been part of the Signal Corps until 1926; by 1938 it was filled with officers who felt that it should be separate from the army entirely. As a result of their rumblings, the Air Corps was reorganized as the autonomous US Army Air Forces in 1941, and would separate completely from the army in 1947 as the US Air Force.

Meanwhile overseas, the rest of the world was preparing—and rehearsing—for war. Italy had seized Ethiopia in 1935 in the face of British opposition and amidst only feeble bleatings from the League of Nations. A civil war had broken out in Spain in 1936, in which Italy and Germany on one side were openly fighting Soviet Russia–backed forces on the other. Japan, having taken over Manchuria in 1931, attempted to seize control of northern China; its gross miscalculation of the nature and endurance of Chinese resistance led to a continuing, bloody struggle to destroy the existing Chinese government. Adolf Hitler, after rearming Germany in defiance of the Versailles Treaty, annexed Austria by a *coup de main* and began to talk about taking over Czechoslovakia as well.

German action against Czechoslovakia occupied the sleepless attention of the chancelleries and general staffs of Europe. Nation was arming against nation as never before. All of this was occurring while the US Army played polo.

Major General Johnson Hagood pointed out that polo, including grooms, stables, and remount stations, was costing taxpayers $1 million a year and that it had no military value. But Major General Frank Parker, the dean of army poloists, pontificated that polo was in fact an *asset* to the army and gathered crack players around him wherever he was stationed.

Sooner or later the US Army would have to face up to the enormous task that loomed ahead, but in 1938 and 1939, polo was sufficient challenge for most.

Second Class Year

As war clouds gathered over Europe at the close of the 1930s, the thoughts of the Class of 1941 were focused on the summer of 1939. The prime thought in *every* cadet's mind for the first two years had been furlough, the only extended leave in the four years, which would be coming in the summer of 1939, at the end of Yearling Year. It would last 77 days. Besides being able to remember the exact number of days, Ben Spiller still recalls that he was "$300 out of debt and *that was money* in the summer of 1939." He went to both World Fairs—the one in New York and the one on Treasure Island in San Francisco—that summer, and he was at the Glen Island Casino the first time Glenn Miller played "In the Mood."

Suddenly, on September 1, 1939, as members of Black '41 returned to the banks of the Hudson after that 77-day furlough to begin their Second Class Year, Adolf Hitler launched his blitzkrieg (lightning war) against Poland. Two days after the invasion, Britain and France declared war. World War II had begun.

The German Wehrmacht (army) and Luftwaffe (air force), in a closely coordinated campaign, attacked Poland with 60 infantry divisions, 14 mechanized and motorized divisions, 3 mountain divisions, more than 4,000 planes, and thousands of tanks and armored cars. The Poles were able to mobilize less than one-third of that strength. Within weeks, Poland fell before the German onslaught.

The US Army of 1939 had been bolstered to 166,000, and there were three organized and six partially organized infantry divisions, but not one approached its combat complement. The two cavalry divisions were at less than half strength. There was not a single armored division in the army, and the total number of men in the scattered tank units was less than 1,500. The entire Air Corps consisted of 1,175 planes and 23,455 men to service, maintain, and fly them. Yet, although the Germans outnumbered the US Army seven to one, the United States had obligations that spanned tens of thousand of miles, from Alaska to Panama to the Philippines.

Although two troop increases, authorized during the summer and fall of 1939, would soon raise the active army at home and overseas to a personnel strength of 227,000, that size was still woefully inadequate to meet the army's commitments, much less to face an army the size of Germany's Wehrmacht.

For those at West Point, World War II came as a shock but not a surprise. "I don't believe I was surprised that World War II started," said Joe Gurfein, "but I remember being very disgusted with British Prime Minister Neville

Chamberlain's politically expedient, but militarily unrealistic, program of delay when an Allied offensive was called for."

Robert "Zeke" Edger observed, "Everyone was emotionally charged against Germany, but there was little or no consideration of military strategy which would influence our lives. Almost nobody thought the United States would be dragged into the war."

"Putting things in perspective, you have to remember that the United States was very isolationist at that time," said Andy Evans, "and there just was not a clamoring for educating cadets in the tactics and techniques of what was going on in Europe. There wasn't much support in the US Army in those days for the belief that aircraft could be used as artillery and could become important in support of the infantry front line."

The official USMA reaction to the war was impartial, but in the long run the war brought the USMA out of the era of horse cavalry and trench warfare and into the age of blitzkrieg, paratroopers, and electronics. After the war started, the cadets were finally allowed to have radios to listen to news.

West Point didn't intentionally isolate itself from the world. The pre-1939 ban on radios was really a ban on "entertainment" to guarantee study periods during the morning and evening. Cadets could read the *New York Times* and listen to radios in the class lounge between 3:00 P.M. and 5:30 P.M. Nonetheless, at West Point the immediate response to the war was one of complacency rather than panic.

"If it were not for the 'saber rattling' of Professor Colonel Herman Buekema," Bradish Smith recalled, "I think we might have regarded the war as merely 'news,' with little concern for its continuing effect on us. Colonel Buekema taught what in those days was called 'Economics, Government, and History.' It covered the whole waterfront. His astute reading of the situation in Nazi Germany before the war opened our eyes much more widely to the threat that Hitler was posing to the world."

To many, the Academy didn't take the war seriously at the start. "Our curriculum and training did not change," said Jack Murray. "For example, we spent untold hours on the back side of horses and none inside of a tank. Except for Colonel Buekema and a few officers in the Military History program, things went on as usual—just as they did in the rest of the country."

According to Pete Tanous, prior to 1939 the USMA perceived itself as a university in uniform, with parades, social affairs, and hops (dances). It was only after Hitler invaded Poland that the atmosphere started to change.

By and large, the cadets had their own concerns to occupy them. Paul Skowronek does not recall that the war made any difference in life at West Point. "Getting through each day was our first, and almost only, real concern. The outside world would have to solve its problems without help from us. Cadets had daily problems of their own."

"It was not until the draft was instituted in 1940 that most of us began thinking about involvement in earnest," said Ben Spiller.

Nonetheless, some of the cadets felt the importance of the events that were taking place in Europe. "When Poland was attacked, Polish radio played a Chopin polonaise each day to prove that they still held out," recalled Bill Hoge. "I remember the keen disappointment when one day we tuned to the proper short-wave channel and it was dead."

After Hitler's conquest of Poland, World War II evolved into the "phony war," a "sitzkrieg" characterized by the stalemate at the Maginot Line through the winter of 1939–40. The world became somewhat apathetic and unimpressed, and the people of the United States lost interest in the war.

On April 9, 1940, without warning, the German Army poured into Denmark and began landing in Norway. A month later, on May 10, Hitler struck Holland and Belgium. By the end of May, resistance in all four countries had been subdued, and four more national flags were superseded by the swastika. The British Expeditionary Force of 200,000 that had been confidently sent to the continent to save Holland and Belgium was surrounded at the French coastal port of Dunkirk. Had Hitler not halted his Wehrmacht to let Herman Göring's Luftwaffe try to destroy the troops from the air, most of the trained and combat-capable elements of the British army would have been captured on the beaches before a boatlift could rescue them.

On June 5, Hitler turned his eye to France, the nation that had held out against Germany for four years in World War I and that now sat smugly behind the "impregnable" line of fortresses named for André Maginot. When Hitler's legions—and the führer himself—strutted into Paris only 10 days later, the world was aghast.

"The conventional wisdom was that the Germans would not be able to crack the Maginot Line and defeat the French and British," said Horace Brown. "I was a little dubious myself, but I had a good friend who'd spent the summer of 1939 in France and Germany. His descriptions of the sloppy and lackadaisical French troops and the sharp and 'on the ball' German troops gave me cause to wonder."

Yet, at West Point, cadets were still using broomsticks to drill, even as Hitler's mechanized Wehrmacht transformed how war was waged.

As Wendell Knowles noted at the time, "There was a certain sense of dismay to see how little was being done by the nation—and the army—to prepare for our inevitable part in the conflict. However, cadet life was so full of our normal and busy routine that we still paid little attention to events in the outside world. Everyone was fed up with the French, especially about the Maginot Line, and the fact that they had surrendered so easily while the British fought on. The British were more highly respected. On the other hand, the mass of armor that the Germans had, their mobility and their firepower, were impressive to the cadets from a tactical point of view."

Among many of the cadets, there was admiration for the German army: They were, at least, tough, seasoned professionals. The abilities of officers such as German Field Marshal and tank tactician Erwin Rommel were impressive. Up until the early spring of 1941, many cadets didn't feel that they would be in a war with the Germans. "By the time we graduated," remembered George Pickett, "we were being mentally adjusted to the fact that the Germans were the enemy, but if we'd been told in the summer of 1940—right after the Germans had overrun France—many of the classmates would have said 'what the hell, the Germans just licked them fair and square.' "

Pickett added, "Nobody really worried about the Germans until the Battle of Britain, when it looked as though Operation Sea Lion [Hitler's codename for the invasion of Britain] would succeed and the Germans would conquer Merry Old England. The British had gun control, and all of a sudden they found that they had no way to defend themselves against the Germans, so they asked Americans for hunting rifles and old Springfields. By the fall of 1940, the average American got used to the idea that he had to help the British."

By the fall of 1940, it would become apparent that the Germans were the enemy.

"Safe to Fire, Captain Timothy"

On June 11, 1940, as the German Army moved within sight of the French capital, the West Point Class of 1940 had graduated and Black '41 had become the USMA's senior class. There was a joke that in his First Class summer, a cadet had more rank than Douglas MacArthur because all the other cadets were more or less intimidated by the First Classmen, who served as squad leaders and company commanders for the entire corps of cadets.

Every summer the senior class traveled to an old World War I–era reservation at Tobyhanna, Pennsylvania, a sleepy resort town in the Pocono Mountains, for three weeks of field artillery instruction. Tobyhanna was the closest place available for the firing of artillery, so cadets went there to practice shooting the old French 75mm cannons that were left over from World War I. The ammunition, too, dated back to World War I, and it left a lot to be desired. It didn't often misfire, but once was once too often. Therefore, a range safety officer always inspected the guns before announcing, "Safe to fire, Captain Timothy!" The regular Battery Commander, whose name was Captain Timothy, would then yell, "Safe to fire!" At this signal, the cadets would jerk the lanyards.

Despite the precautions, one morning, after the cadets had been shooting for some time, the Tobyhanna sheriff appeared at the artillery range. Somebody had put a round *off* the range. The Poconos were a popular resort area, and Black '41 had just shot the roof off the Tobyhanna icehouse in the middle of summer. The safety officer caught hell.

In the town of Tobyhanna, there was a place down by the railroad tracks called Shimco's Tavern that sold beer. It was, of course, off limits to cadets.

Nevertheless, more than a few of them made their way to Shimco's Tavern to, as George Pickett tells it, get liquored up. They then often sang a parody of a popular song about the tavern owner's daughter, whom they called Shimco Sue: "Shimco Sue, I'm sad and lonely."

One night, it was raining heavily as they started home. Antanacio "Tony" Chavez, a cadet from the Philippines, suddenly disappeared from between George Pickett and another cadet. All that was left was his campaign hat, floating on a puddle. But when they reached down to pick up the hat, they found Tony Chavez under it. He was only about 5'3" and had stepped into a rain-filled hole that literally was over his head.

A few nights later, after bedcheck, several cadets made plans for after-taps activities. Jim Sykes was called to the OIC house to answer a long-distance, person-to-person phone call, and while taking the call, Jim noted that Captain Joe Costello, the Tactical Officer in Charge, was in bed and ready to call it a day.

When Jim returned to the barracks, he announced, "The tac has turned in for the night."

All bedlam broke loose.

Pete Tanous, Demo Wilkinson, and Ralph Freese had been fortunate to have met some "debutantes" from Scranton. The young women were waiting in a parked car, well concealed in a wooded area adjacent to the barracks.

The three men dashed out of the barracks to the parked car. Wilkinson and Tanous, unable to decide who should sit in the front seat, decided to flip a coin. As the coin sailed upward, Pete called, "Heads I win, tails you win."

All of a sudden, a hand shot out of the darkness and caught the coin, and a booming voice announced, "I caught the coin, so you *both* lose, misters."

"Yes, sir, Captain Costello, we both lose," the chagrined cadets answered. Freese and the women quickly retreated into the darkness.

After leaving Tobyhanna, the First Classmen went on to the Army Air Corps research establishment at Wright Field near Dayton, Ohio. The man in charge of quarters at Wright Field answered the door and announced when a cadet had visitors. At first there were no visitors, but then one night George Pickett's name was called, and he found a beautiful young woman outside beside a gleaming Packard convertible, waiting for him. It turned out that a large number of young women came calling, and every one of them had been given the name of a cadet. Before long, the cadets had all been paired with a date and were off to a banquet and dance in downtown Dayton. It seems that the Army had set up an extensive social calendar for the cadets to prepare them for their duties as officers.

Although the summer was full of fun and pleasant memories, the cadets learned next to nothing from a tactical standpoint during their summer camp.

Nor were such lessons taught at West Point. The cadets had spent hundreds of hours on horseback, but classes seldom addressed specific tactical problems or how to coordinate a tank attack or infantry attack. As late as the spring of 1941, USMA taught the cadets horsemanship instead of teaching them about tanks.

From the standpoint of tactics, the only real training that had any impact on Black '41 came from Colonel James M. Gavin (Class of 1929), who was an instructor in the Department of Tactics. Gavin taught "sand table" classes in the Infantry Branch Instruction School in the spring of 1941, from which the cadets picked up technical aspects of fighting such as evaluating a tactical situation and issuing orders. Later, during the war, Gavin went on to serve as the Commander of the 82d Airborne Division, and he picked Jack Norton from the Class of 1941 as his operations officer. It was during Norton's First Class Year that the two had gotten to know one another.

Although the cadets received no instruction in tank tactics, the infantry tactics that they were taught came directly out of the Army's experience in World War I. Nor had the USMA faculty updated the infantry branch instruction, so many of the illustrations were of trench warfare in World War I battles. However, Colonel Gavin had researched the German campaigns of 1939 and 1940, and his classes discussed the new German tactics. Somehow Gavin had obtained copies of an English translation of Field Marshal Rommel's book *Infantry Attacks*, and he set up three-dimensional demonstrations of Rommel's tactics on the sand table so that the cadets could better understand them.

First Class Year

The adventures at Tobyhanna notwithstanding, the men of Black '41 survived the summer of Shimco Sue—the summer that France fell—and returned to West Point as aloof and regal as kings. The Yearlings of the Class of 1943 could make the pitiful Plebes of the Class of 1944 crawl like worms, but Black '41 could make the Yearlings themselves sweat.

Ham Avery, who had been called "General" by a special young woman in Baton Rouge when he'd gotten his appointment to West Point in 1936, discovered that rank did indeed have its pleasures. Notable among them were the attraction that First Classmen held for young ladies. Two particularly attractive women—one from Red Bank, New Jersey, the other from Elmira, New York—were especially helpful in contributing to Avery's polish as a gentleman, besides being "downright fun." When the young lady from Elmira came up the Hudson River one weekend on her father's yacht, Avery became an extra in the cast of a never-released Hollywood movie about West Point.

Whatever can be said about the evolution of the Academic and Tactical Departments, football was then, and is still, very important to West Point.

Even those who couldn't care less about the game got caught up in the annual crusade against the Navy goat. As cadets soon discovered, you don't have to love football to hate Navy.

When the men or Black '41 came to West Point in 1937, the USMA had a truly powerful team. Garrison Davidson (Class of 1927) coached it to a 7–2 season that included a 6–0 rout of Navy and an incredible 47–6 slaughter of St. John's. In 1938, William Wood (Class of 1925) replaced Davidson as coach, but the Army mule was still a powerful football symbol as Army trampled the Ivy League schools that made up much of its schedule and kicked the Navy goat back to Annapolis 14–7. Only the 18–20 squeaker against Columbia and the annual loss to Notre Dame (Army *always* seemed to lose to the Irish) marred Wood's first season.

Maybe it was part of the reason that they came to be known as Black '41, but as the men of the Class of 1941 reached maturity, Army's fortunes on the gridiron declined from bad to worse. Bad was 1939. Worse was 1940.

Army's ill-starred 1940 team included just eight First Classmen, and only four of these were in the starting lineup all season: Bill Gillis—now the team captain—at center, Joe Grygiel of Company K at offensive end, Jack Harris of Company A at tackle, and Joseph "Sloppy Joe" Weidner of Company D at guard.

The 185-pound, 6'1" Gillis, who had been eyed for his athletic prowess by Army's coaching staff since he arrived at West Point, had played on the varsity since he was a Yearling and now, in the final season, was an obvious choice for team captain. With his firm jaw, expansive manner, and fearless predilection for pranks, Gillis was remembered as a born leader, but he was also a good student. In addition to being the captain of the 1940 Army football team, Gillis would set two new Academy records during the spring of 1941, just before graduation, in the high and low hurdles. Yet Gillis was unaffected by his athletic awards and glory.

"I didn't like Gillis at first," Bill Hoge said later. "He was typically Texas: very loud and very boastful. But during First Class year I saw more of him, and I thought he was one of the greatest men I ever knew. Maybe he quieted down some after three years at the Academy. . . . Also, he didn't have to boast so much. He'd pretty well proved his points along the way."

Gillis and Hoge would slip out of barracks after taps and go into New York City, then sneak back in just before reveille. "It seems ridiculous now, but we were both a little stir-crazy at the time," Hoge laughed.

Even with Gillis's leadership, 1940 was a season right out of hell for Army fans. The 3–4–2 season in 1939, which included the dreadful 0–10 loss to Navy, was only a preview of the misery of 1940. The season began on October 5 with a game against Williams College in which Army prevailed 20–19. The following week, however, the team suffered its worst defeat in Army history as it fell to Cornell 45–0. Larry Greene, generally cursed with being Bill Gillis' backup at center, got to play in the Cornell game, but it was his only playing time of the season.

On October 16, with Herb Frawley of Company D playing his only game of the season, Lafayette clobbered Army 19–0, its first victory over the cadets since 1893.

The first weekend in November afforded Army the ominous prospect of facing Notre Dame at Yankee Stadium before a crowd of 76,000. Led by Gillis, Grygiel, Harris, and the inimitable Sloppy Joe, the mules blew Notre Dame off the statistic tables. They made 13 first downs to Notre Dame's 4, rushed 174 yards to 62, and completed six passes to one for the Irish. Notre Dame scored in the closing minutes of the first half, but by the closing minutes of the game, Army had outplayed the Irish and had pulled into scoring position. The stands were abuzz with talk of "one of the biggest upsets in football history" as Army marched to the Notre Dame 12 yard line.

Suddenly, the stars stopped shining and fortune no longer smiled. Four yards were lost on running plays, and two passes fell short of their intended receivers. The game ended Notre Dame 7, Army 0.

A week after the Notre Dame heartbreaker, Army met Brown at home for a disappointing 13–9 loss, and on November 16, Pennsylvania buried the mules 48–0.

The November 23 contest at Princeton saw Army at least put some numbers on the board in the cause of its 26–19 loss. These points, however, would be Army's last.

Bill Gillis and his men went to Philadelphia for the annual contest with the midshipmen on November 30, 1940, hoping for a miracle, but their prayers would remain unanswered. Navy scored on its first possession, but it was an even and bitterly fought game until the end of the third quarter, when once again the midshipmen put one across.

The game ended with a valiant Army effort as Jere Maupin ran back a punt 45 yards before he was run down. Army advanced 15 yards on two passes, but a third pass was intercepted with minutes left on the clock and the venerable mule went back to the Hudson scoreless for the second year in a row in the annual crusade.

The Hundredth Night Show was the penultimate event of cadet life, a last big blow-out to mark the milestone of the hundredth night before graduation.

In the lexicon of the cadets, the two months of dark, cold days that began at the end of Christmas leave were known as the "gloom period." To help make the gloom period more bearable, a tradition evolved—marking the end of the gloom with a satirical cadet variety show. Looking forward to the happy separation in June, this event bore the appellation "Hundredth Night

Show," although the program was actually held on the Saturday nearest to the hundredth night.

The Dialectic Society, the cadet organization charged with development of the annual show, typically started the school year with the grandest of intentions, but then, just as typically, found itself on January 2 with the show's script yet to be written.

Black '41's Lew Elder, as the society's president, charged his four would-be producers to "bring back a show from Christmas leave." So it was that in the depth of winter, four cadets closeted themselves at the Hotel Picadilly in midtown Manhattan to complete their mission. Representing the western contingent were Horace "Race" Foster of Company M and Peer deSilva of Company A, who were both from San Francisco. Representing the eastern establishment were two Company F men—Pete Tanous of New York and Bill Vaughn from Upper Darby, Pennsylvania. Their brainstorming sessions lasted all night and into the dawn, as the Picadilly's congenial manager kept them supplied with hot coffee, rolls, and danish. Classmates such as Dave Woods of Company A, who later became the show's director, and Johnny Redmon of Company H, who would soon design the costumes, stopped in to offer suggestions. Others drifted through just to give the four soon-to-be-famous playwrights a bad time.

The creative process began with a consensus that the previous Hundredth Night shows—including the Class of 1940's *Time, Tide & Assembly*—had "failed to express adequately a satire on cadet life." The days and nights at the Picadilly took their toll, but finally, a script emerged "from the debris."

Once the script had been delivered, the set designers started work on the sets, and Bill Gillis, whose talents on the gridiron were matched only by his ability on the ivories, was tapped to write and arrange most of the music.

Mike Aliotta of Company H was cast in the lead. However, Aliotta's role became too much for this "Sicilian reprobate" to handle alone, and Pete Tanous was brought in to play the part of Aliotta's alter ego.

For the cover of the program, Jack Lovett (Class of 1943) submitted a sketch of a seated, beautifully proportioned female in an abbreviated bikini. Unfortunately, Major Richardson—mentor, Officer-in-Charge, and decision-maker—nixed the illustration. Lovett suggested that he cover part of the woman's torso with a black ball. It did not mar the beauty of the cover, and if cadets held some copies up to the light, they could see the original sketch. In a sense, this censorship made the cover more provocative, and perhaps it provides another clue to the origin of the mysterious epithet "Black '41."

On March 8, 1941, the curtain went up for the first of two sold-out presentations of *Malum In Se*, the Hundredth Night Show. The plot centered on the establishment of a second West Point on a balmy south-sea island called Malihony, populated by pretty young women in sarongs intent on making the cadets happy. As Pete Tanous recalled, "Mike Greene, Bradish

Smith and Harry Jarvis—our 'female leads'—looked like a cover on *Mademoiselle*."

Back at gloomy West Point on the Hudson, however, the commandant heard of the idyllic conditions on Malihony and departed with his faithful assistant to whip the cadets back into shape. But when two island belles prevailed upon them to join the leisurely lifestyle on Malihony, the play ended happily ever after. Many of the real-life cadets in the audience, watching the show, didn't realize that they might never be quite as carefree again.

Sandy Nininger, while chairman of the Lecture Committee, did a great deal to brighten lecture periods by introducing a variety of entertainment. He scored a theatrical coup by presenting a Broadway show at the Academy. Academy staff were incredulous when Nininger first proposed the idea, but after he secured permission to do it, he was able to bring the then-current New York hit *Arsenic and Old Lace*, with Boris Karloff and the entire Broadway cast, up to West Point.

There was a serious side to First Class Year, too. This was a cadet's last year, a time when academics counted as they hadn't since the first semester of Plebe Year. This was the year when academic standing translated directly into one's eligibility for postwar assignments. The higher a cadet ranked when he graduated, the better the assignment he got in the Army. So, in the fall of 1940, the sense of excitement was tempered by a sense of how serious everything was becoming. With this came a change in the men's perception of many of the Academy's more immature rituals.

Not all of the men—and certainly fewer as time went on—relished the idea of hazing and arbitrary discipline. During his four years, Ted Reed's impression of USMA had deteriorated from one of great admiration and respect to one of considerable distaste for what he saw as a "behind-the-times outfit, run like a boy's school for the convenience of the authorities. I had attended college one year before going to USMA and resented a number of the indignities imposed upon us in the name of learning discipline," he said. "In fact, they were a detriment to learning during the academic year."

Nearly four years older than his youngest classmates, Ernie Durr was physically tough and agile, an expert at squash, and the mainstay of the "B" squad football team for three years. He also sang in the choir for four years, was a member of the glee club, and sang in barbershop quartets. A tall man with curly hair and dark, sad eyes, Ernie was an expert bridge player, and here, as in his schoolwork, his powers of concentration were remembered by his classmates as "almost forbidding."

"Ernie was good at almost everything," his classmate Paul Gray observed. "He was mature beyond his years, with a fine mind and a personality that built friendships and drew respect."

Ernie had become engaged before he entered West Point. He didn't have much social life except when his fiancée visited West Point—which was not often because she came from California. They were married immediately after his graduation. It was as an honor representative that he tried to change the honor system at West Point.

"Ernie was one of my best friends," said Bill Hoge. "He stood near the top of the class all the time and still had time to help his classmates who were having trouble academically. He was Company M's honor representative, and he took the job seriously."

By the time the Class of 1941 had reached its First Class Year, the Honor Code had become weighted down with a great many petty breaches of conduct and discipline that were considered to be "honor offenses." A cadet often committed unintentional honor violations, and when he realized that he had done so, he was supposed to report himself, whereupon he usually received five demerits for disciplinary offense and was then absolved of his honor offense.

Nearly all cadets tried to conduct themselves in accordance with the rules of the Honor Code, but the Honor Code had become so complicated and the interpretations had become so complex that it was like a civilian legal code. An unintentional violation often resulted from an esoteric interpretation of a minor provision of the code.

"This sad state of affairs resulted largely from laziness on the part of commissioned officers," said Bill Hoge. "They were quick to implement new policies and procedures that were difficult or impossible to enforce and then, to avoid the problem they had just made for themselves, they would decide that any violation would be an Honor offense. Since an Honor Code violation required dismissal from the Academy, a minor violation of a regulation made a cadet liable for the maximum penalty. It was a ridiculous situation that finally came to a head with the cheating scandal over the Electrical Engineering exam in the 1970s."

Ernie Durr had foreseen this debacle during his First Class Year and submitted a plan to the superintendent to remove a great many offenses from the Honor Code and instead have them covered by a Code of Duty. Penalties for a violation of the Code of Duty would be less severe than violations of the Honor Code, and the Code of Duty would cover all unintentional violations. Unfortunately, the chaos surrounding Pearl Harbor and the United States's entrance into World War II six months later resulted in Durr's proposal being shelved for over three decades, before it was adopted in the 1970s.

Graduation Time at Last

In May 1941, a few weeks before graduation, the Class of '41 was taken on a visit to Camp Rodman at the Aberdeen Proving Grounds in Maryland. While they were there, a chicken pox scare broke out and all of the cadets were stripped and examined.

Several, including Bill Hoge and George Pickett, had a "suspect rash," so they were isolated in a ward in the hospital when the rest of the class went back to West Point. Hoge's dad—a lieutenant colonel (later a general) under whom Pickett would serve in the Korean War—came to visit.

"I had known Bill Hoge for four years," Pickett said, "and would never have guessed that he was an Army brat."

Bill and George had no clothes with them except the uniforms they had worn to the hospital. By the time they were ready to be discharged, their white collars and cuffs were black with dirt, so they threw them away and returned to West Point without any—and got away with it. However, another classmate who had been with them in the hospital bought a sport shirt in the PX and wore it back to school instead of his dress coat. "He caught holy hell from his Tactical Officer," Pickett laughed.

Despite his casual attitude toward his classes during most of the preceding four years, academics had come fairly easily for Pete Crow, but his conduct on the eve of graduation almost resulted in dismissal. "Someone got the idea that of the 424 cadets to graduate, I ranked 431 in conduct. I always thought a little fun was in order," Pete said.

During summer camp at Tobyhanna, Crow had ridden a cow into Shimco's Tavern, with Dick Polk twisting her tail. Another time, he had a big trout on the line when reveille sounded. "I just couldn't horse him out," Crow complained. "As the last note of assembly sounded, I was pounding across the area, trout in hand. The Tactical Officer, Joe Costello, shouted out, 'You! Man!' However, he later relented when he realized how good that trout would be for breakfast."

During a break at branch training just before graduation, Crow spotted a big bass just off the rock where the cadets had assembled. He usually had a hook and line in his pocket when he was around water, so he caught a little frog, put him on the hook and tossed him out toward the bass. While he was listening to the instructor, who was explaining water systems, he felt a sharp tug on the line. Turning, he hauled in the big bass. Instruction came to an abrupt stop. Fortunately, the instructor was a reservist and liked bass.

However, the final straw—and one that earned Pete Crow confinement until graduation—occurred when a tactical officer found him asleep on the lawn in front of the headquarters of an ordnance depot when he should have been inside learning the ballistics of a 75mm artillery projectile.

A few days before graduation, Pete was summoned to the superintendent's office. He stood shaking at attention, wondering whether he might be denied graduation because of his pranks. In his quiet, gentle tone, Brigadier General Robert Eichelberger (Class of 1909) asked if Pete's family was coming up for graduation.

"No, sir. It's a long way for them to come," Crow lied. In fact, because he was in confinement, he had been too embarrassed to invite them.

Eichelberger then asked, "Do you have a girlfriend whom you would like to ask to the Graduation Hop?"

"Yes, sir," Crow answered promptly.

"You go call her and tell her that as soon as she reaches Grant Hall, you are released from confinement."

Crow left the superintendent's office mopping his brow, with a smile on his face. He thought back to the scene in his high school principal's office.

"Congressman Starnes, the principal, and my father could now brag just a little," he mused. "Someone from those Alabama hills had *made it*."

June Week, climaxing with graduation day itself, would end the year. Among the events of that spring were the publication of the West Point yearbook, the *1941 Howitzer*, and the Graduation Horse Show, the culmination of a year of training of mounts individually assigned to cadets who intended to select cavalry or artillery as their branch of service upon graduation. To prepare for the Graduation Horse Show, roughly 50 cadets had been given the opportunity in the fall of 1940 to draw lots for those horses which the Cavalry instructor considered the best available for cadet riding classes. During the following nine months, these cadets rode their chosen mounts in riding classes and on weekends.

Paul Skowronek had drawn a handsome horse, one of only two gray mounts in the group. Skowronek decided to keep this high-strung, part-Arabian horse called Jerry, notwithstanding the Cavalry instructor's caution that he might be difficult to train as a show jumper. For months it appeared to Skowronek that Jerry had indeed been a bad choice. He rushed jumps and was generally difficult to control. However, by the end of the year Jerry and Paul were unbeatable, and they easily won the indoor jumping competition held just before June Week to select the cadets who would participate in the Graduation Horse Show.

The Graduation Horse Show consisted of an outdoor course with several cross-country fences and about a dozen more spaced around the polo field. Jerry ran the entire course without a knockdown, winning the blue ribbon and a silver trophy donated by the US Cavalry Association. After commencement exercises, Skowronek's last stop on the West Point reservation was at the cavalry stables, where he and his June Week date said farewell to Jerry.

In addition to Skowronek's interest in horses, he was an avid skier, which led to his being tagged with the nickname "Skibunny." When Skowronek arrived at West Point, skiing was just becoming popular at the Academy, and it soon filled a large share of his recreational time. Skowronek was instrumental in getting the ski club started. As Mike Greene described it, "He just found a slope and started using it." Skowronek won the first downhill race to be held on the newly constructed ski trails. He was elected president of the cadet ski club and captain of the ski team. Greene also recalled that during Second Class Year, when he, Skowronek, and Chuck Willes were roommates, they lived on the first floor of the barracks. "In the middle of winter, once the lights were out, it was Paul's habit to dash outside into the main courtyard, strip naked, find a place where the plows had piled the snow up, and dive in. He found this invigorating."

"I can't believe it's here," Mike Greene said to his roommate on the morning of June 11, 1941. "Even if you had a rough time, or thought you had a rough time, no one really wants to admit that it's about to be over, that they're going to have to start all over again in a different place, that they'd been protected in a way, and now they were going to be out in the big, wide army, left on their own."

There was a lot of apprehension that morning, but also a sense of wonder about where they were going and what it was going to be like. Most of the cadets realized that the United States was likely to be at war soon and that they were all going to be involved. Getting their diplomas was really, in one sense, the beginning of a long journey into the unknown.

Mike and Larry Greene's father (Class of 1913), mother, and granddad (Class of 1887) were all at West Point for graduation. When Mike and Larry entered West Point in 1937, their grandfather had been in Europe, and he had subsequently remarried and moved to New Zealand, but he came back just to see Mike and Larry graduate.

The cadets were sworn in as second lieutenants, and in some cases small ceremonies were held so that fathers who were officers or West Point graduates could pin the gold bars on their sons.

Al Moody—"the studious, serious, and talented type who lived up to the rules and regulations," according to Leon Berger—graduated at the head of the Class of 1941, its valedictorian without a competitor. George Pittman, his roommate, recalled that Moody had superb mind. "He spent more time coaching classmates—and underclassmen too—than he did on his own studies. The final standings of the whole class show him *way* ahead of the number-two man for the four years. He must have established some kind of record in that."

Pete Tanous said that as a cadet, Moody was "the type you knew would succeed. He was not a bookworm, but a real capable person, with a captivating personality."

Moody was nicknamed "Ace," as much for his tennis play as his academic standing. He had also been a crack fencer, serving as captain of the team his First Class Year—cadet officer, instructor, and horseman. A warm and down-to-earth cadet, he was respected and admired by both faculty and fellow students.

Despite its administrative and tactical shortcomings, the United States Military Academy had succeeded in its education and training of the 424 men who went forth in the army on the afternoon of June 11, 1941. On that day West Point graduated some of the best and brightest officers to serve in the front lines in World War II.

The USMA had taken a group of people from diverse backgrounds and had trained them to work together as a group. The cadets had learned how to think like officers and act like officers; in short, to be leaders. They had, for four years, lived together under the same terms in an environment with very little outside influence. They had become a tight-knit group.

As George Pickett pointed out, there used to be an expression at West Point that the only friends you really have are your classmates, and if you don't have any friends in your class, you don't have any. This was the idea of military bonding. This kind of bonding would become all the more important in the tank battalions and regiments in which the men would later fight. "Men who had been through the same thing became a unit," said Pickett. "They were a group of people who were bonded. You knew that the guy next to you had done the same things you had under the same circumstances and that he could be depended upon to respond a certain way. If you had to put your life on the line, you could depend on him, and he on you."

3

Reporting for Duty

Choosing Assignments

As soon as the academics ended in May, the graduation standings for the West Point Class of 1941—the order of merit—was posted and the cadets were called into the auditorium in the East Academic Building. There, seated at an imposing desk placed on the dais, was a representative from the adjutant general of the army. He was flanked by a half dozen assistants and a huge blackboard, on which were listed all the regular army vacancies for second lieutenants that would be in effect on August 1, 1941, when the class was to report for duty.

In ordinary times, the new second lieutenants would have had until mid-September to report for their first assignment. But these were not ordinary times. The graduation leaves had been cut short; the Class of 1941 would get only six weeks.

The cadets filed in, took their places. One by one their names were called, and one by one they chose their assignments. When a cadet's name was called, he would stand and tell the adjutant general's representative where he wanted to be assigned from the remaining available assignments. The next man would then erase that entry from the board.

The top graduates from the West Point class usually selected the Corps of Engineers. The cadets from the bottom rungs of the class were usually assigned to the infantry and quartermasters, because there were more vacancies in those services.

In June 1941, the officer from the adjutant general's office started the selection process with the top-ranked cadet. Al Moody surprised everyone by choosing cavalry (although his unit eventually converted to armor, and Moody later went into a staff job). The rest of the top six graduating cadets—like *all* of the top six in 1940—chose the engineers, as did 17 of the top 25.

Ultimately, however, there would be room for only 49 men from the class in the Corps of Engineers.

When the picking was done, the largest block of men—over a quarter of the class—was in the infantry because it had the most openings. The next largest number of graduating officers—nearly 20 of them—went into the Coast Artillery, because the Coast Artillery had popular seacoast installations such as Fort Moultrie, Fort Monroe, and Fort Barancas at Pensacola. Some graduating cadets signed up for the Coast Artillery because they didn't want to be separated from their sweethearts and future wives. The Coast Artillery had a reputation for being a plush post. It was often said that the unofficial insignia of the Coast Artillery was a baby carriage. The Coast Artillery meant stability.

Nearly as many men—66 and 56, respectively—entered the Field Artillery and the Quartermaster Corps. The branches picked by the fewest men were the cavalry and Signal Corps. The 31 cadets who chose the cavalry had to take their tour on the Mexican border. Those interested in getting into armored units had to select infantry, cavalry, or artillery units. At the time the men of the Class of 1941 chose branches, there were only three or four tank battalions in the entire army.

By the end of the summer of 1941, however, the army had formed two armored divisions. Divisions are the basic building blocks of the army and contain 6,000 combat soldiers plus support personnel. Divisions generally had three regiments, which in turn contained three battalions, each comprising four companies. In the summer of 1941, there were roughly a dozen infantry divisions and two cavalry divisions. At that time, tank and artillery battalions were attached to the infantry divisions, so the creation of armored divisions was a real organizational milestone for the armor advocates. By the end of World War II, there were over 100 infantry divisions and 16 armored divisions. It would be the first time the army had more than a few tank battalions. (Those that existed were still largely equipped with old equipment left over from World War I).

Duty in the Philippines and Panama was traditionally very popular with all the branches because quarters and good food were cheap. A married officer in the Philippines or Panama could employ house boys and cooks at cheap rates. A married second lieutenant at Fort William McKinley in Manila could live better than a major at Fort Benning, Georgia, because of the difference in the cost of living. Beyond this, many men were drawn to the mystique of the Far East. The 15th Infantry was in Tientsin, China, for example, and that was the place men talked about wanting to go.

Perhaps the favorite assignment was the Air Corps, because an officer received pay and a half for flying. While many of the graduating cadets *wanted* to go into the Air Corps—indeed more than chose the infantry as a basic branch—they were required to choose another branch of the service as a "basic branch" so there would be a place for them if they failed flight training. As late as 1941, cadets could not select the Air Corps as a "basic line branch." Of the 188 men who picked the Air Corps as a secondary choice to their basic branch, 98 would pass flight training and never look back to the infantry or the quartermasters.

Some of those cadets determined to fly chose Signal Corps as a basic branch, since the Air Service originally had been part of the Signal Corps and in 1941 its officers were still flying in an observer status or working in communications. George Pittman was commissioned into the Signal Corps. His first choice for a future assignment was the island fortress of Corregidor that guarded Manila Bay in the Philippines. In the meantime, however, he was accepted for Air Corps training.

The Big, Wide World

The Class of 1941 had their cars loaded and their bags packed the moment that the command "Graduation class dismissed!" reverberated across The Plain. Everyone went off in different directions. Some went to get married. There were weddings one after the other—more than a dozen on graduation day itself and another dozen in the week that followed.

"I kept thinking about what I was really going to be doing when I got to be an officer," Mike Greene said to one of his friends. "What was going to be expected of me? I knew this was going to be different than just being a cadet. Was I really ready for this?"

The joy of graduation and the promise of graduation leave was soon clouded by the death of one of the graduates. On June 15, just four days after graduation, Dan Eaton was killed in a car crash in Michigan. As a First Classman, Dan Eaton had lived in the room next to George Pickett. "If you would have told me on graduation day that one of the guys out there would be dead within a week, and had asked who it was going to be, Dan would have been one of the last guys I would have picked," said Pickett. "Dan just wasn't the careless type, not a guy that was going to do something stupid."

Eaton and his mother were driving back to Michigan when somebody pulled out in front of their car. Eaton slammed on the brakes, but he couldn't stop fast enough.

When the class had entered West Point in July 1937, the covers of the major news magazines were graced with the images of society weddings. When the class graduated four years later, the same magazine covers were featuring dictators.

On May 27, 1941, Franklin Roosevelt had told a radio audience of 65 million, in a speech that was broadcast overseas in 14 languages, that the United States would stand up to Hitler: "I have tonight issued a proclamation

that an unlimited national emergency exists and requires the strengthening
of our defense to the extreme limit of our national power and authority."

Roosevelt went on to outline where and when the country would go to
war with Germany. The president firmly asserted the doctrine of the freedom
of the seas and made it clear that he intended to use "all additional measures
necessary" to assure the delivery of supplies to Great Britain. He also declared
an "unlimited national emergency," thus giving the administration somewhat
broader powers in dealing with the crisis.

In his speech to the nation, Roosevelt purposely avoided any mention
of Japan, whose brutal, decade-long adventure in China had cost hundreds
of thousands of Chinese lives. When asked why he had refrained from chas-
tising the Japanese, the president responded that he didn't want to anger the
peaceful faction that he felt existed among Japanese business leaders. Indeed,
Tokyo's stock market climbed when news of the speech reached Japan, and
government leaders there were able to conclude a neutrality treaty with the
Soviet Union in a reported 12 minutes.

In England, Londoners celebrated a three-week lull without a bomb falling on the city, but the Brits withdrew from Crete after a German airborne
attack that left the Wehrmacht in control of a rubble-strewn island. At sea,
however, the Royal Navy was gloating over its success against the great
German battleship *Bismarck*.

Just 11 days after the Class of 1941 motored down the Hudson into
graduation furloughs, Hitler hurled three million men into the arms of Mother
Russia in a brutal invasion that he called Operation Barbarossa and that the
Soviets would come to call the Great Patriotic War. By whatever name, the
invasion of the Soviet Union on June 22, 1941, changed the balance of power
in the world for the duration of the war and ultimately for the next half
century.

More and more during cadet barracks bull sessions—as well as in classroom expressions of concern—the men of the Class of 1941 had been asking
"Why wait and get clobbered by a victory-strengthened Nazi juggernaut?"

"I think the public at large, minus the usual pacifist outcries, began to
realize that we couldn't duck this one," Hamilton Avery said. He began to
realize that the infantry—his basic branch by virtue of class rank—didn't seem
so romantic. Fortunately, he had been accepted for training with the Air
Corps. "If I'm going to tangle with the ruthless, formidable German war
machine," he joked, "let it be from a relatively safe distance of, say, 30,000
feet." Later, in combat against the Luftwaffe, he would live to regret his
remarks.

"The world situation was dark indeed for the democracies in the short
run, but victory would be ours in the long run," said Ted Reed with a confidence not shown by many classmates. "We were well aware of the personal
hardships and dangers we would be called on to face, but in no position to

judge the depth of the threat to the nation. Under the circumstances, the second lieutenant's creed seemed most appropriate: 'Keep your mouth closed, your bowels open, stay away from headquarters, and never volunteer.' "

Said Horace Brown, "My perception was that we were heading toward war, but I was not oriented toward thinking about Japan as much as Germany."

When he mentioned to the commander of his field artillery detachment his interest in the Philippines as a station, the commander's reply was, "Don't do it unless you are prepared to eat fish for three or so years." The possibility of becoming a prisoner of war was something several of the cadets had discussed, and in the end Brown set his thoughts of the Philippines aside.

Most of the cadets, however, although anxious to take on the Germans— and the Japanese, if necessary—were still very inexperienced and had a lot to learn. When Edgar "Ted" Sliney's mother arrived to attend his graduation, Sliney proudly told her that he would be a second lieutenant. "I know that," she said. "But listen to the first sergeant!"

The Class of 1941 had learned a lot, but there was still a lot to learn, and the scope of it was yet to be fully manifested.

The Big, Wide Army

The 1941 army was a "citizen" army as opposed to the "dogface" army of the mid-1930s. By 1941, the army was many times larger and had moved out of its traditional battalion and regimental permanent posts to large training centers of division-sized camps of hastily built barracks and tents.

"We knew the army was being required to mobilize rapidly from a very small to a very large army," said Jim Laney, assigned to the basic artillery school at Fort Monroe, Virginia, "and that we must acquire a vast, detailed knowledge of our basic branch. At West Point we were given a general knowledge of all branches but not much detail of any."

Hugh Foster went to the 4th Signal Company of the 4th Infantry Division at Fort Benning, Georgia. He had received very little Signal Corps training at USMA and, through a fluke in orders, he never went to his branch school. "I went right to troop duty, rather ignorant of my job."

Many of the graduates in the artillery or cavalry divisions noticed that their training seemed to have little to do with training for a modern-day war. Harry Ellis had been assigned to the 14th Cavalry at Fort Riley, Kansas, and afterward he attended the horse course at the cavalry school there. "The army was still playing polo. Polo and horse shows were a part of army life on all the posts where cavalry troops were stationed. I played polo and rode in horse shows," Ellis remembered. "I had no real perception of the task facing the army. In the cavalry, there was little urgency or sense of the impending United States involvement in the war."

Wendell Knowles went to Battery B, 61st Field Artillery Battalion, 1st Cavalry Division at Fort Bliss, Texas. In addition to learning the duties of a second lieutenant trying to train a grossly undermanned unit, he and his unit had to exercise, groom, and otherwise care for 108 horses. Of the men assigned to Battery B, only about half were literate, and many were Hispanic or native American and spoke very limited English.

Knowles' perception in the autumn of 1941 was that the US Army—at least at Fort Bliss—did not have a very well-defined idea of what needed to be done to prepare for the future conflict. It was quite obvious that a great deal had to be accomplished, and soon, but little seemed to be getting done. Most of the 34 regular officers assigned to the 1st Cavalry Division Artillery had applied for transfer to other units. The artillery officers, who had previously fielded a good polo team, had transferred all of their polo ponies and equipment to teams in the cavalry regiments. The 1st Cavalry still had weekly horse shows and gymkhana contests, but polo was finally being phased out of the artillery.

Tuck Brown went to the army artillery school at Fort Sill, Oklahoma, for his three-month "basic course." This involved 310 hours of driving and draft (six horses and three new lieutenants pulling 75mm guns) plus 100 hours of horsemanship (riding up hills and sliding down bluffs) and three hours of communications. "This fully qualifies me to become a communications officer in a truck-drawn artillery battalion?" he asked himself when it was over. Little did he realize that three years later he would be leading mules up and down the steep sides of mountains in Italy.

After the field artillery basic officer course at Fort Sill, Horace Brown and Ted Reed went to Fort Jackson, South Carolina, to join the 8th Division Artillery in November 1941 for the Carolina Maneuvers. Their course at Fort Sill was the same as the other artillery courses, except that 100 hours of equitation and 55 hours of driving-in-draft were *added* to the course because Major General Danforth, chief of artillery, insisted that his regular officers be "familiar with horse artillery"!

Brown's first assignment was as aide to Brigadier General Sloan, the division artillery commander, and he later married the general's daughter, Lucia. "By some quirk, I became for a short time battery commander of headquarters battery, division artillery, for the division artillery training test," Brown mused later. "My communications chief, operations sergeant, and first sergeant explained to me that if I kept out of the way, we would pass. We did.

"From June until the Carolina Maneuvers on November 10, I was really still a student. My perception of the task facing me and the US Army came alive when I reported to the 8th Infantry Division and became aware of our troop and equipment status—a real jolt."

George Pickett requested and received an assignment to the First Armored Division. "When we were coming back from leave in 1939, General George Patton's 1st Mechanized Cavalry Brigade had bivouacked at West Point for the night and I got to see those guys, and there was just something about those tank boots that appealed to me. The same things that appealed to Patton, appealed to Pickett!"

Based at Fort Knox, Kentucky, the 1st Armored Division had been the first armored division in the army, but it was complemented almost immediately by the creation of a companion unit. The flamboyant George Patton was to be in command of the 2d Armored Division. For exponents of high-speed, mechanized warfare, those two divisions were the pride of the army.

Both Mike and Larry Greene—because their father had been a tanker in World War I—wanted to go into tank units. Since the tank divisions were under the control of the cavalry, they had picked cavalry so they could get into an armored infantry unit. Their father, who had been in the infantry, had always been furious that the cavalry had gotten control of the tanks, but ironically, the cavalry quota was full when Larry's time came to choose a branch, so he chose infantry, as did Mike. In order not to compete with each other, Larry had asked for a 1st Armored Division unit and Mike had asked for a 2d Armored Division unit.

All of those who were going into regular infantry had to go to the infantry basic course, while those going into armored units went directly to their units, bypassing the basic school. Larry Greene went to Fort Knox, where the 1st Armored Division was being organized, and Mike went to Fort Benning, where the 2d Armored Division was headquartered. When Mike arrived, he met several of his classmates who were there to attend the infantry school basic course prior to being assigned to a unit for duty.

The Greenes' father was Regimental Commander of the 67th Armored Regiment in the 2d Armored Division. Mike was assigned to a tank platoon in Company G of the 66th Armored Regiment in the same division.

On August 1, Mike Greene reported to First Lieutenant McCaskill, Company G, 66th Armored Regiment. McCaskill looked at the youthful Greene and muttered, "You must be kidding!" It was immediately apparent that the youthful West Point graduate was as green as a young sapling.

"I was told to report here and that I was going to be a platoon leader," Greene informed McCaskill.

"You're going to be a platoon leader in *this* company?"

"Yes, sir," Mike said.

"Just stand right where you are for a minute." He then went to the door and shouted, "Sergeant Plourde, come in here!"

The sergeant came in, and McCaskill said to him, "You see that lieutenant?"

"Yes, sir."

"Sergeant Plourde, this is your new platoon leader."

Greene recalled that the sergeant's face fell in despair.

"Now," Lieutenant McCaskill added, "what you do, Sergeant Plourde, is you take Lieutenant Greene to the motor pool and show him every one of our vehicles, and when he is thoroughly familiar with every one of them, you come back here and tell me that Lieutenant Greene is now fairly familiar with all of our vehicles."

Lieutenant McCaskill then told Mike Greene, "You'd better learn what the hell to do with your platoon, because we're getting ready. There's a war on you know, lieutenant."

"Yes, sir."

"Dis-missed!"

Mike Greene went to the motor pool and learned to drive just about everything—trucks, jeeps, everything but motorcycles. He soon discovered that at Fort Benning, as elsewhere in the army, there was a pervasive sense of urgency that transcended all the ranks.

Everyone also knew that the 1st and 2d Armored Divisions would be in the thick of things if there was a war, yet the divisions were improvising because they had little equipment, and what they did have wasn't in very good shape. For military exercises, they were still using jeeps to simulate tanks and trying to perform tank maneuvers with the jeeps.

The army had a long, long way to go if it was to be prepared to fight the Axis.

The Air Corps

The US Army Signal Corps had bought its first airplane from Orville and Wilbur Wright in 1907. Yet at the commencement of World War I, the army's aircraft strength of 266 airplanes of all types made the United States, in terms of airpower, the most ill-equipped major force in the world. During the war the strength of the US Army Air Service, founded in May 1918, grew to 7,889 airplanes and 195,023 personnel. After the war, though, the Air Service, like the rest of the army, declined sharply. In July 1926, the army put the Air Service on par with the Signal Corps and the Corps of Engineers by creating the US Army Air Corps, but the strength of the new corps did not keep pace with the appellation. It was not until after the men of the Class of 1941 were in their Plebe Year that the army started taking its Air Corps seriously.

When Henry H. "Hap" Arnold (Class of 1907) took over as Chief of the Air Corps on September 29, 1938, the situation began to improve, and by January 1939 America's Air Corps ranked sixth in size in the world.

Eight months later, the Germans put on an aerial display as the Luftwaffe stunned Poland with its brutal efficiency. The Luftwaffe was instrumental half a year later against Scandinavia and Western Europe as well. When France fell on June 22, 1940, and Germany began the Battle of Britain, the role of airpower in modern warfare was no longer in question. American attitudes toward it changed virtually overnight.

When Arnold took command, personnel strength of the Air Corps stood at 21,089. Two years later, that had more than doubled, and by December 1941 it had increased nearly seventeen-fold to 354,000. During the same period, the number of air bases more than quintupled, and training facilities proliferated as did (with generous new tax advantages) aircraft factories in the private sector.

On June 20, 1941, nine days after the Class of 1941 graduated, President Roosevelt signed the executive order under the War Powers Act of 1941 that permitted Army Chief of Staff General George Marshall to create the US Army Air Forces—an autonomous, if not independent, air arm—with Arnold as chief. Six months later, Arnold's position was upgraded to Commanding General USAAF, and he gained a seat on the Joint Chiefs of Staff and the Allied Combined Chiefs of Staff.

Tactical and air defense units in the continental United States were divided geographically into four numbered air forces, with the first and second in the northeast and northwest quadrants of the country and the third and fourth in the southwest and southeast. Separate commands, similar to the commands of today's US Air Force, were established to oversee such duties as training and maintenance, while the ferry command supervised delivery of lend-lease aircraft to Britain. Among the other subsidiary commands of the USAAF were regional commands in United States possessions in the Pacific. The Hawaiian Air Force was established in November 1940, and a year later the Far East Air Force was set up in the Commonwealth of the Philippines.

Over 150 men from the West Point Class of 1941 entered the US Army Air Forces for flight training in the autumn of 1941. Those who went through Gulf Coast training started at the Spartan School of Aeronautics in Tulsa, Oklahoma, for primary training with the old Fairchild PT-19 Cornell monoplanes, before being assigned to Randolph Field near San Antonio for basic training. Finally, the men destined to become fighter pilots would go on to nearby Ellington Field or Kelly Field for advanced training—and ultimately their solo flights—in North American AT-6 Texans. Those assigned to fly multiengined aircraft went through advanced training in the twin-engined, five-seat Cessna AT-8 and AT-17 Bobcats.

Among those from Black '41 assigned to Tulsa were Richard Kline, Wayne Rhynard, Andy Evans, Paul O'Brien, Jim Walker, Ben Mayo, Dick Polk, and Pete Crow.

Pete Crow and Dick Polk rented a little house in Tulsa, and as Pete said,

"I suspect we put evening entertainment ahead of flying. Even so, we made it through basic at Spartan. Soloing was worth all the drinks we had to buy for everyone."

Just as they all were about to leave Tulsa, Ben Mayo and Joyce Kirby were married. As had been the case with the evenings of preceding months, the ensuing party was a drunken affair. A carload of Black '41 classmates, straggling out from the party, crashed into a bank on a dead-end road. Pete Crow happened along as the police were pulling them out of the wreck and throwing them into a paddy wagon.

"What are you doing?" Crow yelled at the police. "Can't you see those boys are hurt?"

The next thing he knew, Pete himself was in the wagon with them. At the station, the police told him that he could come to get them in the morning when they had sobered up. This he dutifully did. They were released without being fined, but he was fined $25 for using "profane and threatening language in speaking to an officer of the law."

Pete Crow, along with the others, was sent to Randolph Field for flight training. Not long after he arrived, he happened to be looking through the section of the San Antonio newspaper that covered horse shows when a picture of a striking blond on horseback making a jump caught his eye. Her name was Tulah Dance. The next horse show found Crow in his best tweeds, smoking a pipe. Although Tulah was only 16, Pete found himself falling in love. While he was stationed at Tulsa, he saw her at every opportunity. On one such occasion, returning from her house after a few drinks, he found a tree in his way. "I spent the night with classmate Hank Boswell at Fort Sam Houston, while a wrecker hauled my car away."

When they finished the course at Tulsa, the '41 classmates went on to Little Rock, Arkansas, where Jim Walker was best man at Ben Mayo's wedding. Pete Crow continued to correspond with Tulah as much as possible.

Hamilton Avery was assigned to a civilian primary flying training unit at Jackson, Mississippi, a few hours' drive from his family's home in New Orleans, under the "baleful eye of regular Air Corps officers." As soon as he arrived in Jackson, however, Avery's world narrowed to the width of the cockpit of the PT-13 and the desk in the classroom. For the next four months, he was "too damned busy trying to keep from washing out as a pilot to think about anything else!"

Another classmate, Zeke Edger, after wrecking one airplane and having another catch fire during his training, decided "there was a message in this" and opted to go back to the Ordnance Corps. He later served as an ordnance company commander in Australia.

George Pittman was sent to the Gulf Coast for flight training. He and his bride—the daughter of the West Point Hospital executive officer—shared a house on Sylvan Beach in LaPorte, Texas, with classmate Clinton Ball and his wife. The two couples paid $35 a month rent for a completely furnished cottage with a fishing walk pier that extended 1,000 feet out into Galveston Bay. To save money, George and Clint took turns using one car while their wives drove the other car to visit friends, play cards, and participate in fashion shows. Since each couple was paying only $17.50 per month in rent, they had plenty of money for steak, liquor, and phone calls. At one point, Clint talked his cousin Lucille Ball, the soon-to-be-famous actress, into coming to visit them.

The Louisiana Maneuvers

In September 1941, the Army scheduled the Louisiana Maneuvers, the largest massing of troops in North America since the Civil War, and the largest by the US Army anywhere since World War I. The maneuvers were to pit General Walter Krueger's Third Army, with 270,000 men, against General Ben Lear's Second Army, with a strength of 130,000. While the size of these two forces was small in comparison with what the Germans had in the field, the exercise was a milestone for the army.

In assessing the importance of this exercise, General Dwight Eisenhower would later write that the beneficial results were incalculable. "It accustomed the troops to mass teamwork; it speeded up the process of eliminating the unfit; it brought to the specific attention of seniors certain of the younger men who were prepared to carry out the most difficult assignments in staff and command; and it developed among responsible leaders skill in the handling of large forces in the field. Practical experience was gained in large-scale field supply of troops. No comparable peacetime attempt had ever been made by Americans in the road movement of food, fuel, and ammunition from railhead and depot to a constantly shifting front line."

Another important by-product of the maneuvers was to destroy the distinction between draftees and regular army. When the men got back from maneuvers, they *all* looked like a bunch of old pros. Their attitudes were different. They were one team, one unit.

Despite the serious backdrop against which they were held, the maneuvers were not without their lighter moments. At one point during the Louisiana Maneuvers, some infantrymen found that a small backwoods bridge on a dirt road had collapsed. The platoon leader, Joe Gurfein of Black '41, was instructed to form a platoon of 25 men, get a bulldozer and other necessary equipment, and rebuild the bridge. That involved cutting down nearby trees, tying them together, and then laying them across a 20-foot span. Fifteen feet is the normal span for a wooden bridge. Typically, a 20-foot span required special concrete supports. However, Gurfein was able to design a bridge without supports, which would span the 20 feet.

The men cut the trees and built the bridge. When it was finished, Gurfein told them to drive across it. One of the men looked at him and said, "You designed the bridge, *you* drive across it."

Gurfein did—and it held.

Mike Greene went to the Louisiana Maneuvers as a platoon leader of Company G of the 66th Armored Regiment of the 2d Armored Division. When the division was activated, it didn't have any vehicles, and the men assigned to it found they would have to make do until, bit by bit, the tanks started coming in.

Paul Skowronek's first assignment had been to a horse cavalry regiment at Fort Riley, Kansas. His regiment was sent almost immediately to participate in the Louisiana Maneuvers. "These maneuvers created serious doubts about the survivability of horse cavalry on the modern battlefield," he said afterward, "doubts about my chosen branch of service."

"Horses were literally ridden to death trying to accomplish impossible flanking actions, or being sent on all-night marches to be in position for dawn attacks," he observed. "The high mobility of mechanized units and truck-mounted infantry riflemen showed that horse troops were no longer able to move fast enough or far enough to be effective. Trying to do so accounted for countless horses going lame or developing incapacitating saddle sores, particularly the pack horses that were carrying machine guns and ammunition."

By the conclusion of the Louisiana Maneuvers, it had become clear that tanks, trucks, and jeeps could run rings around horse troops and do it with fewer supply and maintenance problems. There were other questions about horses in modern combat that were never satisfactorily solved as well, such as what to do with a horse when the rider was in a foxhole taking shelter from aerial or artillery bombardment and how to get horses overseas when available shipping was already overloaded.

At the beginning of World War II, when the German panzer divisions almost effortlessly destroyed the Polish horse cavalry—reputed to be the best in Europe—with surprisingly few casualties, the survivability of horse-mounted troops on the modern battlefield was shown to be in serious doubt. Further successes of the German blitzkrieg had only confirmed that mobile firepower and shock action, which traditionally had defined the role of cavalry, would henceforth be carried out by tanks.

While the German Army in 1939 and 1940 was demonstrating in combat the tremendous advantages of its highly mobile tank forces, the shift to tanks had barely begun in the US Army. The last gasp of horse troops came during the Louisiana Maneuvers.

After the maneuvers, horse cavalry regiments rather quickly became cadres for rapidly forming armored divisions. When the war started, the army's only horse cavalry division, the 1st Cavalry Division, would leave its horses in Texas and go to fight as infantry in the Pacific.

Meanwhile, contrary to this logical trend, an all-black—except for its officers, who were all white—horse cavalry division had been activated in south Texas. It appeared to Paul Skowronek at the time to be an effort to use the army's excess inventory of horses and to find a place for thousands of conscripted blacks. "It was an ill-conceived plan and doomed to failure," he recalled. "After numerous training failures, the hopelessness of the project was recognized. The division was shipped to North Africa to be deactivated and broken into port battalions and service units. Most of the white officers moved into combat staff positions or transferred to other branches of service."

The first peacetime draft in American history was originally established by the Selective Service Act on October 16, 1940, with the goal of training 900,000 men for a period of one year. The first 10,000 were inducted on November 25. Every man between 18 and 45 was subject to military service, but only for one year.

Regular army officers were more than a little bit concerned with what would happen if the draftees would be allowed to leave after their training. George Pickett, for example, had four regular army soldiers and over 30 draftees in his company. If the draft in fact expired in October 1941 and the drafted soldiers went home, his entire company would consist of only one officer and 27 men. (Pickett was the only regular army officer in the company.) At the last minute, the draft was extended by Congress by a one-vote margin. "We came within one vote of having a whole army disintegrate in October 1941," Pickett noted later.

Pickett's first realization that the United States was going to get into the war—that war was imminent—came when he found himself in the midst of the Louisiana Maneuvers.

"Everybody took a break for about three weeks, the tanks were all broken down—we had to practically rebuild those things. We didn't even have guns, we had broom handles with 'machine gun' written on them. We'd set wooden models up in a field and they were supposed to be antitank guns. Can you imagine being on maneuvers with your tanks and your tracked infantry vehicles, when all of a sudden the enemy runs out waving broom handles and tells you that they've wiped you out with those broom handles? That was the type of army we had then."

An umpire had just declared Pickett's platoon out of action in the maneuvers when a major happened by. He said, "You all don't have to be in a maneuver any more, but use your time the best way you can, because in a couple of months we won't be in a maneuver, we're going to be in the real thing."

4

Infamy

America's Wake-Up Call

The first wave of 189 carrier-based aircraft from the battle fleet of Admiral Chuichi Nagumo was detected by United States forces at 7:02 on a sleepy Sunday morning; it was officially deemed not to be a threat. The planes arrived over the green hills of the Hawaiian Island of Oahu at 7:55 A.M. They were America's wake-up call.

By the time that Nagumo's warplanes were finished 110 minutes later, the United States had suffered the worst military disaster in its history. The navy lost 3,077 men, with another 876 wounded, while the army had 226 killed and 396 wounded. Nearly 100 aircraft were destroyed on the ground—most of the force present in the Hawaiian Islands.

The most strategically important loses were those suffered by the US Navy at the Pearl Harbor Naval Base. At 7:30 A.M. on December 7, 1941, the United States was one of two important naval powers in the Pacific Ocean. By 10:30 A.M., Japan was, for all practical purposes, the *only* naval power in the Pacific Ocean.

Three destroyers, three cruisers, and five auxiliary ships were lost, along with eight battleships. The USS *Arizona*, the USS *California*, and the USS *West Virginia* were sunk. The USS *Nevada* was "beached in a sinking condition," while the USS *Pennsylvania*, the USS *Tennessee*, and the USS *Maryland* were severely damaged and put out of commission indefinitely.

In 1941 the battleship was, as it had been since the turn of the century, the principal means by which a nation could project military power in the geopolitical ring. Battleships were what nuclear deterrents became in the latter half of the century. At 7:30 A.M. on December 7, 1941, the balance of power in the Pacific in terms of battleships was 11–9 in favor of Japan. At 10:30

A.M. the ratio was 11–1. Two entire battleship divisions had virtually evaporated from the US Navy roster.

In Washington, the War Plans Division of the War Department had assumed that the United States would enter World War II—perhaps in response to the sinking of a single ship, as had been the case in World War I—in approximately April 1942. However, events had overtaken the planners, and they were far worse than anyone's wildest projection. The United States was suddenly at war!

Fate dealt a violent blow that galvanized a nation, yet the full extent of the withering losses was so far-reaching and debilitating that the American public could not be told for nearly three years how bad, how *really* bad, the destruction at Pearl Harbor had been. The fact of the raid, however, became painfully clear to the nation almost instantly. Few people who were alive on that fateful Sunday will ever forget where they were when they learned of it. Even 50 years later, the mere mention of the words "Pearl Harbor" conjures up memories as vivid as any in American history.

As Walter Mather of the Class of 1941, who was in Hawaii and who survived the attack, reflected later, "All our way of life came to an end on December 7, 1941."

Peacetime Garrison Conditions

Although Mather did not then think it significant at the time, four engineers from the Class of 1941 had been assigned to Oahu and had arrived in early October 1941. Three of them—Curtis Chapman, Allen Jensen, and Mather himself—became company commanders that month in the newly activated 34th Combat Engineer Regiment. It was their first real assignment in the Army, just four months after graduation. The fourth classmate, Elmer Yates, became the assistant adjutant of the 3d Engineer Battalion of the 24th Infantry Division. All four classmates were initially stationed at Schofield Barracks, north of Honolulu, with work assignments spread throughout the island of Oahu.

There had been a series of alerts in November, but the last of them had been lifted in early December, and peacetime garrison conditions prevailed. On the night of December 6, Elmer and Nat Yates, along with Walter Mather and his wife Midge, were guests of the Chapmans at their rented beach house at Lanikai, near his company's work site on the Ulupau Peninsula close to Kaneohe Naval Air Station. Only later did the Chapmans learn that their landlord, a German named Bernard Otto Kuehn, was a spy for the Japanese.

After leaving the party, the Mathers stopped at a lookout point in the hills above Honolulu, from whose height on that beautiful, clear night they could see all of Pearl Harbor beneath them, the ships brilliantly illuminated from bow to stern for the Christmas season. It was around 1:00 A.M. At that same hour, the Japanese carriers were in their final preparations for launch of the attack.

The morning of December 7, Terry Chapman heard the snarl of airplane engines and ran to the window, where she watched with disbelief as five Japanese planes, flying at eye level along the Lanikai shore, attacked the seaplane base at Kaneohe, destroying all the aircraft except the two planes out on patrol and sending great columns of black smoke high in the air. Curt was gone, not to be seen again for two weeks.

The enemy planes were so close that Terry still remembers the pilots' leather helmets and goggles, the fixed landing gear, and the sensation that the two-man crews, seated one behind the other, were watching her as she spotted them.

That afternoon, Curt sent her a pistol and a note: "Practice shooting this into the ocean. You may need it." For the first time, Terry was genuinely afraid.

Radio silence was almost immediate, but at infrequent intervals the radio broke silence, reporting possible Japanese landings at specified sites. Yet no one seemed to know where those sites were.

During the night of December 7–8, amid the terrible confusion, Nat Yates and Midge Mather pulled close to one another for comfort in a hastily blacked-out room of the Mathers' rented beach cottage at Makuleia on the North Shore. They watched the flickering lights offshore, some approaching the beach and then receding. No one knew whether the Japanese were landing, had landed elsewhere, or had left. Not until the next morning did they learn that the lights were those of Hawaiian fishermen going about their business just as if nothing unusual had happened the day before.

In the wild first blackout of the war, with rumors and broadcasts of Japanese parachute and amphibious landings everywhere, Curt Chapman sat on the hood of his jeep with a flashlight as his company slowly drove in blacked-out conditions the 50 miles back to Schofield Barracks. There, over the next few days, his company as well as Jensen's and Mather's used trench diggers and bulldozers to fill Schofield Cemetery with hundreds of wooden coffins of navy and army dead. Everyone waited for the next attack. The state of army communications and intelligence was such that it would be several days before they realized the Japanese had withdrawn.

The only heavy earth-moving equipment on Oahu available to the military was that of the 34th Combat Engineer Regiment, so it was this equipment that became indispensable in those early days for the reconstruction and replacement of destroyed facilities. Weeks later, as more men and equipment arrived or were requisitioned from the civilian economy, the engineers mass-produced and placed hundreds of concrete pillboxes at almost every intersection in Honolulu and installed antiaircraft guns around Pearl Harbor using—in many cases—navy five-inch guns recovered from ships sunk in the attack.

Evacuations of all dependents, except those holding "indispensable" government positions, began shortly after the Pearl Harbor attack. By the middle of December 1941, many of the wives of the senior officers had left. Thus, on December 17, and until the following April, Elmer Yates and Walter Mather—though they were only second lieutenants—were jointly assigned to grandiose senior officer quarters on Engineer Loop at Schofield Barracks. The huge house was completely devoid of furniture, except for a massive dining room set and army cots. It was here that Nat and Midge learned to cook meals in coffee cans, often saying, "Oh, this tastes just like my mother's," and pressing the button under the dining room table to call the maid, who, of course, never came. They took hula lessons, practicing at night in the blacked-out living room, while waiting for Elmer and Walter who, every few weeks, would manage a few hours' visit with their brides.

This Is Real, This Is Not Orson Welles

"If you had asked me who the enemy would have been," George Pickett said later, "I would never in a thousand years have said the Japanese. We never heard a damn thing about the Japanese. There may have been a lot in the press, but we were in the woods for 90 days on the Louisiana and Carolina Maneuvers, and I don't think I saw a newspaper seven times in those 90 days."

The 1st Armored Division had gone straight from Louisiana to South Carolina and returned to Fort Knox on December 6, 1941. They were unaware that there would be war the next day, so two-thirds of the division had gone on leave before dark. The captain took off, and there were only three second lieutenants left for duty as senior Second Lieutenant Pickett became the de facto company commander.

The next morning, Sunday, December 7, Pickett's mother came up from Alabama in his car, which he had stored in Montgomery while he was on the maneuvers. It needed antifreeze, so they drove down to Elizabethtown, Kentucky, to find an open gas station. Across the street was a drugstore, and they decided to go over to have a sandwich and a soda.

When they walked into the drugstore, people were laughing and joking. Music was playing on the radio. Suddenly, the radio announcer said, "We interrupt this program for an important announcement. The Japanese have bombed Pearl Harbor . . ."

A man from the 27th Field Artillery at the soda fountain said, "Oh, hell. This is just another Orson Welles extravaganza," referring to Orson Welles' broadcast of H. G. Wells's *War of the Worlds*, in which Welles pretended to be reporting the "invasion" of New Jersey by creatures from Mars.

"This seems hard to explain today, but in 1941, the army was a very small, parochial thing," Pickett said later. "As we saw in 1990 with the Persian Gulf situation, this could never happen today, with television, and it could never have happened after the Korean War because the army started teaching people in the war colleges to be more politically savvy, to take more of an interest in world events and listen to what was going on, to vote and keep up with the issues. In 1941, there wasn't any Information and Education (I&E). You reported for duty, you did your job, and that was it."

Pickett said good-bye to his mother and reported back to the post almost at once. When he reached Fort Knox, officers and men, on leave after the maneuvers, were coming in by droves. Others had started calling in for instructions, and George Pickett had a hard time finding anybody with any authority who knew what to do. Colonel Virgil Bell, the regimental executive, told him, "I can't get any instructions from anybody. Pickett, if they call in and ask you what to do, tell them to give you a phone number or an address, and we'll get in touch with them if we want them back."

The News Sinks In

George Pittman and Jesse Thompson were at the house they shared, along with their wives, in San Antonio while in Basic Flight Training at nearby Randolph Field on December 7. At breakfast, George casually got up from the table and turned on the radio, catching the first news flash in the middle of the broadcast. The two men quickly dressed in their uniforms and reported for duty at Randolph Field.

Pete Crow was also at Randolph Field, visiting Tulah Dance, when the news came in. That evening was a somber one. Tulah's father, Dupre R. Dance (Class of 1922), who had resigned from the army in 1925, decided to volunteer. He was assigned to West Point as Assistant Professor of English, reporting in March 1942.

Lynn Lee was driving south on US Route 301 from Annapolis (where his younger brother was a Plebe) to Langley Field when the news came on his car radio. Lee was a bit scared. His transport battalion had been ordered overseas, and his father, Lamar Lee, a navy lieutenant, was on a ship tied up to the submarine base at Pearl Harbor. He later learned that his father had survived the attack.

Horace Brown had just completed the maneuvers in South Carolina and arrived at Fort Jackson, South Carolina, in early December. On Saturday night, December 6, he attended the Saint Barbara's Day Ball. (Saint Barbara is the patron saint of artillery.) "It was a great party and the occasion where I met Lucia, my future wife."

He called on her the next day, when Pearl Harbor was attacked. "I still called her even though the world had changed."

"Although we were not yet prepared for war, we had all been mentally geared that way and saw it as inevitable," Brown said. "The army was expanding. The National Guard and the Reserves had been called into service, and we all knew why. Pearl Harbor was just the fuse that set things off."

When Brown called Lucia on the morning of December 7, they talked about Pearl Harbor and he made arrangements to come over. She mentioned that her father, a regular army officer, had already been called to Fort Jackson for a meeting. The couple enjoyed a short dating period of about four months in a time of great stress, then eloped and were married in Lexington, South Carolina, on April 12, 1942.

December found Ham Avery in Augusta, Georgia, at basic flying school. On December 7, he and a group of officers and flying cadet trainees were being entertained at the home of their flying instructor, Captain Kammerer. Kammerer was a former fighter pilot in the German Luftwaffe who came to the United States to get away from the Nazis. He was such a good instrument pilot that all the American airlines wanted him, but he wanted to work for the Air Corps instead. The army had compromised a possible security situation by letting him train combatants.

The conversation at Captain Kammerer's turned lively after the radio announcement. "We have to survive the next few months of training to get our wings now!" one of the trainees said.

Nobody remembers Captain Kammerer's full name. He rarely discussed his background except while in the air, when he would say in a cool, quiet voice, "For that crass maneuver you would have been grounded by the Luftwaffe!" After the United States declared war on Germany, Kammerer was transferred to a less sensitive job, and Avery never saw him again.

Black '41 classmate Wendell Knowles was on duty with the 1st Cavalry Division Artillery at Fort Bliss when Pearl Harbor was attacked. His immediate reaction was one of surprise, consternation, and disbelief. "How could our armed forces in Hawaii have been caught so by surprise?" he asked rhetorically of one of his friends. "How could our national leadership have been so lacking in perception of what the Japanese could do?"

The division commander called an alert and ordered that the troops set up and occupy defensive positions in and around Fort Bliss in the event the Japanese extended their attack on Hawaii to West Texas. Knowles and his 61st Field Artillery Battalion dug in and set up a perimeter defense around Biggs Airfield, which then consisted of nothing more than a dirt strip and a balloon hanger. "We stayed there for several days until it became apparent that the Japs had no immediate interest in Fort Bliss," Knowles recalled. "At that time calmer heads prevailed, and we went back to our barracks and to horse exercises and grooming."

All across the country, soldiers were put on alert. Jack Murray, stationed at Fort Lewis, Washington, was having breakfast with his wife at the home of another officer when they heard the news about Pearl Harbor on the radio. An advisory was issued for all military personnel to return to their units. When Murray arrived at Fort Lewis, no one had received any orders. That evening, the battalion's previously scheduled formal dinner party was held at the Officers Club. Then at 10:30 P.M., everyone was confined to the post. Jack Murray found himself in the best-dressed battalion that ever went to war. Since most of the men lived off post, their wives had to go home for their uniforms.

"It looked a bit scary, but our president assured us that this was not our war, didn't he?" Murray quipped. "At least he got reelected on that theme in 1940. Professionally, we were better prepared to fight with Napoleon or Grant than against Germany. I can't recall a single person suggesting that we might be fighting Japan."

Paul Skowronek was attending the basic cavalry officers' course at Fort Riley, Kansas, and recovering from a broken leg at the post hospital. The fracture resulted from being kicked by a horse during advanced equitation (fence jumping) training. The declaration of war shortened Skowronek's convalescence, and he went back to classes on crutches.

Mike Greene, a platoon leader in Company G, 66th Regiment, 2d Armored Division, was at his parents' house when he heared the news. They had just returned from Mass and were sitting down when the news came on. "Before the attack on Pearl Harbor," Mike Greene reflected, "I hadn't thought about the Japanese at all. I had been following what they were doing to the Chinese but had never focused on the fact that *we* would ever be involved."

General Greene's immediate response was, "The 2d Armored Division will probably be one of the first units committed. We'd better get ready."

Both General Greene and his son thought they were going to get into the war rather quickly. In fact, Mike would not go overseas until 1944, and his father never did.

Tuck Brown was with the 38th Field Artillery Battalion of the 2d Infantry Division at Fort Sill, attending classes during the week and training during the weekends. At about 7:00 P.M. on December 7, the 2d Division artillery general called everyone together and said that the Japanese had attacked Pearl Harbor and that training would get serious now. Since the 38th Field Artillery hadn't had a day off for weeks, Brown wondered how the general could expect more of the men. He would soon find out.

On Monday, December 8, as the lamps in Washington, D.C., were lit for the long night of planning, Henry L. Stimson took up his pen to record his thoughts. He was already thinking ahead, not so much of the tactical disaster in Hawaii but of what Pearl Harbor would mean, in the strategic sense, for the crusade of defeating the Axis power.

"When the news first came that Japan had attacked us," Stimson wrote, "My first feeling was of relief that the indecision was over and that a crisis had come in a way which would unite all our people. This continued to be my dominant feeling in spite of the news of catastrophes which quickly developed. For I feel that this country united has practically nothing to fear, while the apathy and divisions stirred up by unpatriotic men have been hitherto very discouraging."

5

In the Philippines

The Pearl of the Orient

The Philippine Islands, snatched from the Spanish in 1898, were America's outpost in the Orient, the largest and most distant colony that the United States possessed during its short flirtation as a colonial power. Given limited autonomy in 1916, permitted a legislature of its own in 1934, and given commonwealth status in 1935, the Philippines were scheduled for full independence in 1946. Indeed, when he was governor general of the Philippines in 1928–29, Henry L. Stimson helped to draft the instruments that would lead to independence.

The defense of the Philippines was the responsibility of the US Army and, to a lesser extent, the US Navy, which had bases at Subic Bay and Cavite on Manila Bay. The Army retained sizable forces in the Philippines that were principally assigned to Luzon, the largest of the islands. Metropolitan Manila, a city of 684,000 in 1941 known as "The Pearl of the Orient" was less than half a day's drive from most of the US Army posts.

During most of the first four decades of the 20th century, the Philippines had been a prized assignment in the career of any Army officer. In the Philippines, a second lieutenant could live as well as a colonel could in the States, and a colonel could "live like a king."

Because the Philippines lie 7,000 miles and two weeks' sailing time from San Francisco, they were really a world apart from the United States. Thus, military commanders were, by necessity, given a level of autonomy that they would not have enjoyed at home. Life was easy, and there was no sense of military urgency.

One of the first things that Manuel L. Quezon did upon being inaugurated as the Commonwealth's first president in 1935 was to ask his old friend General Douglas MacArthur to help him form an army. MacArthur, who was retiring from his job as chief of staff of the US Army, agreed, and President Franklin Roosevelt officially designated him military advisor to the Commonwealth. When MacArthur arrived there, Quezon made him a field marshal in the new Philippine Army.

MacArthur did what he could within the context of the limited financial resources of the Philippines, and by July 1941, the Philippine Army numbered about 20,000, with provisions for training 75,000 more by March 1942. These forces were in turn backed up by 2,552 US Army troops and 7,921 Philippine Scouts—Philippine nationals serving in units organized and commanded by the US Army—although no formal operational relationship existed between the US Army and the Philippine Army.

For months, General MacArthur had hounded Army Chief of Staff George Marshall to take a more active interest in possible Japanese aggression; the Japanese occupation of Indochina in July of 1941 was the catalyst that started the wheels to slowly turn.

On July 26, the same day that Roosevelt froze Japanese assets in the United States, the War Department created the US Army Forces in the Far East (USAFFE), which consisted of the Philippine Department, those military forces of the Commonwealth ordered into active service for the period of the emergency, and "such other forces as might be assigned." At the same time, MacArthur was recalled to active duty, effective on July 26, with the rank of major general. With the establishment of USAFFE and the simultaneous induction of the military forces of the Commonwealth government, the two separate military establishments, which had existed in the Philippine Islands since 1935, were placed under one command for the first time.

On July 31, General Marshall approved a proposal by the War Department's War Plans Division to reinforce the islands' defense "in view of the possibility of an attack." The next day, MacArthur was informed that he would receive substantial reinforcements, and Marshall told his staff that "it was the policy of the United States to defend the Philippines."

During the next few months, MacArthur's requests for men received almost instant approval. The first reinforcements disembarked in the Philippines at the end of September with 108 M3 tanks and a coast artillery regiment with a dozen three-inch guns. By the middle of November, the War Department had approved MacArthur's requests for 1,512 officers, 21,015 men, and 25 nurses. Among the 1,512 officers were five men from the West Point Class of 1941: Ira Cheaney, Jr.; Robert Kramer; Sandy Nininger, Jr.; Robert Pierpont; and Hector Polla.

Bob Pierpont was the first of the class to arrive in the Philippines. Having graduated from the seventh instructor course at the engineer school at Fort Belvoir, Virginia, on September 6, he embarked for Manila on October 4 aboard the Army transport ship USS *Tasker H. Bliss.* Both Sandy Nininger and Hector Polla had been at the infantry school at Fort Benning, Georgia, when MacArthur's requests for troops were being processed, and their orders quickly came through.

Sandy Nininger had only three days at home with his parents before sailing. When they bade him goodbye, his mother said to him, "Sandy, suppose the war in Europe brings this country into it?" He took his mother in his arms and said, "Remember, you are the mother of a soldier."

Few of his classmates had perceived in Sandy the kind of firm dedication that is suggested in these words to his mother. They saw him as a quiet, self-contained boy who would rather take a good book and a canoe and go off by himself than join his classmates at a football game. Pete Tanous described him as "one of the mildest fellows in the class. He was the type of guy who wouldn't squeeze a grape or step on an ant."

Shy and studious, Sandy loved poetry more than aggressive, competitive sports, and it was hard for his classmates to picture him as an infantry officer, but there was another, radically different, side to Second Lieutenant Nininger that would soon show itself in the heat of battle.

Ira Cheaney's marriage to Lillian Jackson shortly before he finished training at the infantry school at Fort Benning drastically changed his priorities. When he was ordered to Manila, Cheaney knew that his wife could not accompany him. At the same time, his friend Tom Cleary was unmarried and had orders to Honolulu.

With a great deal of hesitation, Cheaney asked Cleary if he would exchange assignments. Being footloose at the time, Cleary readily accepted. They sent telegrams to the War Department and awaited a reply. Before the exchange could be approved, however, Cheaney went to Cleary again and stated, "It would be challenging fate to make the change and would be unfair to you." Without further discussion, Cheaney called Washington to cancel his request and have the original orders reinstated.

Pre-war Manila was an exciting place, full of the sweet smells of exotic spices and lumpia—a popular dish in the Philippines—sizzling in the open air cafes. Ships came and went in the harbor, loaded with copra, coconuts, and supplies for MacArthur's growing army. Somehow, the idea of impending war was not uppermost in the minds of the people of Manila. True, the *Manila Post*—carried by the Americans in their crisp, tropical whites who came and went at the Yacht Club—was filled with war news and discussions of MacArthur's military buildup, but no one really expected the Japanese to actually attack the Philippines. There may have been some scrapes between ships at sea, and planes may have been seen, but if a crisis came, it would come and go out of sight of Dewey Boulevard and downtown Manila, so there was really nothing to be too put out about.

Bob Pierpont was assigned to the 14th Engineer Regiment of the Philippine Scouts and placed in command of a platoon in B Company. It was here that Pierpont was reunited with his former classmate Bob Kramer, who was also assigned to the 14th Engineers. The 14th Engineers were part of the US Army Philippine Division, the backbone of MacArthur's ground forces and the tactical unit that in October 1941 contained practically all US Army and Philippine Scout personnel not assigned either to the Harbor Defense Artillery or the Air Corps.

It took two months from the end of July for the War Department to begin a significant flow of men and matériel to the Philippines, but after a sluggish start and a certain reluctance to divert resources from the potential war against Germany, things began to move well. In the four months after General MacArthur's assumption of command, the flow of men and supplies increased tremendously and all preparations for war had been pushed actively and aggressively. Time was running out rapidly, but at the end of November, many War Department planners still thought it would be several months before the Japanese struck. The month of April 1942 was commonly accepted as the critical date, and most plans were based on that date.

Upon their arrival in the Commonwealth, Sandy Nininger, Hector Polla, and Ira Cheaney were assigned to the 57th Infantry Regiment of the Philippine Scouts, one of only two infantry regiments in the Philippine Division. The 57th was assigned to Fort William McKinley, just south of Manila, which was the official headquarters of the Philippine Division, although the division was scattered throughout the Commonwealth—mostly on the island of Luzon—and operationally it functioned more as a group of regiments than a division. It was a matter of too much area to cover with too few troops.

While they were stationed at Fort William McKinley, Polla, Cheaney, and Nininger occasionally ran into Bob Kramer and Bob Pierpont because the 14th Engineers provided instructors for two engineer schools at the fort. However, aside from an occasional chance contact and a few beers in the Officers Club, the classmates had little contact in the increasingly hectic atmosphere of late November and early December 1941.

For Pierpont and others, working with or trying to train the Philippine Army (as opposed to the Philippine Scouts) presented numerous problems. In many units there was a serious language barrier, not only between the American instructors and the Filipinos, but also *among* the Filipinos. The enlisted men of one division may have spoken the Bicolanian dialect, while their Filipino officers usually spoke Tagalog, and the Yanks spoke neither. In the Visayas, the problem was even more complicated, since most of the officers were Tagalogs from central Luzon and the men spoke one or more of the many Visayan tongues. Transfers were made to alleviate the situation, but no real solution to the problem was ever found.

Meanwhile, Nininger, Polla, and Cheaney drilled their assigned platoons and prepared for possible war. Part of this preparation involved the backbreaking chore of storing supplies, and Fort William McKinley was designated as one of six supply depots. Yet despite the activity of preparation, there remained an air of optimism. The men were looking forward to their first Christmas in the tropics. The War Plans Division had given them until April—plenty of time—to prepare for war.

On Friday, December 5, 1941, Chief of Staff General George Marshall sat down at his desk in the War Department in Washington to draft a letter to General MacArthur in the Philippines, detailing what he had done—and was currently doing—to build up the US Army Forces in the Far East. "Reinforcements and equipment already approved," he said, "require over one million ship tons."

Fifty-five ships had already been obtained, and approximately 100,000 ship tons of supplies were enroute, with twice this amount ready for immediate shipment to ports of embarkation in the continental United States. Requests for equipment for the Philippine Army (except those for M1 rifles) had been approved, and other items were being sent as rapidly as they could be assembled and loaded on ships.

"Not only will you receive soon all your supporting light artillery (130 75mm guns), but 48 155mm howitzers and 24 155mm guns for corps and army artillery. I assure you," Marshall closed, "of my purpose to meet to the fullest extent possible your recommendations for personnel and equipment necessary to defend the Philippines."

It was late in the day when Marshall finished his letter. A cold, grim winter evening was closing in on Washington, and the weekend beckoned. He would mail his letter on Monday. Marshall opened his desk calendar to Monday, December 8, 1941, put down his pen, and turned off the lights.

War Comes to the Commonwealth

Across the international date line in the Philippines, the weekend passed quietly. As the first bombs were falling on Pearl Harbor, it was almost 2:30 A.M. on Monday morning in Manila.

Marine Corps Lieutenant Colonel William T. Clement was the officer on duty at the US Navy's Asiatic Fleet Headquarters in the Marsman Building in downtown Manila when the radio operator intercepted the message, "Air raid on Pearl Harbor. This is no drill." Recognizing the technique of the sender, the operator brought the message to Colonel Clement, and within half an hour it was in the hands of Admiral Thomas C. Hart, Commander of the US Navy's Asiatic Fleet.

MacArthur's chief of staff, General Richard K. Sutherland, got the news via a commercial shortwave broadcast from the United States, and he immediately contacted MacArthur on a specially installed hotline to MacArthur's apartment in the penthouse at the Manila Hotel. The two men concluded that a state of war now existed, and MacArthur told Sutherland to alert the troops to prepare for battle.

At Fort William McKinley, Nininger, Polla, and Cheaney received word of the Pearl Harbor attack before breakfast. By now the men were also receiving the disturbing news that the holocaust in Hawaii was not an isolated jab. The "day of infamy," as President Roosevelt would characterize it in his speech to Congress, had dawned in many places. This had happened despite the fact that few American or British military analysts believed the Japanese were either willing to undertake, or capable of, a multitude of coordinated strikes throughout the Far East.

The number and diversity of the Japanese attacks took the Allies completely by surprise. During the early morning hours of December 8, Japanese naval and air forces struck almost simultaneously at Kota Bharu in British Malaya, at Singapore, across the border in Thailand, and at Guam and Hong Kong. At Wake Island, landing operations had begun almost immediately. By dawn, Japanese forces were in possession of Shanghai. Even as the first bombs were dropping on Pearl Harbor, Japanese troops were on their way into Hong Kong. By the end of that day they were but a few miles from Kowloon. Attacks on the Philippines—once considered impossible, recently considered improbable—suddenly seemed imminent.

Words can scarcely capture the abrupt change of mood. A somber pall settled over the Commonwealth. There was little panic, but nervousness was endemic, and rumors spread like quicksilver. Had the Japanese landed? When would they attack?

The answer to the latter question would have already been answered had heavy fog not rolled over the island of Formosa (now Taiwan) in the middle of the night, closing the Japanese air bases situated there. The Japanese plan had called for an air strike against American installations in the Philippines to take place at dawn and to be coordinated with the other attacks throughout the Far East, but the fog intervened.

MacArthur's Air Corps commander, General Lewis H. Brereton, had 35 heavy bombers based at the sprawling complex at Clark Airfield 50 miles north of Manila. The big bombers easily had the range to assault the Japanese bases on Formosa, and Brereton advocated an immediate strike.

At 7:15 A.M., Sutherland told him to stand by for further orders, and a few minutes later he got a phone call from the Air Corps chief, General Henry H. "Hap" Arnold. Realizing that many of the aircraft based at Pearl Harbor had been destroyed on the ground, Arnold told Brereton to get his precious B-17s into the air to avoid a repeat disaster.

Since MacArthur's chief of staff still hadn't authorized an air raid against Formosa, Brereton ordered his B-17s into the air *sans* bombs. By 8:30, all the heavy bombers were airborne, along with many of the 107 (some sources say 90) Curtiss P-40 interceptors based in the Philippines. Their job was to patrol the skies and to watch and wait for the Japanese.

It was about 10:30 that morning when Brereton finally spoke to

MacArthur directly, and reportedly they agreed that an air strike would be mounted against Formosa by late afternoon. An hour later the B-17s were back on the ground being loaded with bombs, and the P-40s had returned to refuel, having found no Japanese aircraft to intercept.

Their timing couldn't have been worse. Just after noon, a flight of 27 Japanese bombers from Formosa came over Clark Field at 22,000 feet. They had been sighted by spotters in northern Luzon, but the teletype warning was never received at Clark. Had there been no fog in Formosa that morning and had the Japanese taken off on time, they would have reached Clark when the B-17s were on patrol. As it was, the entire armada was lined up on the apron amid truckloads of bombs and aviation fuel.

With several hours of warning having been afforded the Americans, the Japanese had expected the planes to have been dispersed. But there they were—sitting ducks on the ground. What the earlier attack on Pearl Harbor had done to the US Navy, the attack on Clark Airfield did to the US Army Air Forces. Only 55 Americans were killed, but half the B-17s—along with 56 fighters, 30 other aircraft, and much of the Clark Field complex—were destroyed.

The calamity at Clark Airfield was devastating to both the morale and the fighting capability of USAFFE. In a sense, it was a worse disaster than Pearl Harbor because it was an avoidable mistake, because the losses represented a higher proportion of available forces, and because it happened to American forces 5,000 miles farther from home.

No amount of postwar finger-pointing—and there was a great deal of that—can erase the sense of hopeless isolation that the men and women of the US Army in the Philippines felt as they sat down to dinner, a *late* dinner, on the night of December 8, 1941. Only the day before, they had eaten their Sunday ham with a sense of security. Now, in less than 24 hours, they had lost half of an already small air element and were hearing reports of smaller attacks that the Japanese were making all around them. The enemy seemed to be everywhere. Could the Philippines be the next to fall?

Over the next several days, the Japanese bombers struck American positions throughout the Philippines on a daily basis. Nichols and Neilson airfields were hit, as was the naval base at Cavite on Manila Bay. The US Navy's Asiatic Fleet was forced to withdraw. The surviving B-17s were moved south to Del Monte Field on the island of Mindanao, but by December 15 a shortage of parts and supplies forced them to be withdrawn to Australia, leaving a mere handful of obsolete fighter aircraft out of an air force that had totaled 277 planes barely a week before.

There was a sense of despair in the United States that more could not be done. Despite his having been one of the architects of the "defeat Germany first" policy, even Henry L. Stimson wrote in his diary on December 14 that the United States "could not give up the Philippines in that way; that we must make every effort at whatever risk to keep MacArthur's line open and that otherwise we would paralyze the activities of everybody in the Far East."

On December 19 an order was issued by which all the junior officers in the Philippines were advanced by one rank, with the pay and allowances attendant to such a promotion. For the second lieutenants of the Class of 1941, it meant trading their gold bars for the silver of first lieutenants, but frankly, there was too much else on their minds for the event to be accorded the kind of celebration it would have received a few weeks before.

The first Japanese forces had landed on December 10 on the north coast of Luzon and, meeting scant opposition, established a secure beachhead. Because northern Luzon is characterized by rugged mountains and Manila Bay was heavily fortified, it was assumed—correctly—by MacArthur's staff that the primary thrust would come at Lingayen Gulf on the west. A foothold there would allow an invader to approach Manila and the population centers on Luzon across a flat, level plain rather than over mountainous terrain.

Over the course of ten days, the Japanese made numerous small-scale landings throughout the Philippines, while MacArthur's forces braced for what they knew would be the main offensive. There was even an instance when the 21st Field Artillery of the Philippine Army shot up some Japanese reconnaissance ships in Lingayen Gulf, which in turn led to the widely circulated rumor that the Japanese invasion had been repulsed.

On December 22, Japanese General Masaharu Homma's 14th Army put ashore at Lingayen with 43,100 men. The initial landings were met with stiff resistance, but Philippine and American forces were too thin to prevent the Japanese from consolidating their beachhead within 24 hours. The 57th Infantry was not among the units assigned to the North Luzon Force that was to oppose the Lingayen landings, so Sandy Nininger, Hector Polla, and Ira Cheaney listened to the radio and talked to the truck drivers who were hauling supplies and bringing out the wounded. They waited, listening to the reports, left to speculate about the rumors.

Sandy Nininger's last communication with home had been a radiogram to his parents sent on December 19, 1941, and picked up by an amateur station in San Diego, California. The message said "Well, Merry Christmas. Sandy."

By Christmas Eve, the Japanese 14th Army had begun its relentless march south toward Manila. In the capital, as at Fort William McKinley, a queasy sense of being encircled began to prevail.

MacArthur realized that his meager forces, having been unable to stop the Lingayen landings, would be unable to prevent the Japanese from capturing Manila. He decided to implement War Plan Orange, a pre-war contingency plan that called for a holding action, followed by a declaration that Manila was an open city and then a complete withdrawal to defensive positions on the Bataan peninsula to the west, across Manila Bay from the capital. MacArthur had pre-positioned enough supplies on Bataan and the adjacent island fortress of Corregidor to sustain his force for six months. This, War Plan Orange assumed, would allow enough time for a rescue party to be sent out to the Philippines from the United States.

War Plan Orange went into effect on Christmas Eve, and MacArthur began withdrawing his staff to Corregidor. Simultaneously, two Philippine Army Divisions went to Bataan to begin preparation for the defensive actions there.

The North Luzon Force, defending the 120 miles that separated Lingayen Gulf from Manila, based its plan of withdrawal on five delaying positions or defensive lines that had been reconnoitered before the war, and that were separated by the distance that could be traveled in one night. Each line took advantage of natural terrain features such as rivers, swamps, and high ground. Each would be held for as long as possible to buy enough time to organize a successful defense of Bataan.

Christmas passed with little notice as the Japanese 14th Army overwhelmed the First North Luzon Force defensive line. There was a nervous attempt at Christmas dinner by the 57th Infantry, but it was eaten on the run as the last preparations were being made to move everything to Bataan. In Manila, department store windows were lavishly decorated with tinsel and piles of packages, but an overall sense of foreboding pervaded the city. Men in red and white Santa Claus outfits presented a jarring counterpoint to the sandbagged air raid bunkers.

By the day after Christmas, the headquarters of the 57th Infantry had been relocated to Bataan, and Manila became an open city. The bombing stopped, and an eerie sense of peace prevailed after the lights came back on.

New Year's Eve was a bizarre spectacle as nightclubs, hotels, and cabarets became jammed with thousands of revelers dressed in evening gowns and tuxedos who danced and drank the night away in an almost orgiastic frenzy. As the night wore on, bartenders adopted what they called a "Scotched earth policy," which involved breaking all unconsumed bottles of booze to prevent them from falling into the hands of the onrushing Japanese forces.

As 1941 slipped away at midnight, General Homma's 14th Army was practically within earshot of the capital. The North Luzon Force abandoned their fourth defensive line on December 27, and the fifth and final line was broached on New Year's Eve.

At four o'clock in the afternoon on January 2, 1942, General Homma finally entered the city. The Commonwealth government was taken over the following day as Japanese troops, assisted by Japanese civilians who had been living in Manila before the war, fanned out to round up American and British nationals still in the city.

Across Manila Bay on Bataan and Corregidor, the men of the 57th Infantry dug in to await the reinforcements that they were certain would soon arrive from the United States. MacArthur was already planning how he would integrate the expeditionary force into the Bataan force when it arrived. The men had abundant supplies, a good defensive position, and help was surely on the way.

Digging In

No rumor was more widespread during those early days on Bataan then the phrase "help is on the way." Everyone heard it. Everyone repeated it. And *almost* everyone believed it.

"Help is on the way from the United States," General MacArthur had said. "Thousands of troops and hundreds of planes are being dispatched. The exact time of arrival of reinforcements is unknown, as they will have to fight their way through."

Declaring that no further retreat was possible, he asserted that "our supplies are ample" and that it was imperative to hold until aid arrived. Though the message carefully stated that the date of arrival was not known, men eagerly hoped that it would come soon.

Lieutenant Henry Lee of the Philippine Division expressed the mood of the men when he wrote: "MacArthur's promise is on everyone's mind. The time is secret but I can say that swift relief ships are on the way. Thousands of men and hundreds of planes—back in Manila before the rains!"

Back in Washington, serious efforts to reinforce the Philippines, which had begun in earnest in early fall, had now ceased. The Japanese naval blockade around the Philippines was so complete that only submarines could slip through. The only way that a convoy of supply ships could penetrate it would be with an escort of a battleship division, and the backbone of the Pacific Fleet's battleship force had been sunk at Pearl Harbor. The real impact of the disastrous losses at Pearl Harbor was so far-reaching that its true scope remained classified. There were no ships to break the Japanese blockade of the Philippines. Help was *not* on the way, nor would it be any time soon.

Several attempts were made to get freighters in from Australia, and three did arrive in the southern Philippines at Cebu and Mindanao, but except for a handful of barges that managed to sneak through the blockade in the first six weeks of the siege, the men on Bataan were essentially on their own. Submarines were able to get through, but they were not designed as supply ships, and throughout the siege they brought in a total of only 53 tons, enough food to feed a single meal to 60 percent of the defenders of Bataan.

Despite its isolation, Bataan provided a particularly good defensive position. A peninsula 25 miles long and 12 to 20 miles wide, it has virtually impenetrable mountains in the center; any invader would have to approach it by way of the relatively level terrain on the east or west coasts of the peninsula. Because these approaches were narrow and thickly jungled, they were ideally suited to a defender who lacked air support. In this sense, it was not unlike the triple-canopy jungle that so favored the Vietcong a quarter century later.

The defenders divided themselves into two corps areas—one to defend each of the two coastal approaches—and dug in. The first, or I Corps, guarded the west coast facing the South China Sea, with about 22,500 men. The second, or II Corps, with 25,000 men, guarded the east coast facing Manila Bay.

Six divisions were assigned to II Corps, with three of them committed to the main battle line—named Abucay after a village in the region—a line that paralleled the meandering, mosquito-infested Balantay and Calaguiman rivers. These divisions were the 52d and 41st Divisions of the Philippine Army, and the 57th Infantry, commanded by Colonel George S. Clarke. After the frustration of having to watch the war from afar, Cheaney, Nininger, and Polla at last had their chance to fight.

On January 7, the 57th Infantry dug in at the village of Mabatang on the only main road on the east side of the Bataan Peninsula. At night they could look across Manila Bay and see the lights of the capital and of Fort William McKinley. Manila looked strangely peaceful. It was hard to imagine the Pearl of the Orient was occupied by the Japanese.

Things were oddly quiet that first week on Bataan as they waited for the assault that didn't come. "Maybe they're going to let us sit here and rot," one man said hopefully.

He was quickly rebuffed. "They can't do that. The guns on Corregidor control the entrance to Manila Bay. The Japs now own the best harbor in the Far East, but they can't use the damned thing as long as we're here."

The troops were on full rations. The division had brought 10 to 25 days' supply of food with them, and under War Plan Orange, six months supply for 43,000 men had been pre-positioned on Bataan. There was, for example, nearly three million pounds of canned salmon on Bataan. There also had been some serious foul-ups. The transfer of rice from one province to another was prohibited by Commonwealth regulations, so bureaucratic red tape had denied the men access to a half-million tons of rice stored at the Commonwealth warehouse at Cabanantuan.

On January 9, the men of the 57th Infantry had their lunch—canned salmon and settled down in the muggy heat of the afternoon to oversee the road that crossed the Calaguiman River north of Mabatang and to wait. Flies buzzed about. Men joked and even dozed in their foxholes. Several Japanese patrols had been sighted over the previous few days, but there was still no sign of a major attack. The 57th Infantry, both the US Army officers and their Philippine Scouts, were as ready as they would ever be.

At 3:00 P.M., there was a sudden roar like thunder and the earth shook with the explosion of 75mm artillery shells. The II Corps line was being attacked, and the road guarded by the 57th Infantry was the prime objective.

General Akira Nara had planned his attack on the assumption that the Americans and Filipinos had been exhausted by their withdrawal into Bataan and that resistance would be weak. He thought the whole campaign would be over in a couple of weeks.

Nara sent Colonel Takeo Imai to the 141st Infantry Regiment to capture Mabatang, but Imai ran into a withering fusillade by the 24th Field Artillery supporting the 57th Infantry. Imai never even made it to the Calaguiman River.

The next day, Imai's 141st Regiment struck again. The men at the 57th Infantry outpost at Samal succeeded in decimating the lead units of the advance before falling back to the Abucay line on the Calaguiman River. The Japanese kept up the pressure throughout the night, but it was not until midday on January 11 that the first of Imai's troops managed to cross the Calaguiman.

Around 11 o'clock that night, a battalion of Imai's 141st Infantry appeared 150 yards away on the other side of a sugar cane field from the 57th Infantry's line. Earlier, many of the cane fields had been burned and cleared to open fields of fire for heavy machine guns, but the 57th Infantry had decided against clearing this particular field under the assumption that the 75mm guns of the 24th Artillery could knock out any infantry that attempted to broach it.

The cane stalks gleamed silver in the moonlight as the 57th Infantry sentries spotted the Japanese soldiers rippling through the field. Moments later, the cane field erupted in a blinding sheet of fire and smoke as the 24th Artillery opened fire, filling the air with the smell of gun powder and cordite and the stench of burning cane. But the Japanese infantry charged the Americans in a horrifying human wave. Hundreds of screaming Japanese ran headlong into the American and Filipino perimeter like madmen.

It was the 57th Infantry's first banzai charge. It would not be its last.

The first wave was cut down in the face of intense .50 caliber machine gun fire, but another wave took its place, and then another. Human beings who had once knelt quietly on tatami mats amid shoji screens to drink tea in ceremonies precisely prescribed by centuries of tradition now were heaps of mutilated corpses. Men who had, a few months before, driven down the Hudson River Valley in '34 Chevys while listening to the exploits of the Red Sox and the Yankees on the radio were now squeezing the triggers of M1 Garands until the barrels were too hot to touch.

Company I of the 57th Infantry was the first to be overwhelmed, at about midnight. Japanese infantrymen who had been an unreal spectacle moments before were now in the trenches, bayoneting Americans and Filipinos in a frenzy.

On Company I's right, Company K began to crumble, but Company L, the battalion reserve, arrived in relief, and the position held. Colonel Clarke sent in more reserves; the line stabilized, and the Japanese withdrew.

In the first battle of the Bataan campaign, the 57th Infantry had taken the full force of the attack, and, although it had suffered gravely—losing Company I—it had held the line.

A few hours later, the sky over Manila began to grow light. The morning star glowed brightly as the others faded, and the horizon grew bloody red in anticipation of the dawn. The exhausted men of the 57th Infantry, many of whom had now been without sleep for 24 hours, could see the awful carnage before them. The cane field was charred and blackened, and more than 300 Japanese bodies lay about on the ground and draped over the barbed wire.

In the dawn of January 12, 1942, company commanders realized that, while the main brunt of the assault had been repulsed, the Japanese had infiltrated the line where Company I had been overrun. There were dozens of Japanese snipers hidden behind the lines and perhaps a company-strength Japanese unit in Company I's sector as well. If the line was to hold, this sector had to be retaken. An assessment of the size of the Japanese force was needed.

Colonel Clarke came forward to discuss the problem with Major Yeager when Sandy Nininger of Company A approached Yeager.

"Fred, I know of a good approach," Nininger told the major, pointing to an irrigation ditch. "That leads into the area. Give me ten good men and I'll try to pick up enough information so that we can figure out a plan of counterattack."

Yeager agreed, and Nininger set out with a patrol.

Colonel Clarke later wrote, "His attitude struck me as a soldier who at last was doing the job he had been trained to do . . . Sandy received permission to go forward in the 3d Battalion sector. He was loaded down with grenades and with a Garand rifle slung over his shoulder. He carried under his arm a Japanese 'Tommy gun' . . . Reports of his action . . . were reported to me by the company commanders of L and M companies, as well as other corroborated reports from men in Company K.

"Sandy shot his first Jap out of a tree, and as the body fell at his feet he was so excited he stood up in the face of terrific rifle fire and yelled like a school boy . . . He threw grenade after grenade. Men of Company K counted 20 Japs killed by his grenades. Our counterattack was succeeding, and their artillery laid down a fearful barrage. Many reports of further action by Sandy were made by the second in command of the 2d Battalion making the counterattack to regain Company K's position. Sandy apparently had used up all his ammunition . . ."

Nininger was wounded and staggering from loss of blood when three Japanese charged toward him with bayonets. He killed all three before falling at last from loss of blood, exhaustion, and weakness.

As Colonel Clarke said of Nininger in a letter to Sandy's father, "I cannot tell you how many of the enemy Sandy accounted for, but this I will say: his personal actions at this particular time cannot possibly be evaluated. Suffice it to say his actions acted like a tonic on the men around him, and added greatly to the success of our counterattack . . . He exemplified 'Duty, Honor, Country,' and reflected great credit on his regiment, his alma mater, the army, and his country."

And so Sandy Nininger became the first of the West Point Class of 1941 to die in battle in World War II. Major Yeager later wrote of him, "He wanted to be regarded, above all other considerations, as a man fulfilling West Point's guiding motto of Duty, Honor, Country. In the evenings Alex [as Yeager called Nininger] and I would usually reminisce about old times. Of all his cadet activities, the one he must have enjoyed more than all the others was the Lecture Committee. He used to tell me of the various celebrities who came to speak at West Point—of his meeting with them—of his impressions of them . . . this seemed to be one of his underlying pleasures . . .

"The whole defense position of all units then fighting on Bataan would have been made to collapse, unless prompt measures were taken to locate the exact position of the enemy, and the strength of the Nips, so that immediate measures could be taken to eliminate them.

"It is my honest opinion that had not this counterattack, based on Alex's findings, been made successfully, the entire Bataan campaign would have ended in January instead of three months later. These three months, I believe, saved Australia and enabled us to end the war many months before it otherwise would have. In my own mind, [Sandy Nininger] will always be a shining example of what an officer and man should be. He was the most fearless and most courageous officer or soldier I have ever seen."

Sandy Nininger, who became the first soldier in World War II to receive the Congressional Medal of Honor, served not only as an inspiration to his classmates and other troops on Bataan, but—as word of the events of January 12 reached the United States via shortwave—to Americans preparing to meet the Axis throughout the world.

The tenacity of the 57th Infantry's heroic counterattack forced Colonel Imai to completely rethink his battle plan. An easy dash down the highway through Mabatang and Abucay was clearly out of the question for the time being, so he began probing west toward other parts of the II Corps line for a possible breakthrough. Even as he did so, the 57th Infantry became part of a counteroffensive that pushed the Japanese back far enough on the east that by January 15 their advance along the coast had been completely stalled.

Despite this setback, Imai managed to overwhelm, or flank, the Philippine 51st and 53rd divisions to the west and, on January 23, it became necessary to withdraw from the Abucay line.

It had been nearly a month since Christmas Eve. Hector Polla and Ira Cheaney remembered the breathless rush of the yuletide season, a push to move supplies—as much as possible, as quickly as possible—from Fort William McKinley to Bataan. Hector Polla and Ira Cheaney had seen death, they had smelled death, and they had seen one of their own classmates fall. They had rapidly grown to manhood leading their platoons against the enemy in the cane fields and jungles of Bataan.

There had been a certain optimism when the 57th Infantry had dug in along the Abucay line, a sense that it could hold the Japanese there. "Defensive positions are always favored in any battle," they'd learned at West Point. Now, as they trudged south toward Pilar, they tried to suppress a feeling of despair.

A new line was formed at Pilar. The only road across Bataan intersected the coast road there to traverse the Bagac on the west coast, forming a natural line. Would the Pilar-Bagac line hold? What if it didn't? Where would the next line be? Help was on the way, though. Wasn't it?

Bataan had been ". . . saved for another day," wrote Lieutenant Henry Lee at the time in a poem. "Saved for hunger and wounds and heat. / For slow exhaustion and grim defeat. / For a wasted hope and sure defeat." Lee later died in Japanese captivity, but his notebooks, hidden by him before his death, were recovered after the war.

Having punched through the main American-Filipino lines and forced a withdrawal from roughly one-third of Bataan, General Nara decided on a bold flanking move rather than a new assault on the newly consolidated Pilar-Bagac line. This action involve amphibious landings on the rugged southwest coast of Bataan, which was characterized by tiny bays, steep cliffs, and heavily forested hillsides. Landing a full-scale force was impossible here but, by the same token, a small-scale infantry landing would be difficult to dislodge. Nara also realized that American reserve units would be heavily committed to covering the withdrawal from the Abucay line.

Nara's plan went badly at the beginning. The invasion barges became separated in the dark, and an American PT boat from Corregidor stumbled across two of the barges and sank them. Only 300 Japanese troops finally managed to get ashore. Once there, however, they dug in and were able to fend off the initial American counterattack.

MacArthur decided to pull trained infantry units off the main line of defense forming between Bagac and Pilar to fend off the Japanese incursion. Among these units were elements of the 57th Infantry. One detachment went to Longoskawazan Point, while the 45th Infantry, reinforced by Company B of the 57th, attempted to retake Quinauan Point.

When Ira Cheaney and the rest of his company arrived on Quinauan Point on the evening of January 28, they found the 45th Infantry and an ill-assorted array of Air Corps ground crewmen and Philippine policemen stretched through dense jungle. Visibility was poor, and the Japanese positions were well concealed.

Major Dudley Strickler, Commander of the 45th Infantry, told Cheaney and the others of the situation this way: "The enemy never make any movements or signs of attack, but they just lay in wait for us to make a move, and when we do, we take casualties and we *still* cannot see even *one* enemy. We have used machine guns to shoot up the trees where we thought the snipers were most likely to be. We blasted these trees from top to bottom, but still we only made 10 or 15 yards all day at some points. That's why I had to call for reinforcements."

Strickler's 45th Infantry, with Cheaney and the rest of Company B at his right flank, moved out at first light on January 29. It was a dreadful day. Even without the Japanese bullets buzzing about like angry wasps, progress would have been difficult. The forest was thick with vines that stung the men's faces and dense undergrowth that snarled their feet. Mortars were useless and machine guns nearly so because of the density of the jungle. The entire 45th Infantry counterattack came down to each man and his M1 Garand.

The Japanese were neatly camouflaged in their defensive positions, and it was often impossible to see or detect them farther than ten feet from their rifle barrels. It was a brutal kind of warfare that would be seen again and again—on Bataan, in New Guinea, and later in Vietnam. Men were killed by unseen snipers firing single shots and by machine gun nests that they stumbled into at point blank range.

Almost half the force on Quinauan Point was killed. Major Strickler disappeared without a trace on February 1 while on a personal reconnaissance of the front lines. An intensive search led to the conclusion that he'd been shot by a sniper and his body swallowed by the heavy undergrowth.

For Company B of the 57th Infantry, the toll included Ira Cheaney, who was shot down on January 30 by one of the many unseen snipers. He received, posthumously, the Distinguished Service Cross. As Tom Cleary would write in Cheaney's obituary in the *Assembly* four decades later: "Ira distinguished himself as a man and a soldier, and his spirit is now a part of the marvelous fabric of dedication, courage, and sacrifice woven by the men of West Point on the battlefields and in the skies wherever our nation's interests and security have been threatened."

On February 4, tanks were sent in and the Japanese were pushed back to the beach, but it was not until four days later that the last Japanese resistance was finally neutralized. The toll to the American and Filipino forces was devastating.

While the battle on Quinauan Point reached its crescendo, the Japanese again attempted further landings on Bataan's southwest coast, this time at the beaches around the mouths of the Silaiim and Anyasan Rivers, which were about a thousand yards apart and about two miles north of Quinauan. By the morning of February 1, all of the 57th Infantry—except for Cheaney's Company B—had been pulled in to defend the Silaiim-Anyasan area. The 57th Infantry now had a reputation as a crack unit and one with high morale and *esprit de corps.*

Hector Polla was part of the force designated to push the Japanese back. Artillery was ultimately ineffective because of the difficulty of communication between forward observers and the gunners. The 57th Infantry had machine guns, but they were used sparingly because the heavy ammunition had to be carried in by hand. Thus the Silaiim-Anyasan fight—like that on Quinauan—came down to each individual rifleman with his M1 Garand.

Advancement was painfully slow until February 9, when the 2nd Battalion of the 57th Infantry spearheaded a spirited foray against the toughest Japanese point on the line. During a coordinated attack under heavy fire, Hector Polla brought an adjacent unit up into position, enabling the the 2d Battalion to hold a position that could not have been held otherwise.

The following day, the 57th Infantry retook the Japanese position, and within a week the last of the enemy had been killed or captured. Polla finished the action with his life and a Silver Star for his heroism. Of the three classmates in the 57th Infantry who had witnessed the start of 1942 while evacuating Manila for Bataan, only he had endured the first month on this Godforsaken peninsula.

February was a good month for the defenders of Bataan. They had turned the enemy back from their landings on the southwest coast and stopped their advance at the Pilar-Bagac line. Nevertheless, the strain was beginning to show. Each small action, however successful, exacted irreplaceable casualties and weakened those who remained in the line.

In January and February the daily food ration had averaged 30 ounces, compared to the peacetime garrison ration of 71 ounces. Canned milk and fruit had started to become scarce, and corned beef and bacon had disappeared entirely. The men had taken to hunting pigs and water buffalo, but these had to be eaten quickly due to the lack of refrigeration. Some of the men had tried dogs and monkeys for protein, and the 26th Cavalry had slaughtered their 250 remaining horses by the middle of March. Scurvy, malaria, and amebic dysentery became the rule rather than the exception, and vitamin deficiency was virtually universal. It was later estimated that men in the line units were expending up to 4,000 calories daily, but by March they were barely consuming 1,000 calories a day.

On March 11, General MacArthur turned his Corregidor headquarters over to General Jonathan Wainwright and left by submarine to Australia, where he announced, "I shall return." His departure was taken by some as a sign of abandonment and by others as a symbol of hope.

The Last Stand

A six-week lull in fighting settled over Bataan in mid-February. Planeless airmen and shipless sailors had been organized into makeshift combat units early in the campaign, but the engineers, including Bob Pierpont and Bob Kramer of the 14th Engineer Battalion of the Philippine Division, continued their work maintaining trails and roads and building fortifications. Pierpont and Kramer were among the officers who supervised the installation of the barbed wire and minefields as well as the construction—using whatever materials they could beg, borrow, or scrounge—of obstacles and tank traps and nearly 1,400 improvised box mines. It was not until the first of April that the 14th Engineers joined the other service troops and became de facto infantrymen. By the time that the final Battle of Bataan began at sunrise on Good Friday, April 3, 1942, every man on Bataan who could lift a rifle was an infantryman.

As the first rays of sunlight touched the tops of the palms that framed a cloudless sky, nearly 150 Japanese artillery pieces laid down what was, without a doubt, the most catastrophic barrage of the entire Bataan campaign. Overhead, Japanese bombers blasted American and Filipino positions behind the Pilar-Bagac line. Japanese tanks and troops moved into the attack in mid-afternoon, and within 24 hours the line had begun to crumble.

The 57th Infantry, having been the first unit to face the initial invasion of Bataan three months before, was now in the rear as a reserve unit. The thunder of the enemy's 75mms and 105mms was clearly audible nearby. Easter Sunday found the defenders attending services in what one officer later described as "a serious atmosphere."

On Monday, April 6, the haggard and badly malnourished Americans and Filipinos launched a counterattack—a last, desperate gamble to sustain their position—with both the 57th Infantry and the 14th Engineers taking part. The three surviving West Point classmates fought shoulder to shoulder in what everyone knew would be the final battle in the defense of the Philippines.

General Homma was impatient to resolve his Bataan dilemma. Tokyo had expected him to have conquered Bataan and Corregidor in January. Of all the actions that the Imperial Army had undertaken on December 8, 1941, the campaign in the Philippines was the only one that was yet to be concluded by victory. Guam, Wake Island, and Hong Kong had all been seized in December. Indochina was under Japanese control. Malaya and Singapore had collapsed on February 15. What in the world, the Imperial General Staff asked Homma, was going wrong in the Philippines?

The general was embarrassed, and he was mad.

The American counterattack of April 6 was characterized by several individual success stories, but the American and Filipino troops ran head-on into a Japanese offensive launched the same day. As evening fell, it was clear that the counterattack had failed. The Japanese succeeded in driving a wedge into the defenders' line so deep that nearly all the units faced Japanese troops on two sides.

This counterattack was the last coordinated action of which either I Corps or II Corps of the Bataan force was capable. As the troops fell back in confusion in the oncoming darkness, communications collapsed and it became a war of individual units—some of regimental size, some mere platoons—fighting separate holding actions.

The 57th Infantry found itself withdrawing through an arduous jungle landscape with a great many wounded, who had to be carried by men weak and tired from lack of food and sleep. On April 7, the 14th Engineers became involved in an effort to establish a new partial defensive line at the Mamala River, but the Japanese were now moving so rapidly that the Americans had no time to dig foxholes. By nightfall, the remnants of the 57th Infantry and the 14th Engineers gathered together on the banks of the Alangan River, a brutal two-day hike south of where they had been on Good Friday. Their orders were to once again dig in and hold the line, but air and artillery bombardment, coupled with their own exhaustion, made the construction of any kind of organized defensive position a virtual impossibility.

On April 8, Japanese bombers dropping incendiaries on the dry cogon grass started fires that quickly spread to the bamboo thickets along the Alangan, forcing the men of the 57th Infantry to put their weapons aside and become firefighters in order to avoid being burned out of their positions. They had barely brought the conflagration under control when Japanese infantrymen appeared on the opposite shore.

Meanwhile, the 14th Engineers were taking the brunt of an incursion by two Japanese regiments consisting of tanks and infantry. The engineers were able to cleverly construct a barrier that, by four o'clock that afternoon, had brought the tanks to a halt. But, lacking the antitank guns necessary to destroy the enemy's armored vehicles, Kramer, Pierpont, and the rest of the 14th Engineers were only able to accomplish a holding action. Despite hours of backbreaking work, their brave efforts hardly caused more than a missed step in Homma's relentless offensive.

On the evening of April 8, General Wainwright ordered another counterattack. From his headquarters on the still-untouched fortress island of Corregidor barely four miles from the tip of the Bataan Peninsula, Wainwright knew that the situation was critical, but—because communications were in a state of utter chaos—he had no idea just how grim things had become. General Edward P. King of I Corps and General George M. Parker of II Corps replied that a counterattack was impossible. General King had reached the conclusion that his men were so badly beaten that there was no alternative but to surrender in order to save the lives of the remaining survivors. If they didn't surrender, he estimated that most would be killed within three days. "II Corps, as a tactical unit," wrote King's aide, "no longer exits."

At nine o'clock on the morning of April 9, King put on his last clean uniform and went forward under a white flag to meet the enemy.

At 12:30 that afternoon, silence descended over the Bataan Peninsula as the defenders surrendered unconditionally. Of the five men of the Class of 1941, Hector Polla now stood alone. Nininger and Cheaney had died to protect this ground, which had now been surrendered, and Pierpont and Kramer had disappeared.

The sick, starved Americans, many of whom had not slept more than a few hours in six days, were herded together for transportation to the prison camp at San Fernando, near Manila. General Homma, however, decided that the troops should not be transported. They had embarrassed him. They had angered him. They would *walk*!

Hector Polla was sitting with the men of the 57th Infantry when their Japanese guards passed along Homma's directive. They stood up and moved out, down past the shell-blasted jungle, past the stumps of the palms and piles of Garands and empty ration containers.

Within an hour, they reached the highway overlooking Manila Bay that ran up the east side of the Bataan Peninsula. Only then did the full impact of the situation hit them. The weather was beastly: hot and humid. The pavement on the road was hot enough to fry an egg; many of the men had started the day with no shoes, and others had their shoes taken from them by their Japanese guards. Polla watched in horror as men were shot or bayoneted for seeking shade.

The group of prisoners assembled at Marivelos on the southern end of Bataan and was joined by others who were marched out of the jungle along the way. Soon the long, wearied parade of wasted human forms stretched for several miles. They walked and shuffled through the sticky heat, past the wreckage of their heroic stand. GMC trucks and M3 tanks stood along the road, pushed aside by the motorized column of Japanese troops moving south. At one point, they passed a group of Japanese noncoms feasting on American rations. A Yank captain asked them if he could feed some of the men. He explained that they hadn't eaten anything since the night of April 7. He received a rifle stock across his face and was kicked and beaten unconscious. One of those who moved to help him was shot down, and the rest rejoined the line as it continued to stagger north.

Many prisoners were beaten or gunned down by Japanese troops within sight of their officers. In one two-and-one-half mile section of the highway, 62 bodies were left along the embankment. The Japanese consistently refused to allow their captives to refill their canteens or to get drinks from roadside wells. At one point, 30 men were permitted to leave the march to fill their canteens in a paddy field, only to be machine-gunned just as they reached the edge of the water.

As Hector Polla hiked slowly north, he saw again the positions that he and his comrades had occupied during the 98-day siege. They passed Pilar, the anchor of the Pilar-Bagac line that they had tried to defend. Men continued to be killed for asking for food when they passed through Abucay. Then they reached Mabatang and the place where Sandy Nininger had died, the place where the 57th Infantry had held the line . . . so long ago.

For nearly four days the Bataan Death March wound its way across the Bataan and Luzon Peninsulas. The officers and men of General Homma's 14th Army continued to hand down death sentences for such crimes as asking for water, refusing to bury unconscious men in shallow graves, and not walking fast enough. Their brutality equaled—and frequently exceeded—that of the Nazis. Hector Polla finally reached the military prison at Cabanantuan, which would be his home for the next two-and-one-half years.

The Japanese forces regrouped and prepared for the final assault on Corregidor. After intense bombardment, they invaded on May 5. General Wainwright was forced to concede that "no possibility" of resistance existed, and he surrendered to General Homma on May 7. The American defense of the Philippines was over.

Bob Pierpont, Bob Kramer, and eight other men evaded capture and attempted to reach Panay, which was not at that time occupied by the Japanese. On April 18, all of the party except one, whose leg wound had forced him to surrender, departed from Bataan in a small boat and, by paddling all night, crossed the mouth of Manila Bay to land at Cavite. They continued their journey, losing members along the way, toward southern Luzon, where they split into two groups. In August 1942, Pierpont and his three companions surrendered voluntarily to the Japanese because of threats of reprisal against Filipinos who had aided them.

Thereafter Pierpont was confined at Cabanantuan and in Manila. He withstood the rigors of prison life, an achievement his father later attributed largely to his training at West Point. His consistent good humor under adverse conditions was an inspiration to his fellow prisoners.

Listed as missing in action, Bob Kramer and his companions eventually reached Batangas province, where they melted into the jungle. They established a camp and remained there well into 1943, making several raids against Japanese outposts. They also organized a group of Philippine natives into a band of guerrillas, with whom they operated for two years.

General Masaharu Homma was indicted in 1946 for his part in authorizing the Bataan Death March in violation of the Geneva Convention. He was tried, sentenced, and executed.

The Bataan Death March marked the closing chapter of America's first campaign in World War II. It was America's Holocaust, a living nightmare for those who experienced it, a rallying point for those who did not. It left little doubt as to the kind of adversary the Japanese would be.

6

Awakening the Giant

The Rainbow

The United States on the morning after December 7, 1941, has been characterized as a sleeping giant awakened with righteous anger. The same could be said of the US Army—except that it was not a giant.

By comparison to the army of a few years before, or even a few months before, the US Army in the waning weeks of 1941 was indeed a growing force, expanding in men and materiel. The War Plans Division had placed on General George Marshall's desk a grandiose plan—The Rainbow Plan—for defeating the Axis, but the army, navy, and air corps had no forces capable of launching an offensive against Germany or Japan.

The Rainbow Plan called for the defeat of Germany first, followed by Japan. Marshall's War Plans Division was faced with the task of figuring out how to implement the plan. Initially, it involved reinforcing Australia and England, which were only a step away from the Axis forces. The territories of Alaska and Iceland also had to be held because they were only one step away from the continental United States.

The Philippines had to be written off. There was no way to save that doomed garrison. The Panama Canal, however, had to be protected at all costs. Without it, the task of fighting a global war would verge on the impossible.

Despite the desperate need to simply hang on, the War Plans Division went about the task of developing a grand strategy for implementing the initial counterattacks that ultimately led to victory. The first counterattack would be entirely symbolic. USAAF Colonel Jimmy Doolittle, renowned as a fearless and skilled test pilot in prewar civilian life, would lead an air attack against Tokyo using USAAF B-25 Mitchell bombers launched from the aircraft carrier USS *Hornet*. It was an idea that looked crazy and even suicidal at first glance, but if it worked—and Doolittle was confident that it would—the psychological impact on both sides would be profound. However, the psychological impact would only be of finite value unless it could be followed up by concrete action. Even as early as January 1942, a blueprint for the invasion and reconquest of continental Europe began to take shape.

President Roosevelt and Prime Minister Churchill, along with their military chiefs of staff, met in Washington between Christmas 1941 and January 14, 1942, to discuss the global situation and the enormous task that lay before them. This conference, code-named Arcadia, realistically accepted the premise that the Anglo-American allies had to prevent *losing* the war before they could think about *winning* it. The key result of this meeting was the formation of a strategic partnership, which resulted in an unprecedented level of cooperation between the two nations. Roosevelt and Churchill agreed to create a formal, combined chiefs of staff organization in which the chiefs of staff of the armed services of the two countries would work together to plan and implement both the global strategy and specific tactical operations to win the war.

As Roosevelt said at the time, in comments drafted for him by the hand of Henry L. Stimson, "Our joint war plans have recognized the North Atlantic as our principal theater of operations should America become involved in the war. Therefore it should now be given primary consideration and carefully reviewed in order to see whether our position there is safe." The first essential was "the preservation of our communications across the North Atlantic with our fortress in the British Isles covering the British Fleet."

Winston Churchill was urging Roosevelt to take action as soon as possible, and Roosevelt was pushing Stimson. Stimson was prodding Marshall, who in turn was exhorting his commanders to move quickly and expeditiously. However, it would all take time. Mobilization was less than a year old, and there was much to be done. Weapons had to be produced. The supplies and ammunition to keep the weapons functioning had to be manufactured. The ships to transport the weapons, supplies, and ammunition had to be built and, above all, the people upon whose skills the outcome of the global war would depend had to learn those skills and then be able to teach them to others.

Tuck Brown was with 36th Division in Camp Blanding, Florida, from February through June 1942. When he and his wife drove to Daytona Beach on weekends, they could see the large globs of oil from oil tankers sunk by the Germans. The so-called "brown outs" that reduced lights in cities near the beach did not keep the U-boats from getting a good silhouette of the tankers and other ships.

All in all, it was a gloomy time in which the United States' ability to wage a successful war seemed doubtful.

Desperate Days

Within a week of Pearl Harbor, Ben Spiller's antiaircraft regiment was ordered out of Virginia to Washington, D.C., to take up the defense of the nation's capital. They had to contend with shortages of all kinds of equipment. "It was well known that we had wooden machine guns—cut to look like the real thing—sitting on office buildings," Spiller said later. "In the event of an attack we would have fired what we had, which were pretty good long-range three-inch, and later 90mm, guns. The main limitation would have been against low-flying aircraft. However, our intelligence considered it unlikely that either German or Japanese airplanes could get that far inland because of a lack of support—bases to support a sustained, coordinated attack."

Meanwhile, though the German "fifth column"—saboteurs—came to be feared by some, most people never saw them as a real threat. There were few, if any, proven cases, although there were suspicious events, such as the time that George Pittman was called in to help investigate an accident involving a Martin B-26 Marauder medium bomber that had crash-landed in a Texas pasture one night. The pilot would have succeeded with a forced landing except for a large rock in the middle of the field. In the wreckage was the wing main spar, marked with the words, "I hope the SOB who flies this airplane goes to hell!"

"There were many unexplained events," Pittman said, "but I sort of doubt whether much of it was really sabotage or fifth column activity."

Getting It Together

As winter turned to spring in 1942, the men of the Class of 1941 who had been so dismayed at the army's indecision and lack of preparedness saw things change dramatically. They watched the US Army as it—to use Pete Crow's words—"exploded from a small nucleus to the greatest fighting force the world has known."

After Pearl Harbor, it was necessary for the United States to build an enormous army quickly. The more established units were required to send officers and noncommissioned officers to newly activated units as a core around which the new units could be built. This procedure was repeated over and over again. Training became intense, and 10- to 16-hour days were fairly typical. There was only time to train in essentials. Things such as close order drill, ceremonies, and "spit and polish" were de-emphasized.

New equipment and weapons were produced and issued. In September 1941, for example, there were only six 105mm howitzers in the entire US

Army; by February 1942, Tuck Brown's 36th Division became fully equipped. "We also got the new helmets in February 1942."

Classmate Ted Reed could see similar changes from the perspective of the 43rd Field Artillery. Matériel was rapidly upgraded, training methods and training improved and intensified, facilities sprouted like mushrooms after a rain, and morale, though burdened by a realistic appraisal of the task, ran high. The introduction of the 105mm howitzer to replace the old 1914 French 75mm howitzer as the workhorse of the artillery exemplified the changes. There were improved supplies, better training, and soon a pervasive feeling that "this is our war and we *are* going to win it."

Reed found the training to his liking. "West Point said it was training cadets to be generals, but I first had to learn to be a second lieutenant. I taught and *was* taught; I led and *was* led; I performed, and I loved it. The variety of tasks completed in the eight months I was with the outfit provided enough experience for four years of peacetime. We had 11 officers and two warrant officers when the requirement was for 37 and two." At one point, Reed had seven jobs, some in Battery C and some at the battalion level.

Joe Gurfein commanded one of the first racially integrated units in the United States Army in 1942. At the officer candidate school at Fort Belvoir, he was asked to take over Company C, which had one black platoon and two white platoons. The rationale for selecting Gurfein was that he was Jewish and could "handle" blacks. Gurfein worked with everyone and had a black sergeant running a white platoon. There was only one incident of a soldier saying, "I won't take any orders from him." Gurfein wouldn't stand for it.

In May 1942, Horace Brown was the first man to sign into the 88th Infantry Division, which was to be activated in July 1942. The 88th was a draftee division: All the men were draftees except for a small cadre of officers. The 88th Infantry became the 24th division to go overseas and the first selective service division sent into combat.

When Brown reported to his battalion, the battalion commander was the only regular army officer in place. When the battalion commander was transferred to Division Artillery Headquarters, Brown became the sole regular army officer in the battalion. The new battalion commander, who had training with the National Guard and was promoted within the battalion, was not as well qualified technically as the previous commander, but he turned out to be a superb combat commander. "I would have gone with him anywhere and undertaken any mission," Brown said.

Shortages of supplies continued. For example, in order to get nose drops at the dispensary at Fort Leonard Wood, Walter Mather remembered, a man had to bring a Coke bottle with him because there were no bottles for the medication. Winter clothing was, in Mather's words, "a sham. We had local tailors cut olive-drab wool blankets into short coats, bought Sears-Roebuck long johns, and wore rubber overshoes large enough to fit over our shoes and canvas leggings."

A number of the graduating Class of 1941 had already gotten married. Following the attack on Pearl Harbor and the preparations for war, many more decided to get married.

George Pickett's OAO, Beryl Robinson, was not quite 18. George had met her while he was still at the Academy, and shortly before Pearl Harbor they had become engaged. "Maybe you should finish college," George told her, "and then we'll marry when the war is over."

Beryl didn't like that idea, and George really didn't care for it either. The idea of waiting didn't make any sense, and Pickett had just heard a rumor that his 1st Armored Division was going to be shipped out to Hawaii in January 1942 to help defend the islands. There were rumors about everything. There were even rumors about sending the unit out to join Douglas MacArthur in the Philippines. Beryl had planned to finish art school in New York City, but when the war started, she packed up, came to Fort Knox, and told George, "There's no reason for us to wait until the war is over. We'd better get married before the division gets shipped out."

On December 27, just three weeks after Pearl Harbor, the two were married, and Beryl remained at Fort Knox. After New Year's, George received notice that he was to attend the gunnery instructors' course at Fort Knox, so he and Beryl rented a place at Tip Top—a Sunday School classroom in a church—and set up housekeeping. It was the only placed they could find to live in those days. They used the bathroom facilities at the church and ate breakfast—consisting of Cokes and cookies—at the gas station across the street. Beryl drove George into Fort Knox. They ate lunch and dinner on the base before driving back to the church to spend the night.

George Pickett's experience in 1942 is a typical example of how quickly the Army was expanding and changing. Late in January 1942, Pickett was notified that he was going to stay with the 1st Armored Division, but as soon as he finished the gunnery instructors' course he was ordered to Camp Polk, Louisiana. As it turned out, there were a lot of second lieutenants from the Class of 1941 in the 1st Armored Division, as many as 18 to 20 at any time. Both Hill Blalock and George Pickett were assigned to the division's 6th Armored Infantry Regiment.

One morning, the second lieutenants were called in and told that they were being reported for reassignment. Colonel Virgil Bell, the commanding officer of Pickett's 6th Armored Infantry, told them, "I don't have anything against you young men, but look at it from my standpoint. Do I want to keep a bunch of second lieutenants who are still wet behind the ears, or do I want to keep experienced majors?" When the time came to break up the 6th Armored Infantry Regiment, Blalock went to Camp Chaffee, Arkansas, and Pickett was sent to Camp Polk, Louisiana, with the newly formed 7th Armored Division.

Pickett drove down to Louisiana with his wife on February 14, 1942. He was told that the 55th Armored Infantry Regiment to which he was being assigned wouldn't be organized until August. "In the meantime, we'll put you to work with another outfit that has just been organized." Ten minutes later, Pickett ran into Colonel Bell, who had also been ordered to report to the 55th Armored Infantry Regiment.

In August 1942, Blalock showed up at Camp Polk as the regimental supply officer of the 55th Armored Infantry Regiment of the 11th Armored Division, where Pickett was now the adjutant. The two classmates would stay together throughout the war, until the middle of April 1945.

SAAD Days

Despite its general success, the rapid expansion of the Army also brought problems, not the least of which was substandard instruction.

At the advanced flying school at Ellington Field, Texas, the instructors were, according to George Pittman, "goof-offs and failures." As a result, Pittman and other officers from the Class of 1941 found themselves reorganizing the ground training at Ellington even as they were being taught to fly! George Pittman inherited the job of reprogramming the courses.

Because of the war and the frenetic schedule of flight training, members of Pittman's flight training class were transferred out on March 7, 1942, one month ahead of schedule. All the members, that is, except for George Pittman. Upon graduation, wings were mailed to those who had graduated. But through a mysterious fluke in the War Department's Personnel Department, Pittman's name had been omitted from the list of the graduates. The personnel people at Ellington expected his orders to come in "momentarily," so Pittman found himself unassigned. He went to Captain Blanchard, who was in command of the training squadron and who later became a three-star general and Vice Chief of Staff of the US Air Force. Initially, Blanchard was no help, as he had to have "new" personnel orders to put Pittman on flying status. Pittman became a tower officer and ground school instructor but was prohibited from flying. He was grounded except for whatever flight time he was able to bootleg.

Pittman's father-in-law, Colonel Paul Crawford, inquired at the War Department why his son-in-law had been passed on. The only explanation Crawford received was that the secretary typing names of those to transfer to the Air Corps must have "gone to lunch just before typing Pittman's name and skipped it when she came back."

Pittman was finally given orders transferring him from Ellington Field to Duncan Field. By this time, however, he had already been transferred to *Kelly* Field, where he was able to fly as a pilot in navigation school to make up for the time he had lost. Thus, when he received his Duncan Field orders, he was already well-established in the Kelley Field navigation school and, in fact, was on the next promotion list.

Orders were orders, but when Pittman reported to Duncan Field, he was summarily taken off the promotion list, because he had left Kelly Field. This "Catch-22" action followed him for years and effectively put two grades behind most of his classmates. It also put him a couple of months behind the others in being transferred to the Air Corps from the Signal Corps.

At Duncan Field, Pittman was assigned as a flight test engineering officer for the SAAD—the San Antonio Air Depot. Under the chief test pilot, Captain Victor Anderson, a former American Airlines pilot captain, Pittman learned to fly everything that came into the depot—often the hard way. After copiloting an aircraft for two or three hours and two landings in the left seat, he was checked out and flew test flights in the plane with any and all pilots who followed him into the system. It was a case of, as Pittman put it, "read the operating handbook, sit in the cockpit long enough for familiarization, then fire it up and fly it away."

In twin-engine aircraft such as the Douglas C-47, the Lockheed Hudson, or the Douglas B-18, Pittman got one ride as copilot, one takeoff, and one landing in the left seat, and then he was on his own. The same thing happened with the North American B-25 and Martin B-26 medium bombers. Such a procedure would have been unheard of in 1939 or today, but in 1942, everything was telescoped. Some of the men from the Class of 1941 had left flying school early to form new squadrons and commanded flights across oceans with pilots who had had less than ten hours in their aircraft.

After a few months as a SAAD test pilot—flying literally night and day, seven days per week—Pittman was called in by Colonel Ott, the maintenance chief of the San Antonio Air Depot, and given the job of straightening out quality control for the depot. Too many unsatisfactory reports were coming in on SAAD–overhauled or SAAD–manufactured items, including aircraft, engines, propellers, and instruments. Pittman asked for and received blanket authority to pick technicians to convert into inspectors as well as the authority to require that base technical training schools set up whatever quality control classes were needed to train or reorient the workers. Ott also agreed to let Pittman hire and fire people as needed to do the job.

One SAAD project involved stripping B-24s and rebuilding their tail sections because some had lost their tails over North Africa. By this time, Pittman had flown so many different B-24s that he could tell whether the Consolidated Vultee Aircraft Company (Convair) had built the plane at San Diego or whether it had been built under license by Douglas at Tulsa, by North American at Dallas, or by Ford at Willow Run, Michigan. He once mentioned to a Convair representative that they should go to Tulsa and find out why Douglas-built B-24s flew so much better than those built in San

Diego by Convair, the B-24's designer and original builder. A few months later, Pittman flew a B-24 that flew like a Tulsa aircraft and was surprised that it had been built by Convair in San Diego. It turned out that the representative had reported his conversation, and Convair evidently did what Pittman had suggested.

One day, while walking through the hangar, Pittman happened to lean on an AT-6 wing and discovered by accident that it moved too much. He had the inspection plates pulled and found that one wing's attaching bolts were missing. A check of the records revealed the wing had been changed at Moody Field, Georgia, just before the flight to the depot at Kelly Field. Only the small bolts holding the skin of the wing to the skin of the center fuselage section had held the airplane together.

A short time later, the depot was readying 59 Douglas A-20B Havocs for the North African campaign. The pilots flew the aircraft out night and day, seven days a week, whenever the aircraft became ready. In one particular instance, however, Kelly Field had been closed due to weather "below minimums" (heavy fog) for nearly a week. There were several A-20s that were backed up, awaiting test flight before they could be delivered to combat crews who were on their way to North Africa. Once airborne, the pilots could not see straight ahead, but they *could* see what was below them, so the chief test pilot set up a procedure so that the planes could be flown in a long, elongated racetrack pattern above the right side of the railroad to Poteat and Divine, Texas, from Kelly Field for testing. Fortunately, the test pilots all knew the area, and the testing went without a hitch.

About a week later, Pittman asked the civilian flight test maintenance crew chief of one of the A-20s to prepare the aircraft for a test flight. It was routine for the crew chief to fly on these test flights, but after filing his flight plan, Pittman found a different mechanic planning to fly with him in the plane. He told him, "Go get the crew chief."

The crew chief complained of a sick stomach, but there had been several mishaps—including a fatal accident—in this group of A-20s a few days before. In fact, the man was just scared. Pittman told him, "You have two choices: fly with me in your airplane or go to the front gate, collect your paycheck, and prepare to be drafted tomorrow." The maintenance chief flew rather than lose his deferment, and the flight proved uneventful.

Pittman also had an opportunity to fly the "plushed-up" C-87—a modified all-cargo variation on the B-24—that President Roosevelt had presented to Winston Churchill as an executive transport.

Although they were flying dozens of aircraft, the SAAD had no aircraft they could call their own, so they "built" their own Beechcraft C-45 by cannibalizing parts from several wrecked aircraft in the depot "bone yard." After flying the C-45 for many months, they were inspired by Churchill's C-87 to plush it up a bit. Sometime later, a visiting inspector noticed the C-45 and reported it to the headquarters of the Air Service Command at Patterson Field. Soon a message came down to "stop the modification, put it back into condition, and report it, for delivery to a navigation school." Since the C-45 was on no one's official records, the SAAD commander ignored the message. Eventually the inspector was reassigned, the messages stopped, and the aircraft continued to be used for administrative flights until the end of the war.

With help from several older civilian aircraft, engine, and accessories experts, Pittman transformed SAAD's Inspections Department of about 20 clerks and five inspectors to an organization over 500 full-time inspectors known as the Production Inspection Subdivision.

Before Pittman's organization, each foreman in the shops did his own signing off on equipment. Now the authority to sign off was vested in the hands of full-time inspectors rather than in the hands of the foremen. The SAAD quality control organization system of receiving and delivery inspections, and of quality control administration, later became the Air Force standard. As a result of a presentation at the Air Service Command Headquarters at Patterson Field adjacent to Wright Field in Ohio, the 24-year-old Pittman became known as "Father of USAAF Quality Control."

From his job at SAAD, Pittman was assigned to command one of six sea-borne air depots that provided overhaul shops aboard Liberty ships supporting USAAF field units on the islands in the Pacific. He activated and organized the 5th Floating Air Depot at Kelly Field and later moved it to Brookley Field in Mobile, Alabama, for specialized training at a field camp across Mobile Bay. While the 5th was training, Pittman visited the ship daily to be sure that it was being modified properly to provide space for the various shops. A flight deck was built on each ship's fantail for the R-2 helicopters that were to be assigned.

Just before his training at Brookley was finished, Pittman was ranked out of his job by a colonel who was a personal friend of General Hap Arnold and who had extensive helicopter flight time. Since Pittman at this time was a major and the table of organization called for a full colonel, he was reassigned to Tinker Field in Oklahoma City as base operations officer of what was then the busiest airfield in the world.

The Long Wait

As 1942 turned to 1943, many men realized that the war would be a year or two old before they saw combat. Months became years as they trained to go overseas.

Mike Greene was promoted from second lieutenant, platoon leader to major, squadron commander in 30 months. He moved from the 2d Armored Division to the 3d Armored Division, to the 7th Armored Division, to the 11th Armored Division—all in the States—while being trained and equipped for combat. His brother Larry was with the 1st Armored, which went overseas in 1942. Larry stayed with the 1st Armored until 1945, advancing from a first lieutenant to a battalion commander. Mike, however, did not go to England until September 1944.

In December 1941, shortly after the war broke out, Mike Greene and his father were both part of the 2d Armored Division. Mike received a call from the brigade commander, General Crittenberger, informing him that in January he would be transferred to Camp Polk, Louisiana, where the Army was organizing the new 7th Armored Division. "General Devers, the armored force commander, doesn't believe that father and son should be in the same division," Crittenberger told him.

"Well, sir," Mike asked innocently, "why don't you transfer my father?"

General Crittenberger didn't find the remark very funny.

Mike Greene went to Camp Polk in January 1942, where he was assigned to the 3d Armored Division, pending activation of the 7th Armored Division. In March, his father, a brigadier general, became regimental commander of the 67th Armored Regiment of the 2d Armored Division. Mike, now a captain, was once again in the same division as his father. Headquarters had inadvertently transferred them both.

It was not until August 1942, when Mike was moved to the 11th Armored Division, that he and his dad went to two different divisions. Initially, Mike was with the 42d Regiment of the 11th Armored Division as company commander, but that lasted for only a week before he became Regimental S-3.

Douglass Greene later went on to serve as a combat commander in the 7th Armored Division. As a major general, he activated and trained the 16th Armored Division, then was given command of the 12th Armored Division, which was due to go overseas. However, when he arrived at the port in New Jersey, he was informed that General Patton would accept only division commanders who had already seen combat, so Colonel Greene became a commander at an infantry training center at Camp Gordon, Georgia. He retired as a major general in 1946.

If a man was in a job that called for a higher rank, he could get promoted to that rank as long as he'd been in the job for at least six months. As a result, Mike Greene made captain in September 1942 and major in April 1943 as regimental operations officer.

Black '41 graduate Hugh Foster was pulled from the 4th Infantry Division in March 1942 and sent to Harvard and MIT for graduate studies in electronics and radar. While he was still with the 4th Division, however, he took part in one of the most ingenious innovations in the history of army communications.

Seventeen Comanche Indians had been recruited and assigned to the 4th Division for the purpose of transmitting voice messages over telephone and radio in their native tongue. It was correctly assumed that the enemy would not be able to understand them even if the conversations were monitored. The Comanches have no written language, and their vocabulary does not include words for many common military terms. Hugh Foster was assigned responsibility for training them, which for the most part consisted of enlarging their vocabulary to include selected military terms. He did this in such a way that not even other Comanches would understand what the 17 were saying when they used the new words and phrases. He created a vocabulary of about 250 military terms and phrases, and met with the Comanches three times a week for an hour or so. The sessions would begin with an English word, which everyone would discuss for a few moments. Then one of the Comanches would tell Foster, in Comanche, how they had agreed to say it. He wrote their Comanche word phonetically in his pocket notebook and read it back to them for verification. For instance, the Comanche word for "road" is "puy." An overpass became a "puy-puy-uba-ike-avee" and an underpass became a "puy-puy-tu-kike-avee."

The Comanches eventually went to England and were part of the Normandy invasion. Then Foster was sent to northwestern Africa, and, in 1944, to Italy. He never had a personal opportunity to see the Comanches in action, but 14 of the original 17 did considerable work in France. In 1989, the French government belatedly recognized their services by presenting individual awards to the three surviving Comanche communications men and to the Comanche Nation.

Early 1942 found Hamilton Avery in an antisub patrol in the Gulf of Mexico, along with classmates Thomas Corbin and Herb Frawley. They were in different squadrons and too busy with the checkout procedures for the monstrous B-24 to share much time together, but Avery and Corbin did manage to fly a shakedown cruise together to Albuquerque, New Mexico. Frawley was killed on May 18 in an accident while returning from a patrol during severe weather.

Other West Pointers were assigned together, at times, as well. Six Class of 1941 men were assigned to the 20th Fighter Group, all trainees learning to fly the Curtiss P-40. Long days of hard study and difficult flying, seven days a week, were interrupted occasionally by good times at parties in the evenings. Bill Mitchell, Mickey Moore, and Fox Rhynard were together for 30 days at Wright Field, Ohio, where they service-tested the new Republic P-47B Thunderbolt fighters that were first delivered to the USAAF on December 21, 1941.

"We sometimes pushed the aircraft to failure," Rhynard said, "which was the object of the testing—to determine failure potential and failure rates. [It was] mostly grinding, boring flying, but punctuated with an occasional failure, which quickly became an emergency. We survived and gained invaluable experience and confidence."

Between March 1942 and December 1942, Fox Rhynard and his new wife moved 13 times as Rhynard transitioned from AT-6 trainers into fighters, building up his flight time and his flight skills at bombing and gunnery. As Rhynard put it, "The Army was going balls out—as the country was—to get equipped, trained, and able to fight. By the end of 1942, we weren't there yet, but were starting to make a show in Europe and the Pacific."

Upon finishing his flight training at the Ryan School in Bakersfield, California, in March 1942, West Point 1941 graduate Mac Home, after a brief assignment to a bomber squadron, was assigned to Key Field in Meridian, Mississippi, to fly Lockheed P-38 fighter planes. The twin-boomed, twin-engined Lockheed P-38 Lightning, which entered service with the Air Corps in 1939, was the best fighter in the USAAF when the United States entered the war. But it was soon superseded by the North American P-51 Mustang, which ironically was first ordered by the British as the top American fighter of World War II and was perhaps the best piston-engined fighter ever to see combat with any air force.

The first Mustang had flown in 1940, and by the summer of 1942 its successors were starting to arrive in the USAAF inventory in sufficient quantity to make the P-51 fully operational. Some of these began coming into Key Field shortly after Mac Home was assigned there. The Mustang was like a dream for him, and soon he was acting as the base engineering officer and a self-taught P-51 expert. This fact was not lost on the harried USAAF hierarchy. Home found himself promoted to captain and detailed to the Air Service Command Headquarters at Patterson Field, Ohio, to act as a test pilot on P-51s before he left for North Africa in February 1943.

After graduation, Bill Gillis had turned his eyes back to Texas, where he violated West Point custom and tradition by marrying his hometown sweetheart, Lenore Riley, the "prettiest secretary that Governor Coke Stevenson ever had." Gillis and his wife moved first to Fort Benning and later to Camp Rucker, Alabama, where their daughter Georgia was born in 1943. After infantry school at Fort Benning, Gillis joined the 2d Infantry Division at Fort Sam Houston, Texas, for a short period, until being assigned to the infantry training center at Camp Roberts, California.

In June 1942, Bill Hoge was sent to Nome, Alaska, to prepare for the arrival of his unit, the 81st Field Artillery Battalion, which was scheduled to follow about a month later. The mission of the 81st was to defend Alaska against the Japanese invasion that then seemed inevitable. Bill Hoge's job was to select bivouac areas for the battalion and make arrangements for a temporary shed for each battery to use as a kitchen, mess hall, and orderly room.

Military service in Alaska at the beginning of World War II consisted almost entirely of prolonged, hard labor under harsh climatic conditions. All the combat units were doing the same sort of work: unloading ships, hauling supplies to wherever they were to be used or stored, and erecting a variety of buildings and other structures. Bill Hoge's jobs at Nome included supervising work on the docks, running a narrow gauge railway, and supervising the construction of various types of prefabricated buildings. Most of the work was the sort of thing that might have been assigned to transportation units or engineer general service units if such units had been available. There was very little time for tactical training. What training there was came during periods of intense cold, when prolonged outdoor activity was impossible.

As Hoge put it, "We weren't properly equipped, the battery was understrength, but the men we had were good. It was bitterly cold in December. Each morning we went out into the forest and hid until the sun came up and the USAAF flew a patrol over the area, then we returned to the post."

The situation in the early fall of 1942 was grim. Hoge personally anticipated that the coming winter would be much like the Continental Army's winter at Valley Forge. "We were living in tents. We didn't really have enough Arctic clothing, and I didn't believe the stocks of fuel and food were sufficient for the winter. The Bering Sea freezes in late October or early November, so ships could not get to Nome. Air transport from Fairbanks would still be available, but with the constant bad weather, we knew they wouldn't be able to deliver much."

Suddenly, in early September 1942, a fleet of supply ships arrived carrying nearly everything that was needed. The men worked feverishly around the clock to get them unloaded so the ships could get south of the Aleutian chain before the winter freeze-up. By early December, the troops were reasonably comfortable and well fed.

In March 1942, Bill Kromer was transferred to the infantry training center at Fort McClellan, Alabama. By December 1, he had made captain and was named as Regimental S-3. Pamela Hamlin Kromer, Bill and Jane's first daughter, was born on December 28, 1942 at Fort McClellan.

In June 1943, while his promotion to major was pending, Bill was ordered to Fort Benning to take the Advanced Infantry Course. Because of his age, his promotion was withheld, and at the completion of the Advanced Course, he joined the 345th Infantry Regiment of the 87th Division, which was then at Camp McClain, Mississippi. He took command of Company A of the 345th Infantry, and soon the men in Company A had begun calling themselves "Kromer's Kadets." Acting as battalion executive officer, he guided the 1st Battalion through two months of maneuvers in Tennessee, after which the division moved to Fort Jackson, South Carolina, to wait for deployment for England.

In the summer of 1943, when Henry L. Stimson came to Fort Jackson to review the 87th Division, Kromer was assigned as his aide for the day. For the first time since Stimson had handed Kromer his diploma on the field at West Point, the secretary crossed paths with one of the men of Black '41 whom he had sent into arms.

Henry Stimson had many foremen, not only men like Kromer and his classmates, but throughout the army and society. He also recognized the respective roles within the macrocosm, as few others did. He wrote in his diary in January 1944, shortly after he and Bill Kromer saw one another again, that the nation had an obligation "to make clear in no uncertain terms the equality of obligation of its citizens . . . The men in war production are not essentially different from the men who are proving themselves heroes in the South Pacific and on the Italian peninsula. They can be more accurately defined as the victims of the failure of the nation to develop a sense of responsibility in this gravest of all wars . . . We must . . . bring home to each of these men the fact that his individual work is just as patriotic and important to the government as any other cog in the great machine of victory."

In the Land of Wind and Smoke

Ultimately, almost all of the men in the Class of 1941 went overseas. For many of them, this deployment did not take place until the massive offenses of 1944. A few, however, shipped out within weeks of Pearl Harbor, and these men often were sent to the now-forgotten corners of the global battle-front where Allied planners feared they would have to make desperate and ill-supported stands.

Buzz Barnett and Lynn Lee had been ordered to Iceland before the attack on Pearl Harbor, but the orders were effective January 3, 1942. Barnett was assigned to the 5th Engineer Battalion, while Lee went to the 21st Engineers.

When Lee left Langley Field, the 21st Engineers had carefully detailed instructions—written before Pearl Harbor—on what equipment to load and strap to which truck. Each man's personal gear was stored in his footlocker, and that in turn was strapped on a particular truck. However, when all of the gear had arrived in Brooklyn ahead of the men, someone had decided, for an unexplained reason, *not* to send footlockers to Iceland. As a result, port personnel had cut all the straps and unloaded all the footlockers. A detail had to be sent in from Fort Dix, New Jersey, to reload all the gear into individual barracks bags and turn in the footlockers to the port's supply office. When he had discovered what was going on, Lee's battalion commander, Captain Brian Rice, insisted that it would require a lot more storage room in the hold of the ship if the gear was not on the trucks. It was only then that he was made to understand that the 21st Engineers would *never* see the trucks again, and if they wanted to see their gear, Rice should stop complaining.

Lynn Lee's battalion was met at the docks in Reykjavík and taken about two miles by truck to Camp Tripoli, the headquarters camp for the US Army Air Forces in Iceland. Soon after they arrived, Lee discovered that his battalion's designation had been changed from the 1st Battalion of the 21st

Engineers Regiment to the 824th Aviation Engineer Battalion. The camp was three-quarters of a mile from the airport, so the 33d Fighter Squadron, flying P-40 Warhawk interceptors, was also based there. Several months later, two of Lee and Barnett's classmates, Andy Evans and Poug Curtis, both pilots, would be assigned to the 33d Fighter Squadron at Camp Tripoli.

The northern tip of Iceland is inside the Arctic Circle, but at Reykjavík, the winter sun rises at 10 A.M., makes a small arc, and sets by 2 P.M. The time between dawn and dusk is about four hours, which is followed by 20 hours of total darkness. On rare nights when it is not storming, the skies in Iceland present a truly awe-inspiring sight, with the aurora borealis appearing almost continuously for six to eight months of the year. The city of Reykjavík ("reykja" means smoky, "vik" means city) is heated entirely by water from hot springs. The smoke in the name, Reykjavik, is the steam from the hot springs, not smoke from man-made fires. The Germans, who were already preparing for their next war long before Hitler came to power, had built the city's heating system for themselves during the late 1920s and early 1930s.

For heat at Tripoli, however, the Americans had inherited British coke stoves. "Coke burns hotter than coal, so it was easier to burn down a hut, and a few were lost this way. But I do admit a cast iron stove glowing red does warm you up physically and mentally," Lee recalls. "On a super blustery night, the wind would suck all the heat out of the banked fire and up the chimney, and the fire would go out at 5 A.M. It was as bad as West Point. A doctor there had insisted the heat be turned off at 11 P.M. and then turned back on at 5:50 A.M., and that a window be opened every night, even in the dead of winter."

When Lee and Barnett arrived in January, all of A and C companies, and most of the Headquarters Company, were being moved to a new airfield under construction at Keflavík, about 35 miles away, so the 1st Battalion was assigned to the Nisson huts being vacated by A Company.

The first couple of air raids made Lee a bit nervous, but after a while he got used to them, mainly because three times out of four, he never even saw a German plane. Sometimes it was because of the cloud cover, but usually there were just one or two German planes, so there really wasn't that much danger. "Iceland is the same size as the state of Pennsylvania," shrugged one of the hands who had been there since mid-1941, "so what are two planes going to do to an area this big?"

Eventually it began to sink in that the planes were not intending to bomb Iceland, but rather were looking for ships, and they only passed over Iceland to get their bearings. Even if a large group of planes were ship hunting, they would send only one over Iceland to get the bearings for them all to fly home on.

Most of the German planes were lone Focke Wulf Fw-200s—long-range bombers—that would, after unsuccessfully looking for ships, fly over and drop their unused bombs on Iceland rather than take them home. They seemed to pick utterly insignificant targets. One bomber flew over half the British Fleet and then dumped its bombs on a sardine cannery, which surely cost less than the plane being risked to destroy it.

On February 4, after an air raid, Buzz Barnett was found dead, killed by a .45 caliber automatic. Lee was convinced that he had shot himself, although the official version of his death was not suicide. While they had been on the ship, Barnett had told Lee his sad story. Just before graduation in June, his West Point sweetheart had married someone else with only a week's notice. Buzz had then married the first woman he met after that, and in New York City, as Barnett was getting on the boat, his wife had told him, "I'm pregnant, but don't worry. It's not yours." Lee felt Barnett never recovered from these two events, and the isolation in Iceland was too much for him.

When Lee and his men had arrived at Keflavík, they set up a rock and gravel quarry between the two airfields—Meaks Field and Patterson Field— that the 824th Aviation Engineer Battalion was building. As the quarry neared completion, two squads of GIs were assigned to build huts, and a third group stayed in the quarry as "powder monkeys" to drive the railroad and to direct the trucks that came to pick up gravel.

Lee's first powder monkey was getting far less than a ton of rock per pound of explosive. Industry standard called for an average of just over one ton per pound. After a month, a young draftee in Company C approached Lee's top sergeant and said he thought he could double the current break per pound. When Lee talked to him, the draftee explained: "My father was an explosives contractor who got paid by the tonnage broken loose by his blasts, and I was one of his powder monkeys since I was 12."

On that particular day, the quarry crew was taking out dirt from the top of the rock face and removing it from the quarry. The young draftee looked up and shrugged. "Y'know, this is all unnecessary."

Lee was taken aback by his audacity, and asked him what he meant. "If you drill five or six holes in the floor of the quarry, load and shoot them, when the dust settles, you'll have blown good-sized pits in the floor of the quarry practically under the railroad tracks. Then the dirt can be dumped directly into the cars and the blown out rock could be used as extra gravel."

Lee thought for a minute and agreed to do just that. It worked. The kid had a job. With the new teenage powder monkey at work, the pound per ton yield went up and the *total* yield went up as well, because there was much less oversize rock in each break.

Because of the occasional German bombing, Lynn Lee was required to set off explosions *exactly* at midnight—to the second—so people would know what the big boom was. The quarry ran 24 hours a day, as did the grading and paving on the two airfields, so Lee was at the quarry for the midnight big bangs two or more times a week, and no one expected him to keep routine hours. He slept and ate based on what was happening at the quarry.

One day Lee was still in the sack at 7:30 A.M. when the weekly staff meeting started. Lieutenant Colonel Sid Spring, who had taken over the 824th Aviation Engineer Battalion from Colonel Dave Morris, noticed his absence. Displeased with this affront to his authority, he decided to devise an appropriate punishment.

The battalion had just received an E-27 concrete mixer, which was full of dried concrete. Spring gave Lee the job of getting it cleaned out, to help remember his meetings. The mixer was delivered to the quarry, and Lee put his two smallest jackhammer operators in it to try to chip it out. After an hour they reported no real progress. He and his top sergeant looked inside and saw that only small holes had been drilled. Further examination disclosed that this was not a single load of concrete that had been allowed to set up but was, in fact, more than 50 thin layers, each about an eighth- to a quarter-inch thick, that formed an arch from one 16-inch paddle to the next. No paddle had more than one inch of its height still showing. No jackhammer was ever going to get the mess out.

The young powder monkey came to the rescue. "If you give me a day to run a few tests, I think I could blast it out."

Lee looked at the machine, which cost five or ten years of an Army engineer's salary, and sighed. "I sure don't want to owe Uncle Sam 10 years of pay!"

"You won't," the kid assured him. He set one-ounce blocks of TNT in 20 different cracks and crannies around the quarry in the next 24 hours, watched each tiny explosion, and sometimes went so far as to fit the pieces back together. About noon the second day, he got inside the cement mixer and drilled three holes, all close to the paddles. Finally, he put the two smallest Icelanders back inside the drum and had them drill a total of 20 additional holes. He chose the location, the slope, and the depth of each hole. When the last hole was drilled that night, he said, "I've decided to sleep on it before I load the TNT."

"Are you getting scared?" Lee asked.

"Not in the least, just being careful. You've heard the expression, 'There are bold pilots and there are old pilots, but there are no old, bold pilots.' Well, that goes double for powder monkeys."

At ten o'clock the next morning, the powder monkey was ready to shoot. Meanwhile, rumors of what he was about to do had floated all over camp. The idea seemed so preposterous that none of the higher ranking officers believed it.

Lee decided to postpone the shoot until 11:30, when almost everyone had left the work site for lunch. Then an idea—more of a feeling or an urge—

struck him, and he told one of the sergeants to get a load of materials and be ready to pour concrete. This was done, and at 11:35 Lee gave the plunger a full twist.

The explosion was not too loud, but the mixer disappeared in a cloud of dust, and a bolt flew overhead.

Lee and his men ran for the mixer. As the dust settled, the mixer looked unharmed. Examination showed only one missing bolt and a four-inch slit. After ten strokes with sledge hammers by a couple of big GIs and ten minutes of welding, the mixer was working once again. All but 20 of the 4,000 pounds of hard concrete could be removed. "The important thing for public relations," Lee told his men, "is to get it mixing *right now*."

A half-hour later, when Lee walked nonchalantly into the mess hall, a shout went up: "You blew up the mixer!" There were a dozen witnesses who had seen it disappear in a cloud of dust, most of whom had rushed back to camp to spread the word that the crazy young West Pointer had really done it this time.

Lee answered with feigned indifference, "Where did you hear anything that ridiculous? It's pouring concrete right now."

The story of how the mixer was cleaned out with TNT traveled around the world. When Lee got back to the States six months later and ran into some friends, they asked about the mixer even before they asked about his health. It took 20 years for the story to finally die out, and, as Lynn Lee was to say later, "Of course, it wasn't me who should have been famous, but the 19-year-old powder monkey from a strip mine in Ohio. All I did was have nerve enough to say 'Ah shit, let's try it,' and then give the handle a hard twist."

Late in September 1942, Lynn Lee received a letter announcing, "First Lieutenant Lynn Cyrus Lee will proceed to room number '99X99' in the Pentagon by the first available transportation." It listed five different kinds of transportation he could use, and closed with "by order of General George C. Marshall, the Chief of Staff, US Army."

The next afternoon about 3:30, word came to USAAF Headquarters that if Lee could be at the dock before 5 P.M., he could get on a transport home.

The ship was the US Army Transport *Chateau Thierrey*, a regular army transport that had been on the run from Brooklyn to the Philippines while Lee was in Honolulu in the mid-1930s. Lee had just stepped onto the deck when he ran into classmate Russ Gribble, who had come to Iceland with him nearly a year before. From him, Lee found out at last why he had been lucky enough to be on his way home. Gribble said that Iceland had seven percent regular officers and the army as a whole had under one percent, so some expert in Washington wanted to level things out. Altogether, there were eight or nine men being returned on the same orders.

The ship normally carried over 2,000 passengers, and with only 150 on board, it felt lonely. At most meals in the officers' mess they only used two or three of the 40 tables, and they had a different waiter every meal. All the "walking wounded" on Iceland—about 125 men—whom the doctors wanted sent back to the United States were also on the ship with Lee and Gribble. None of these wounded had been hurt in enemy action. Truck wrecks were mainly the cause, but there were also a dozen injured Merchant Marines on board.

The *Chateau Thierrey* sailed first to Londonderry on December 13, 1942, to load over 200 wounded enlisted men, most with casts. On the morning of the third day back at sea, they awoke to find themselves in the middle of a 60-ship convoy, with only two destroyers and six corvettes for escort. The slowest ship set the speed, so they were doing a top speed of only seven knots, at which speed it would take over 12 days to reach New York City.

The captain of the *Chateau Thierrey*, who had just made a round-trip to India without an escort, was furious. A couple of passengers who could read wigwag said he was fussing at the British commodore to be set free so he could run at 15 knots, which was faster than a German U-boat could go. Because of the more than 300 men in casts, the commodore refused his request.

On the fourth day out, they ran into a powerful North Atlantic storm, and for the next seven days they made less than 100 miles. Most of the time the ship's engines were turning just enough to hold the bow into the wind. On the seventh day, the men woke up to a clear sky and a fairly calm sea, but there was not a single ship in sight.

With the all the banging around that his patients had endured, the ship's doctor wanted to x-ray every broken bone and change every cast. He had neither the skills nor the equipment to attempt it, so he tried to persuade the skipper to get to a port as quickly as possible.

On Christmas Eve about 8 A.M., a few corvettes were spotted on the horizon, and by ten o'clock 30 or 40 ships were in sight. Suddenly, there was a large flash on the horizon, followed in 30 seconds by several bangs. U-boats! All but two of the escorts headed for the area. About 12 miles away, Lee saw a ship up-end and slide beneath the waves. The battle went on until 1 P.M., and for an hour the booming of the depth charges dropped by the escort ships was nearly continuous.

On the third day after the battle—two days after a truly miserable Christmas—the *Chateau Thierrey* sped up, swung right, and headed out of its place in the formation. Two corvettes soon appeared on the western horizon. When they were alongside, they did 180-degree turns and escorted the transport due west. By evening, the rumor was that the doctor had won his point, and they were being taken to Halifax, Nova Scotia. Thirty-six hours later, Lee spotted the coast of North America.

Lee and Gribble got to Boston and then jumped onto a train for New

London. They arrived at the New London railroad station about 8 P.M. on New Year's Eve, in time to catch Lee's sister and her husband before they went out for the evening.

Because of the time of year, they hadn't even tried to get leave orders issued, so they had to travel "immediately" to Room 99X99 at the Pentagon as their orders said. When they got there, the room was occupied by a section that had nothing to do with personnel nor the Corps of Engineers. The men in the room had no idea who had been there 120 days earlier when the orders had been issued or why Lee and Gribble had been wanted, so the two men went to the department of adjutant general who had signed their orders. The adjutant general's office fussed for a few days and finally sent Lynn Lee to the engineer school at Fort Belvoir for five months, where he was assigned to teach "traffic circulation in the division area." The response to his inquiry as to why the army was not using his experience, was that "the department that teaches airfield construction has no vacancies."

Finally, Lee asked General George Hume Peabody, Sr. (classmate Hume Peabody's father) to get him transferred, and in May 1943, Lee finally moved to the USAAF School of Applied Tactics in Orlando, Florida, where they put him to work teaching "airfield construction in a theater of operations." The USAAF tried to run it as though it were in a theater, but as Lee said later, "This effort was a B-minus because supply time was too fast for any realism. And, of course, it ran close from 8 A.M. to 5 P.M., and dependents accompanied all ranks. Nonetheless," he added, "it sure felt good to be back in the United States again!"

7

A Grand Strategy

Festung Europa

The Anglo-American Allies adopted, as the focal point of their strategy to defeat Germany, the invasion and liberation of the nations of western Europe and Scandinavia, which Hitler had corralled into his *Festung Europa* (Fortress Europe). This was to be no small task, as the Germans had fortified Europe north of the Alps into a single vast citadel.

Anglo-American cooperation was later expanded to include the Soviet Union. Stalin was fighting the Germans on his own soil, and he made it perfectly clear to Roosevelt and Churchill that he wasn't going to be satisfied until they opened a second front. Although Roosevelt and Churchill agreed to this goal, the problem remained of where and when to implement it. Invasion routes through Norway and Spain were considered. Churchill favored an attack through Greece and Yugoslavia. A route through Italy could be taken, but the Alps provided such a formidable natural obstacle that it would be impractical to attempt an invasion of Germany this way. In the final analysis, the obvious choice was the only choice. Great Britain would be transformed into an immense base camp for a cross-channel invasion of France.

Having selected the location, Allied planners turned to timing. It was simply impossible, given the strength of the German forces in France *vis-à-vis* Anglo-American capability, to launch a cross-channel invasion in 1942. On the other hand, *something* had to be done during 1942. By the end of that year, the United States would have been in the war for a year without having undertaken a major offensive against the Germans. Stalin could not understand this. He watched as Russian blood was spilled on Russian soil every day. He wanted a second front, and he wanted it now.

The Combined Chiefs of Staff met in London and Washington in April and June of 1942, with both Roosevelt and Churchill participating in the latter conference. Operation Bolero, the immediate build-up of American forces in Britain for the eventual cross-channel invasion, was confirmed. Operation Sledgehammer, the 1942 cross-channel invasion, was considered and superseded by Operation Roundup, a 1943 cross-channel invasion plan, which would itself eventually be supplanted by Operation Overlord in 1944.

In the meantime, in order to keep the Germans occupied, the British and American leaders agreed on Operation Torch, the invasion of north-western Africa. The British Eighth Army was now pushing the Germans westward from Egypt, so the primary objective of Operation Torch was to be to place Field Marshal Erwin Rommel's Afrika Korps in a vise between the two Allied commands.

Known as the "Desert Fox" because of his tactical brilliance in the North African campaign, Erwin Rommel appeared to be virtually unstoppable. However, in July 1942 at El Alamein, the British had blunted Rommel's advance toward Cairo, the Suez Canal, and the Middle East oil fields. Rommel held his own line, and it was not until November 1 that the Eighth Army was finally able to go on the offensive. On November 8, Operation Torch got underway as the American forces, under General Dwight David Eisenhower, landed simultaneously at Casablanca, Morocco, and at Oran and Algiers in Algeria.

By November 13, subsequent landings put the Yanks ashore as far east as Bone near the Algeria-Tunisia border. Morocco and Algeria were French possessions under the control of the pro-German Vichy government, so the Operation Torch landings faced resistance—albeit half-hearted—from French troops. These troops ultimately fell into line with the Anglo-American and Free French forces under General Charles de Gaulle when a way was found for them to do so without losing face.

Roosevelt, Churchill, and their Combined Chiefs of Staff gathered in newly liberated Casablanca in January 1943 to plan the next steps in the series of events that would lead to the eventual invasion of Europe. It was decided to step up the ongoing strategic air offensive against Germany and to consolidate Allied control of the Mediterranean region. The latter involved defeat of the Afrika Korps in North Africa and the invasion of Italy. Control of the Mediterranean and the defeat of Italy were seen as essential in the grand plan to prevent a possible Axis flanking action when the cross-channel penetration finally took place. In addition, the Allies agreed to accept nothing short of unconditional surrender from the Axis.

In late 1942 and early 1943, the US Army was still ascending the learning curve. For example, as Richard Kline pointed out later, the United States had produced a lot of gasoline-powered tanks, which proved unsuitable in the North African battles with Rommel's diesel-powered tanks. Meanwhile, the United States aircraft industry was turning out airplanes faster than Kline and the USAAF could train crews to man them.

As preparations for war continued, more of the West Point Class of 1941 fell in the line of duty.

On October 5, 1942, only three weeks before Hume Peabody was to move on to Gibraltar, he shared a beer with his old West Point roommate, George "Bizz" Moore, in Cleethorpes, England. They talked about their cadet days, of the relay record that Hume had helped hang up on the wall of the West Point swimming pool, and of running the quarter-mile track at the North Athletic Field during Yearling Year. Each marveled at what a long way they had both come in the 16 months since graduation. When the two men parted that night, they slapped each other on the back. Then Peabody, now a captain, strode down the platform and stepped onto his train.

Bizz Moore went on to see his first action as the Forward Observer in the 1st Armored Division during a medium tank attack in southern Tunisia during Operation Torch. He was detailed to the 9th Infantry Division, which was attempting to reach an infantry company that had been cut off. Walking across an open valley to reach his company, Moore was wounded in an artillery barrage, which landed him in a hospital in Oran for two months.

Shortly after Peabody arrived in Gibraltar, he took off from the British airfield in a British de Havilland Moth on a routine reconnaissaince flight. A few minutes into the flight, he had a mid-air collision with another plane. No one survived.

Everyone who knew him was shaken by his loss, perhaps no one more so than Lynn Lee, who in 1943 was transferred to Peabody's father's command. Hume and Lee had graduated from high school together and were roommates at prep school. "In my whole life to that point this was as near as death had come to me," Lee said. "He was my best friend for over seven years. He was his father's son and blindly dedicated to flying. Nothing except the Air Corps was real to him."

Back in the United States, preparations for war continued at a harried pace. Black '41 classmate Hugh Foster's experience epitomized how chaotic the process could be.

By December 1942, Hugh Foster had completed his postgraduate studies in radar at Harvard, MIT, and the army radar school at Camp Murphy, Florida, and was now a fully qualified radar officer. In late January 1943, he was ordered to report to Captain Jay in "C" Stage at Drew Field, Florida. ("C" Stage was the final staging for overseas movement.) When he reported, Captain Jay told him, "As of this moment, you are Commanding Officer of Company E, 560th Signal Aircraft Warning Battalion. Company E is an augmentation company, scheduled to join the 560th, which is already overseas at an undisclosed location."

Jay went on to tell Foster, "Company E has an authorized strength of 12 officers, 5 warrant officers, and 297 enlisted men. There is a company headquarters with 2 officers and 12 enlisted men, plus 5 platoons, each with 2 radar officers, 1 warrant officer technician, and 57 enlisted men."

"How many men are present at this time?" Foster asked.

"There are no enlisted men and no other officers yet assigned," Jay answered crisply. He went on to explain that the company was a "virgin" unit, and Foster soon realized that no barracks had been assigned and no equipment had been issued. Just as these facts were sinking in, Jay added that "Company E will be going to the Port of Embarkation (POE) in ten days, and it *better be ready!*"

Those 10 days proved to be 18 days of utter chaos. Fifty to 60 soldiers were assigned over the 18-day period, only to be transferred out a few days later because they were physically unfit for overseas service. Finally, Foster had a skeleton crew in his command, and Company E left Drew Field by train for the POE, still not knowing its ultimate destination.

As the train was about to leave, Foster discovered to his dismay that the company was short five enlisted men! He hadn't even gone overseas and he'd already lost five soldiers. Insisting that the railroad hold the train at the station for half an hour, he raced back to Drew Field, where the five additional men were hauled out of bed and rushed to the station.

Each soldier had two barracks bags, an "A" bag that he was to take on the train with him and a "B" bag that had gone to the boxcar on the day before departure. Private Alvin Smith, a 45-year-old volunteer, had complete upper and lower dentures, and he packed them in his "B" bag, which went into the boxcar, so Foster had to arrange for the cooks to give him food that he could eat *without teeth* for the three days of the train trip.

At one rest stop, a lieutenant came to Foster and asked if he could go into the station to buy candy and magazines for the enlisted men in his car. Foster granted permission, and the lieutenant went to the depot snack bar. The lieutenant had selected two dozen magazines and two dozen candy bars, but when he attempted to pay for them, the young lad behind the counter said, "Mister, don't you know there's a war on? You can have *one* candy bar."

The lieutenant argued with the clerk to no avail, and as voices got louder an old man came out from the back room and asked what was wrong. The boy behind the counter explained. The old man looked at the lieutenant, then finally turned to the boy and said, "Able, he's an officer. Let him have *two* candy bars."

When the train left Washington, D.C., on the last leg of the journey, the roadbed was so rough that at speeds faster than 15 mph, the baggage/kitchen car bounced around unbelievably. When breakfast was late on that last morning, Hugh Foster went forward to investigate. When he finally reached the kitchen, he just stood in the doorway of the car—braced against both sides—and watched utter chaos.

The ranges had large, square skillets with three-inch sides, yet it was not possible to keep eggs in them long enough to cook. The car jerked so much that the eggs slopped over the sides. Milk and water cans were rolling around like battering rams, and the cooks had to grab stanchions and jump over the cans as they came by.

Eventually the officers and men of Signal Company E settled for a breakfast of *very* roughly made sandwiches. Upon their arrival at Camp Kilmer, New Jersey, they climbed into trucks and were taken to their assigned barracks. Hugh Foster had no idea what to expect, but he soon realized that the company was missing numerous items of equipment—both personal and unit—which he had been assured would be issued at the POE. The men did get their individual shelter halves, poles, and ropes. However, tent pins were not issued at Camp Kilmer. Company E was expected to already have them. There was an Army regulation that said no unit going to a POE could leave the "C" Stage location lacking any authorized item that was available for issue at the "C" Stage post.

When he had time, Foster got together with a couple of experienced sergeants to figure out the best way for the soldiers to pack their gear to transfer it to the ship. They finally settled on a long blanket roll folded double, with its ends tied together with the tent rope. A soldier's head and one arm would be inserted through the loop thus formed, and the blanket roll would be worn diagonally over one shoulder, along with a standard infantry pack of World War I vintage, plus a cartridge belt, a first aid packet, a canteen, and a trenching tool.

At Camp Kilmer, Company E was temporarily issued a single two-and-one-half ton truck and one jeep, which were kept busy drawing supplies and taking Foster to planning meetings. All commanders were told they would receive maps of their destination area once at sea. After five days at sea, when Foster asked for his maps, there were none. "That information is aimed at assault landing forces only," he was told.

Their radar equipment—of a type that Hugh Foster had *not* studied—was to be issued at their still-unnamed overseas destination. Each of the five radar sets was shipped as 284 separate barrels, boxes, crates, and bundles of steel girders. This gave Company E a total of 1,420 separate items, which were scattered individually throughout an 80-ship convoy. The ships landed at numerous ports along the Atlantic and Mediterranean coasts of northwest Africa, and the company never did recover all the pieces. It was later learned that the engineers had commandeered many of the steel girders to build bridges.

As he was leaving Drew Field, Foster was finally promised that Company E's "tentage" would be "issued at the overseas destination." Of course, it never arrived. Everything of that kind that Company E obtained was filched off the docks in Oran. They were told that their vehicles would be issued "at destination," which they were, although the process took five months.

Airpower in North Africa

In 1942, despite the reorganization that accompanied creation of the USAAF, the *Army Field Manual* still called for air units to be under the operational control of the ground force commander. The inherent fallacy in this directive, which would not be fully recognized until the USAAF got into combat, was that it disallowed the need to gain air superiority over the theater of operations. It can be said, as indeed it has, that the fundamental basis for the organization of airpower in a theater came out of the first American experiences over the sands of North Africa in the fall of 1942.

At that time, the Twelfth Air Force, which was organized on August 20, 1942, as an air support command, was assigned to the American II Corps to cover the invasion of northwest Africa. Egypt-based units of the RAF and the USAAF were not coordinated with those in northwest Africa. The Luftwaffe, on the other hand, despite serious supply problems, was a well organized and serious adversary. Because they were attempting to provide ground support before gaining air superiority, Allied squadrons were easy targets for the Germans and consequently failed in their ground support mission as well.

Losses, both on the ground and in the air, were severe, and Allied air commanders went to work to formalize a new command structure, which was presented at the Casablanca Conference in January 1943 and approved by Roosevelt and Churchill. The cornerstone of the new structure was the creation of a single unified air command for the entire Mediterranean theater. Designated the Northwest African Air Force, it incorporated the Twelfth Air Force, which followed Allies from North Africa into Sicily in July 1943, and into Italy less than a month later.

Among the men who flew with the Twelfth Air Force were several West Point 1941 classmates: Dave Taggart, 37th Bomb Squadron; Mac Home, 86th Fighter Group; Jim Walker, commander, 428th Bomb Squadron; and Richard Aldridge, operations officer, 428th Bomb Squadron. Of the group, only Aldridge, who was captured and spent two years as a POW, survived the war.

At the time of Operation Torch, the 37th Bomb Squadron was one of the first USAAF units into North Africa. The desert campaign was rough, and the high attrition inflicted by the combat-seasoned Luftwaffe, together with Taggart's leadership qualities, soon found Dave Taggart commanding the squadron.

On January 15, 1943, Taggart led the squadron in an attack on a bridge near Sax, Tunisia. Rommel's panzers had mauled the fledgling US Fifth Army, and their logistical backup was flowing across that critical bridge. The weather was clear and the bomb altitude was 10,000 feet, making the B-26s a good target for German antiaircraft gunners. Nevertheless, Taggart got his Marauders through the withering 88mm fire, and they managed to obliterate the bridge.

On the way home, flying on the deck, Taggart's formation was hit by German fighters. One German fighter pilot, diving out of the masking glare of the desert sun, put his ring sight on the lead ship, and Dave Taggart went down.

Created on November 12, 1942, the Ninth Air Force joined the RAF Middle East Air Force in Egypt to form the Allied Middle East Air Forces. The role of the Ninth was a mirror image for the eastern Mediterranean theater of what the Twelfth did in the west. Among the men of Black '41 who served with the bomber force of the Ninth Air Force were Richard Travis of IX Bomber Command Headquarters, Ham Avery of the 98th Bomb Group, and Joe Silk of the 586th Bomb Squadron.

After graduating from advanced flying school in March 1942, Ham Avery had been assigned to the 44th Bomb Group at Barksdale Field, Louisiana. Suddenly, he had to fly the four-engined B-24 Liberator, and it was like learning to fly all over again.

Along with several other "new men," Avery was flown down to McDill Field, Florida, to assimilate the new aircraft. The instructor was Colonel— later Major General—Alfred Kalberer, an airline pilot who had flown five million miles with Royal Dutch Airlines (KLM) in four-engined aircraft before the USAAF grabbed him. Kalberer was on his way to inspect air units in the North African desert. His presence on the scene at McDill was purely accidental, but it was providential.

In July 1942, Ham Avery and three other junior officers had been transferred from the 44th Bomb Group at Barksdale Field to the 98th Bomb Group as replacement pilots. In addition to becoming intimately familiar with the B-24, Avery learned to cope with squadron assignments in the 98th Bomb Group as a new second lieutenant and to fly copilot to the commanding officer on antisubmarine missions out of his hometown of New Orleans into the Gulf of Mexico.

Shortly thereafter, the 98th Bomb Group followed Kalberer to North Africa by sea. Since the aircraft and crews were already stationed in North Africa, these officers were destined to accompany some 8,000 ground personnel on the transport ship *Louis Pasteur*. It was a 30-day voyage from New York to Port Said on the Red Sea, with only one stop at Durban, South Africa.

Most of the men of the 98th Bomb Group contingent were relatively green troops, much in need of preparation for combat-zone training. Avery was the only West Pointer aboard.

There was much resentment by everyone at being in close proximity and being fed mutton for 30 days. Added to the crowded accommodations was the constant threat of submarine attacks on the lone ship—which was later reported sunk off the west coast of Africa—and the 100 degree heat below decks. Against this backdrop, Avery and the three other regular army officers decided to impose a disciplined schedule of rigorous training. The four men quickly became the most unpopular people aboard, and several months later their tent was burned down while they were out flying a mission.

"I assumed the no-longer greenhorns got their revenge," Avery said. "However, when German and Italian sabotage units began landing on the North African shore near our base, those same thankless bastards at least knew how to handle weapons and act like soldiers. We lost only one B-24 on the ground."

In June 1942, shortly after he arrived in North Africa, Ham Avery was involved in one of the most extraordinary—albeit unsuccessful—missions of the war. Avery was assigned to fly one of three B-24s acting as part of a diversionary feint during a raid on Tobruk, Libya, intended to capture Field Marshal Erwin Rommel. The joint British-American operation was, however, based on faulty intelligence, which stated that on a particular night the "Desert Fox" was to meet his staff at Tobruk prior to the battle of El Alamein, to plan his Blitzkrieg into Egypt. Unfortunately, he left for Germany the day before on sick leave.

The British had three destroyers and 1,800 marines to assist in the kidnap phase, while the 98th Bomb Group supplied the three aircraft—laden with 1,000-pound bombs—to act as a diversion to the commando landing. Avery's three bombers were to confuse the local antiaircraft batteries—not to mention local security troops—by remaining over the target for an abnormally lengthy period. In addition, each plane was to randomly drop, one at a time, the five bombs at progressively lower altitudes, starting at 30,000 feet.

When the B-24s got down to 200 feet on their fifth run, the master searchlight suddenly caught Avery's bomber in its beam. Seconds seemed like hours as the German 88mm guns started hammering at the bomber. Meanwhile, Avery's tail gunner hadn't paused. He returned fire, pouring a river of tracers into the glowing mouth of the light. Seconds later, it blinked out in a hail of broken glass and Avery was able to scoot out to sea.

"How all three aircraft survived that operation, especially each other's random bomb-dropping, is beyond me," he said when it was over.

The raid didn't succeed. Even if Rommel *had* been present, German intelligence had already gotten wind of the plan, and this prevented the element of surprise. The British lost one destroyer and quite a few troops. Furthermore, due to its security measures, the United States' involvement was never officially reported. Hence, there was no mention of it in USAAF dispatches and certainly no recommendations for decorations, no matter how well deserved.

While pilots and crews of the 98th Bomb Group were flying combat missions from bases in the Libyan desert, they were offered six days leave in Cairo, Egypt, every six weeks. Transportation for the 700-mile trip was provided by a rickety, barely airworthy B-24, which was the product of cannibalism from combat wrecks and was appropriately dubbed *The Mongrel*. All of the pilots in the 98th Bomb Group took turns flying it.

One day, Ham Avery happened to be at the Heliopolis airdrome near Cairo preparing for takeoff in *The Mongrel*. The plane had a full load of "goodies" for those back at the base who were not on combat status, as well as a passenger load of 17 officers, in addition to the crew of five.

At the last minute, as Avery was revving up the engines, a jeep raced up. It was General Kalberer. The officer asked if there was room for him and his aide.

"Silly question," thought Avery.

Since Kalberer had been Avery's instructor-pilot for his checkout in the B-24, Avery insisted on moving over to the right seat as copilot and letting the general take the controls.

"There was method in my madness," he said later. "I wasn't completely confident that the lumbering cripple of a crate could even take off with that overload, let alone be able to clear the minarets of the famous Blue Mosque just off the end of the runway."

As *The Mongrel* lumbered down the runway, the general "poured on the coals," while the rest of the crew sweated, alone with their thoughts. Avery and Kalberer had barely raised the wheels when suddenly the right outboard engine sputtered and quit! The big plane had begun to stall out.

"All pilots are schooled in the procedure for a failed engine on takeoff," Avery explained, "but the next few seconds could mean not only whether or not we crashed and burned, but also whether we would take out the better part of the Blue Mosque—the Egyptian Taj Mahal—causing a Jihad within a war."

General Kalberer, with all his five million miles of multiengine experience, remained as cool as the proverbial cucumber and "fire-walled" the throttles of the good three engines, yelled at Avery to feather the prop of number four, and resisted the temptation to turn away from the obstacle looming ahead. *The Mongrel* shivered and shook as the wheels came up, and it cleared the last minaret by inches.

"To this day, I am not sure I would have made the correct decision [*not* to abort the takeoff] and the right emergency moves in the split-second timing required that 'saved our bacon' and an unholy fracas," Avery said many years later.

There were also fighter pilots from the Class of 1941 who served with the Ninth Air Force in North Africa. Fox Rhynard went to Egypt and Libya in December 1942 and joined the 867th Fighter Squadron of the 79th Fighter Group. When he arrived at Kabrit, south of Cairo, he had been a member of the 9th Fighter Wing—a "phantom" organization—and shortly after his arrival, the unit was deactivated and all personnel transferred as "casuals" to different Ninth Air Force units already in Africa.

In the spring of 1943 over the Gulf of Tunis, there was a dogfight between 12 USAAF P-40s and 16 Messerschmitt Bf-109s. In the heat of the battle, the squadron became scattered and Rhynard wound up surrounded by 3 Bf-109s, which took turns driving home, firing, then pulling up to clear the way for the next one to try. Each time an enemy aircraft got within firing range, Rhynard would dive for the water, level off at 20 feet, jerking and turning whenever a Messerschmitt started shooting. The Bf-109s were superior to his P-40 in their speed and ability to dive or climb and in altitude performance, but his P-40 could turn in a tighter circle, so his only real defense was to make as many tight turns as he could to make it difficult for the Messerschmitts to pull in close and get a clear shot at him.

Finally, a Bf-109 tried to turn with him, and he was able to get around behind him and shoot him down. Another one then came down and tried the same thing. Rhynard landed some hits on him, damaging him, and he took off trailing smoke. The third German fighter followed the others home, and Rhynard turned home as well.

Rhynard was later awarded an Air Medal Cluster for the kill and in September 1943, after 84 combat missions, was rotated home. In 1945 he returned to the European theater to fly 52 missions out of England as commander of the 359th Fighter Squadron.

Ploesti

Even as General Jimmy Doolittle's courageous band of raiders were flying their historic April 18, 1942, air attack on Tokyo, the USAAF was in the final stages of planning for another more effective offensive against Japan's capital. The purpose of the Doolittle strike had been to hit Tokyo and hit it quickly. Nothing more. Nothing less.

Nobody in the USAAF had expected Doolittle's bombers to cause much material damage. Up to that point, nothing had yet gone wrong for Japan. The Doolittle raid, which was as much a surprise to the Japanese as Pearl Harbor had been to Americans, dealt a brutal blow to the Japanese morale. Back in the United States, the effect was just the opposite. After Pearl Harbor and the disaster in the Philippines, the American mood was thick with gloom.

Doolittle's raid, despite the fact that it did only minimal damage, was heralded as a major victory.

USAAF General Henry H. "Hap" Arnold and his planning staff realized, however, that Doolittle's attack had been little more than symbolic. Much more would have to be done, and a plan was already in the works that involved sending Consolidated B-24 Liberator heavy bombers to bases deep inside China, from which to launch a sustained air offensive against Japan. As a heavy bomber, the B-24 had an effective range of 2,300 miles with a full bomb load, 1,000 miles farther than that of a fully loaded B-25. Furthermore, the 23 B-24s assigned to the mission could carry nearly 60 tons of bombs, compared to the 16 that Doolittle's B-25 squadron had dropped.

The plan was designated as the Halverson Project (Halpro for short) after the man who was chosen to lead it, the flamboyant Colonel Harry A. "Hurry-up" Halverson.

Shrouded in complete secrecy, the Halverson Project got underway less than a month after the Doolittle raid. Because the Japanese controlled the Pacific and most of Southeast Asia, Halverson led his armada southeast out of Florida, intending to reach China by taking the long way around the world. Loaded with three months' supply of food and equipment, the 23 Liberators reached Khartoum in the Anglo-Egyptian Sudan (today Sudan) via Natal, Brazil, in three days. Halverson had been told to wait there for further orders before proceeding to China, which he did.

A few days later, word came from Washington that plans had changed. The Japanese had captured the bases in China from which Halpro was supposed to commence its air attack against Japan. But there was another mission for the Liberator force. For weeks, it seemed, Allied planners had been anguishing about a target that nobody in Halpro had ever even heard of. Ploesti.

Located in the heart of the most extensive oil fields in Europe, Ploesti was a refinery complex in Romania that provided one-third of continental Europe's gasoline, lubricating oil, and diesel and aviation fuel. Romania was allied with Germany, and as such, was literally fueling the German panzers that were at that moment closing in on Egypt, the Suez Canal, and thus possible control of the entire Middle East, where much of the balance of the world's oil lay.

The Halverson Project was in a unique position to deal with the Ploesti problem. They had the longest range operational bombers in the Allied inventory, and they were now located just south of Egypt, the nearest any Allied forces could get to Romania.

Halpro's mission against Ploesti, the first American air attack on a European target, occurred on June 12, 1942. There was, however, no publicity accorded to the Ploesti attack: The USAAF wanted to keep the Germans in the dark about the capability of the B-24s, to keep them guessing about where the planes were based and what their true capacity for inflicting damage was.

In Washington, Hap Arnold decided to postpone a second attack on Ploesti until the Allies could amass enough bombers to inflict major damage. Any additional small, piecemeal raids would only afford the German forces an opportunity to hone their defensive tactics.

During the winter of 1942–1943, Allied forces turned back Rommel's drive on Cairo and began pushing him back through Libya. This ended Allied fears that the Germans would capture the Suez Canal, and it also provided the Allies with air bases in Libya that were 200 miles closer to Ploesti.

In the meantime, the B-24s that had survived the Halpro mission were augmented by more Liberators flown in from the United States, and this expanded force became the IX Bomber Command of Lewis Brereton's newly created Ninth US Army Air Force. The core of the IX Bomber Command was the 98th Bombardment Group, now commanded by John "Killer" Kane. Known as the "Pyramiders," the 98th was an all-Liberator force that was employed against strategic enemy targets in southern Italy. Although no one in the 98th outside of Kane knew it at the time, the Pyramiders were also destined to be the nucleus of the next, but still secret, hammer blow against Ploesti that Allied planners had appropriately designated Tidal Wave.

Months prior to Tidal Wave, Ham Avery had already completed the allotted combat tour of 30 missions. Instead of rotating him back to the States—the normal procedure—IX Bomber Command, requiring experienced pilots in its Operations Section, transferred him out of the 98th Bomb Group into the Headquarters Group. Undoubtedly, General Brereton needed his Operations Section beefed up to accommodate the B-24 units coming down from England for the then top-secret mission, and Avery was the only eligible West Pointer available in the Command at that time. Also the British had just saddled the IX Bomber Command with the responsibility for coordinating all bombing operations, both American and British, from the Libyan area. (There were two groups of British Lancaster and Wellington bombers in the Benghazi area.)

Through the winter, the IX Bomber Command was further expanded to include the 376th Bomb Group, known as the Liberandos, and the 389th Bomb Group, the Sky Scorpions. For Tidal Wave, these three all-Liberator groups would be augmented by B-24 groups loaned by the Eighth Air Force in England. These were the 44th Bomb Group, known as the Eight Balls, and the 93d Bomb Group, who had arrived at Alconbury, England, on August 19, 1942. Since then, they had come to be known as the Traveling Circus because they had flown missions from so many different bases, both in England and North Africa. Commanded by then-Colonel Edward J. "Ted" Timberlake (Class of 1931), the officer corps of the Circus also contained four West Point graduates from the class of 1941: George Brown, Kenneth "Kayo" Dessert, Harry Jarvis, and Joe Tate. Among them, they had already flown well over 50 missions against heavily fortified targets in occupied Europe.

Less than two years after graduation, each of them had been promoted at least twice and they had collected numerous Air Medals for their heroic tenacity under fire.

George Brown in particular had already made a name for himself as a fearless squadron commander, and at 24 he was already a lieutenant colonel. Both he and Kayo Dessert, who was now a major, were selected to lead elements of Tidal Wave in the assault on Ploesti.

It had already been decided that the Tidal Wave assault would be a different sort of mission than those to which the Liberator crews had been accustomed. The B-24, like the USAAF's other heavy bombers, was designed for high-altitude bombing. Indeed, all the heavy bomber missions being flown by the USAAF were completed at high altitude because the top secret Norton Bombsight permitted precision attacks from 20,000 feet.

Tidal Wave would be flown below 200 feet.

The adoption of this unorthodox tactic was necessitated by the nature of the target. The complexity of oil refineries and the importance of hitting small, key installations demanded a low-level approach. The fact that Ploesti had one of the most highly developed air defense systems in Europe confirmed the need for a low-level raid. The Germans would expect a high-altitude attack. The shells fired by their antiaircraft guns would be fused for high altitude. Their interceptors would not expect heavy bombers at 200 feet, nor would their radar operators. In a low-level attack, the bombers could mask their final approach by hiding from the German radar amid the mountains and canyons that surrounded Ploesti.

It was a daring, controversial plan, but Colonel Jake Smart sold it to the Allied leaders at the Trident Conference in May 1943 by telling them that he had complete faith in the crews who would execute it. To the men who would actually fly the mission, however, the idea of using B-24s at such an altitude seemed more than unorthodox, it seemed suicidal.

The handful of low-altitude missions that had already been flown against German targets had taken heavy casualties. However, Jake Smart was able to sell the idea to the crews by announcing that he himself would be aboard one of the Liberators, as copilot next to Squadron Commander Kayo Dessert.

By the end of May, Brown and Dessert had already started leading elements of the Traveling Circus through practice sessions for Tidal Wave. The specter of dozens of huge B-24s thundering over southern England at 200 feet, frightening cows, raised speculation about the nature of the mission, so the Circus commanders put the word out on the grapevine that their planned target was the German battleship *Tirpitz*, which was known to be cowering in a Norwegian fjord. The Circus even commandeered an unsuspecting Norwegian naval officer and gave him base exchange privileges—who could turn down an unlimited supply of American chocolate, cigarettes, and liquor?— so he would hang around the base and pretend to be busy.

Finally, the time came for the Traveling Circus to proceed to the sprawling IX Bomber Command base that had been built outside Benghazi. They
arrived in the choking heat of early summer. Once there, life alternated between combating the swarms of German fighters encountered on missions
flown in support of the Allied landings in Sicily, and the swarms of vermin
encountered in and around the tents that the men called home.

On July 19, the force that would become Tidal Wave sent 150 Liberators
against targets in Rome. It was the first Allied air strike on the Italian capital,
and it directly precipitated the downfall of Benito Mussolini five days later.
King Victor Emmanuel fired the beefy dictator and replaced him with Marshal Pietro Badoglio, who immediately entered into secret negotiations with
the Allies, which resulted in the surrender of the Italian government in August.

Unaware that they had been the catalyst that brought down one of the
major Axis powers, the five bomb groups terminated operations the day after
the attack on Rome in order to devote their full attention to preparing for
Ploesti. Using color-coded scrap metal as markers, a full-scale "map" of the
refineries was constructed in the desert. To the casual observer on the ground,
it like looked so much rusting junk laying about in the sand, but to Brown,
Dessert, Jarvis, Tate, and the other men flying the repeated dress rehearsals,
the fragments of oil drums precisely mimicked the corners and shapes of the
targets that they would be seeing for real in less than two weeks.

The Traveling Circus was assigned White Targets Two and Three in the
center of Ploesti—the Concordia Vega Refinery and a complex that contained
the Unirea Sperantza Refinery as well as one that had been built by Standard
Oil before the war. The mission was scheduled for August 1, 1943, and on
July 31 the last briefings were held and the final practice sessions against the
scrap metal outlines in the desert—this time with live bombs and spectacular
success—were flown. New engines had been installed to replace the sand-
choked and badly worn Pratt & Whitney R-1830s that had powered the
Liberators. Weather reports came in. They weren't perfect, but they weren't
totally discouraging either.

At the last minute, word was received from Washington that both Jake
Smart and Ted Timberlake, the commander of the Traveling Circus, would
be grounded, as would General Brereton himself. Hap Arnold didn't want
to take the risk of anybody who knew as much as they did about Allied
strategic planning winding up across the table from a Nazi interrogator in
Romania.

Before first light on Sunday, August 1, one of the new Pratt & Whitneys
coughed into life, followed by another, and another, and another. Quickly,
the first of 178 B-24s taxied off the tarmac and thundered down the runway,
with over two tons of bombs. The 29 Liberandos took off first, followed by

the 39 planes of the Traveling Circus, now commanded by Timberlake's deputy, Addison Baker.

Lieutenant Colonel George Brown was at the controls of *Queenie*, followed by Major Kayo Dessert piloting *Tupelo Lass* and Major Joe Tate in his *Ball of Fire, Jr.* Captain Harry "JoJo" Jarvis, the fourth member of the class of 1941 among the 1,763 men in Tidal Wave, flew as an observer and back-up pilot aboard Major Ralph McBride's *Here's to Ya.* Jarvis could have stayed behind, but he wouldn't think of it.

The Traveling Circus was followed by Killer Kane and his 47 Pyramiders and the 37 Liberators of the Eight Balls. Since their aircraft were newer than the others, the 26 aircraft of the Sky Scorpions brought up the rear. The theory was that the newer ships would be better able to keep up with the mass of the armada than would a group of the more battle-worn B-24s. Indeed, of the ten aircraft that had to turn back because of mechanical problems, seven were those of the Pyramiders whose B-24s had been in the desert for nearly a year.

Unfortunately for the men of Tidal Wave, the Germans had intercepted their radio transmissions. They knew that a formidable armada was airborne from Libya, and by the time the B-24s had reached Greece, they were able to make an educated guess as to their destination.

The plan called for the Liberators, which flew most of the 1,000-mile first leg at high altitude to save fuel, to come down to mission altitude about 150 miles southwest of Ploesti, then fly toward the northeast and approach the targets from the north. Everything went well until the Liberandos were about 30 miles due west of Ploesti. Mistaking a landmark, they turned short of their intended turning point and headed southeast toward heavily defended Bucharest and *away* from Ploesti.

Moments later, the Traveling Circus followed the Liberandos into the same wrong turn. By this time, many people were breaking radio silence to alert their comrades of their error. The Traveling Circus then executed a 90-degree left turn—no small feat for two dozen heavy bombers flying at 200 mph, barely 50 feet from the ground. The Circus then reformed into three parallel columns, led by Addison Baker, Ramsey Potts, and George Brown in *Queenie.*

They now found themselves approaching the target from the southwest rather than from the northwest, so all their practiced landmarks were now skewed from side to bottom, and their targets—White Two and White Three—were on the opposite side of Ploesti. So they took aim at White Four and White Five on the south side of the city instead.

The German antiaircraft guns, more than double the number that the men had expected, were ready for them. They had been alerted to adjust for a low-level attack when the B-24s were spotted after taking the wrong turn on approach. The 37mm and 88mm shells tore at the Liberators, ripping fragments of metal and Plexiglas from them as they drove on to the target.

An 88mm shell exploded in the nose of Addison Baker's lead ship, and his hapless bomber began to disintegrate. Nevertheless, Baker continued his course, leading his column with a B-24 that was rapidly turning into a fireball.

A few hundred yards to Baker's right, George Brown saw that Baker could have crash-landed in the fields short of the target, but instead he pressed on, leading his force toward Ploesti.

Three minutes later, Baker took another direct hit while almost directly over the Columbia Aquila Refinery, which was designated as Target White Five and originally assigned to the Eight Balls. Despite the hit, Baker led his column and their tons of high explosives over the center of the Columbia Aquila. Baker then pulled the burning plane up long enough to give his men a chance to bail out. Three made it; Baker did not.

George Brown's *Queenie* led the second column over the Columbia Aquila a few seconds later and released his bombs just as Baker's badly riddled aircraft stalled and began falling to earth, missing *Queenie* by only six feet. Brown glanced back at Baker's ship as it went down. He could see that the cockpit was now engulfed in fire. Baker was probably dead before his plane hit the ground.

Kayo Dessert, in *Tupelo Lass*, came in low over Columbia Aquila, as did *Here's to Ya* with JoJo Jarvis as the observer. Joe Tate's *Ball of Fire, Jr.* was literally surrounded by balls of fire as its bombs plunged into the inferno of what had once been Target White Five.

Everything was on fire now: refineries, oil tank farms, Liberators. The antiaircraft guns were taking a hellish toll on the bombers, firing at nearly point-blank range.

With the death of Addison Baker, George Brown found himself in command of the 93d Bomb Group. He led the main force in a hard left turn over the streets of downtown Ploesti and reformed with Ramsey Potts' detachment that had bombed the Astro Romania Refinery—White Target Four—the multimillion barrel facility that was Europe's largest refinery and the prime target of the entire Tidal Wave mission. Originally assigned to Killer Kane's Pyramiders, it fell to Potts as a target of opportunity because of the wrong turn on the initial approach.

As the Traveling Circus limped away from Ploesti, Brown noted that the formation now consisted of only 15 of the 34 Liberators that had entered the city, and most of these were damaged. Brown himself had crumpled one of *Queenie's* wingtips on a church steeple in their low-level dash across the city.

Heading in a southwesterly direction, Brown pulled the formation together to share defensive fire just as 52 Messerschmitt Bf-109 interceptors dove down out of the clouds. Because Brown had pulled the formation together, the Traveling Circus survivors were able to put up an impenetrable wall of fire, and the German fighters broke out of their attack to concentrate on individual stragglers elsewhere. It was not yet noon, and the Traveling Circus had been over Ploesti for less than 10 minutes. Neither the Liberandos nor the Circus had hit their assigned targets, but they had severely damaged three of the city's major refineries.

The Pyramiders, Eight Balls, and Sky Scorpions had entered the fray just as the first two waves were forming up for the trip home. By now, the antiaircraft barrage was horrendous, but the three groups, unhampered by a wrong turn, bombed their targets, all of which had already been attacked exactly as briefed.

The entire battle took just 27 minutes. Five of the seven targets were in flames, and quick calculations indicated that at least 40 Liberators had been lost. The survivors, many with severe battle damage, found themselves strung out for at least 100 miles across the flats of central Romania as they sprinted south toward the Mediterranean. While most of the aircraft rendezvoused with their groups, at least a dozen had become completely separated from the others and were now in danger of assault by German interceptors.

While the interceptors had managed to make several blistering strikes on the bombers emerging from Ploesti, Kane's Pyramiders were the only group that was intercepted en route to the target, costing them one bomber. Most of their damage had been inflicted by the nearly 300 antiaircraft batteries that defended Ploesti.

The crews breathed easier as they headed south, out of range of the guns and fighters. Suddenly, Brown's Traveling Circus found itself in a rugged mountain valley with antiaircraft on the ridges. The German gunners—they may have been Romanian, but probably weren't—were stunned by the sight of the monstrous Liberators roaring through the valley *below* them. They attempted to lay down a wall of fire, but the gunners couldn't turn their guns fast enough and simply fired wildly, and their shells were avoidable simply by changing altitude.

As they continued south, the major concern for the men of Tidal Wave shifted from the threat of German interception to the viability of their own aircraft. Most of the Liberators had suffered a great deal of damage. Many had lost engines and all were getting low on fuel. There were also instances of midair collisions as the crews fought mountain-top turbulence with bullet-riddled ailerons.

Finally, the planes reached the coast of Greece. Their pilots looked across the Ionian Sea to the last leg of the flight home and experienced a great sense of relief. It was not, however, to be long-lived. The men had been under the vigilant eye of German radar installations, and Jagdstaffel (fighter squadron) 27 at Megara, Greece had been alerted. Ten Messerschmitt Bf-109s took off to greet the Traveling Circus.

The two sides sized one another up for a few moments and then the

Messerschmitts made their move, diving down on the Liberators. The American gunners returned their fire, shredding one of the Bf-109s and losing one of their bombers. The second pass by the Germans cost another Liberator. Brown's 15 survivors now numbered just 13.

By the time the Germans had reformed for a third pass, the American gunners were ready. This time the Messerschmitt formation split up and hit the Liberators from all sides.

Inside the *Here's to Ya*, JoJo Jarvis observed the battle over Ralph McBride's shoulder as the bombers' 50-caliber defensive armament clattered all around him. Suddenly, he saw a Messerschmitt closing on them at incredible speed, 20mm cannon shells streaming from its nose, tracers pouring from its wings. There was a blinding flash, and the mass of steel and flesh that had been *Here's to Ya* cartwheeled toward the Ionian Sea. Captain Jarvis joined the Long Gray Line of West Point graduates who had given their lives for their country and was awarded the Air Medal, the Purple Heart, and the Distinguished Flying Cross "for meritorious service, over and above the line of duty."

The encounter over the Ionian Sea cost the Americans three aircraft and the lives of 22 men, while the Germans lost but two single-seat fighters and their pilots.

The surviving remnants of Tidal Wave finally touched down 16 hours after the mission had begun. Of the 178 Liberators that left Benghazi that morning, only 88 returned. Eight others landed in Turkey, where they were interned by that country's neutral government, and 23 reached safe haven at Allied bases on Cyprus, Malta, or Sicily. Among the 39 B-24s in the Traveling Circus on the morning of August 1, 1943, only a dozen were parked at Benghazi that night. There were, in fact, only 33 aircraft out of the entire Tidal Wave armada that remained battle-ready. Of the personnel, fully one-third had been killed, captured, interned in Turkey, or disabled.

Every man who had flown the mission was awarded the Distinguished Flying Cross. Despite their losses and their failure to destroy two of the refineries, the men of Tidal Wave had won the Battle of Ploesti. The loss of refining capacity took months to restore, and its absence sent ripples throughout the German war machine, leaving the Reich short of vital supplies during months when it needed to expand. Within a year, there would be Allied bases half the distance from Ploesti. The great complex that Winston Churchill had called "the taproot of German might" was never to be the same again after August 1, 1943.

It was the end of the IX Bomber Command, however. The remaining elements of the Traveling Circus, Eight Balls, and Sky Scorpions went back to the Eighth Air Force in England, while the Pyramiders and Liberandos became part of the burgeoning Fifteenth Air Force. The Ninth Air Force itself went to England to regroup to provide air cover for American forces in the great land battles that would occur in northern Europe in the coming year.

The three survivors among the classmates of 1941 returned to England with the 93d Bomb Group. George Brown became its executive officer on August 27. The following May he became executive officer of the 2d Bombardment Division. Kayo Dessert was assigned to the European Theater Command Headquarters, and in 1944 he went to the Pentagon as a staff officer at USAAF Headquarters. Joe Tate, for whom the Ploesti raid was his 26th mission, was promoted to lieutenant colonel and returned to England to take command of the 328th Bomb Squadron of the 93d Bomb Group. In a letter to his brother Daniel L. Tate, a cadet at West Point, he noted that there were cornstalks in the bomb bay doors and added that "I have recently been cured of buzzing anything *forever.*"

On the morning of December 22, 1943, Tate led the 328th on a coordinated attack involving 439 B-24s and B-17 heavy bombers against Münster and Osnabrück in the German industrial heartland. There was a snafu in his pathfinder equipment, and heavy cloud cover, so the mission was not deemed a success. Only two of the bombers failed to return to England. One of these belonged to Joe Tate. The *Ball of Fire* disappeared, and no trace was found. Tate had flown more than his 25-mission requirement but had elected to stay on rather than return to the United States.

"Everyone liked and admired Joe Tate," recalled Bill Hoge. "I don't think I ever saw him without a large grin on his face. I heard that he had completed the normal quota of combat missions required of a pilot and was on orders to return to the States for reassignment. His squadron was short-handed for a mission that they were to fly the next day, so Joe volunteered to fly again. He was shot down on this last mission. Volunteering for the extra mission was typical of Joe Tate."

In writing to Joe's brother Dan, Kayo Dessert said, "He was certainly loved and respected by all his squadron members, his peers, and his superiors. He was a superior leader—willing and able to do anything."

Among Tate's decorations were the Silver Star, two Distinguished Flying Crosses, five Air Medals, two Purple Hearts, and the French croix de guerre with palm. In January 1989, the new ice hockey rink at the sports complex at West Point was dedicated to the memory of Joe Tate and his brother Tony Tate (Class of 1942), who had died in 1944 under similar circumstances.

8

Striking Back at the Empire

The Jungle War

Beginning in 1941 the Japanese had written the script for the war in the Pacific, and for months as 1941 gave way to 1942, events followed their cues with uncanny precision: Pearl Harbor, Wake Island, Guam, Hong Kong, Shanghai, Malaya, Singapore, the Netherlands East Indies, the Philippines.

Then came late spring, and something started to go wrong for Japan. Australia was vulnerable, but the US Navy stopped the Japanese in the Battle of the Coral Sea on May 7 and 8. The Imperial Japanese Navy was, meanwhile, scheduled to quickly capture the island of Midway at the head of the Hawaiian archipelago, but the US Navy decimated the Japanese in the Battle of Midway between June 3 and June 6.

The Japanese insisted on making New Guinea a repeat of the Philippines campaign, with Port Moresby as its Bataan, but the Australian 7th Division, which apparently hadn't read the Japanese script, stopped the invaders 30 miles short of their goal and then counterattacked in September.

The US Army reinforced the Australians, and by January 22, 1943, eastern New Guinea was secure. Meanwhile, the US Marines landed at Guadalcanal on August 7, 1942, and by the end of January, Guadalcanal was secure. It had taken the Allies just over a year, but the Japanese offensive—which had been proceeding like clockwork—ground to a complete halt.

General Douglas MacArthur, who had been named as the Supreme Allied Commander for the Pacific Theater of Operations, was ready to make good on his promise to return to the Philippines, then move on from there to assault and defeat Japan itself. The doctrine of unconditional surrender that was adopted by Roosevelt, Churchill, and their Combined Chiefs of Staff at the Casablanca Conference in January 1943 applied to Japan as well. MacArthur's job was to make that unconditional surrender a reality.

MacArthur's game plan involved "island hopping," in which each phase of advance had an objective that could serve as a stepping-stone to the next advance. His overall strategy was to avoid frontal attacks, with their terrible loss of life, to bypass Japanese strong points, and to neutralize the enemy by cutting its lines of supply, thus isolating and starving its armies on their fortified islands.

For the island-hopping campaign land, air, and sea operations would be thoroughly coordinated in a new type of campaign, three-dimensional warfare that MacArthur called the "triphibious concept." In Europe, ground troops never saw marines and only saw sailors on the trip over. In the Pacific, soldiers fought side-by-side with sailors and marines continuously for four years.

On January 11, MacArthur wired General Marshall to request that a numbered US Army headquarters be established in the Pacific to relieve the burden carried by his own headquarters. He asked for Lieutenant General Walter Krueger, then commander of the Third Army, because of their long and intimate association. In response, Headquarters Sixth Army, commanded by General Krueger, arrived in the southwest Pacific area, and all US Army combat units were assigned to it on February 16, 1943.

Walter Krueger had been war plans chief when MacArthur was chief of staff. "Swift and sure in attack, tenacious and determined in defense, modest and restrained in victory—I do not know what he would have been in defeat, because he was never defeated," MacArthur said of Krueger.

As the Japanese receded from the southwest and northern Pacific, MacArthur and his US Navy counterparts set their sights on the islands of the central Pacific. Tarawa was liberated in November 1943, Kwajalein and Eniwetok in February 1944, Saipan in June, and Guam in July. On each island, the marines and soldiers faced heavily fortified garrisons manned by fanatical Japanese who put up such fierce resistance that 98 percent of the defenders fought to the death, taking a great many Americans with them.

The net result of the island-hopping campaign was that by July 1944, the United States had complete naval superiority in all but the far western Pacific and had established air bases on Saipan, Guam, and neighboring Tinian from which Boeing B-29 Superfortress strategic bombers could launch air strikes against Japan's industrial heartland.

Far East Air Forces

It was in the Pacific—at Pearl Harbor and in the Philippines campaign—that the aircraft of MacArthur's Far East Air Forces (FEAF) fell back to bases in Australia and Java. With its headquarters now in Australia, the FEAF was officially redesignated the Fifth Air Force on February 5, 1942, and the Hawaiian Air Force became the Seventh Air Force on the same date.

The Fifth directed its attention westward toward New Guinea and the Philippines, with the objective of implementing MacArthur's overall plan to destroy forces in these areas and ultimately recapture Clark Field north of Manila, its former home, which was now a major Japanese base.

Tactically, the Fifth Air Force faced a difficult future. The Japanese were on the offensive during most of 1942, and they had air superiority throughout most of the South Pacific theater. This factor kept American fighter units—USAAF, Marine Corps, and Navy—on the defensive most of the time and made offensive bombing missions doubly hard. When it came to this task, MacArthur depended on the Fifth Air Force, and the call went out for the bomber crews then training in the United States: "Visit the scenic south Pacific. Rustic living conditions. No shortage of targets."

Of the men of the Class of 1941 who served with the USAAF in the Pacific, Alden Thompson, Charles Jones, and Paul Larson were the first to arrive. Thompson was assigned to the 38th Bomb Group, while both Jones and Larson went to the 90th Bomb Group. For the 90th Bomb Group, flying North American B-25 and Martin B-26 medium bombers, the primary mission during the dark and uncertain days of 1942 was to support the American and Australian forces then valiantly attempting to cling to, and eventually retake, New Guinea.

The Japanese held a beachhead on the north side of the Owen Stanley Mountains opposite Port Moresby. Any attempt to preserve Allied control of Port Moresby, and to eventually retake New Guinea, depended on pushing the enemy out of Buna and Gona, and that operation depended on the Fifth Air Force. The US Army 32d and the Australian 7th Division would do the pushing, but they could not do it alone.

During the second week of November, most of the medium bombers in the Fifth Air Force had stood down to prepare for the assault. On Sunday, November 15, they sprang into action. The order of the day included silly-sounding places that no man in the 90th Bomb Group had known existed a year before, but which they now studied intensely. On that first Sunday, they would attack Buna and Soputa. On Monday, it would be Buna, Soputa, Giruwa, and Gona. On Tuesday and Wednesday, it would be more of the same.

Japanese troops with mortars and machine guns laid in wait to protect these villages. For four days the battle raged, and the loss of life was high. On Tuesday, November 17, when the 90th Bomb Group limped back from New Guinea, Paul Larson was not among them. He was the first airman from the Class of 1941 to die in the battle. He wouldn't be the last.

By March 1943, Charles Jones, Paul's classmate, was commanding the 19th Bomb Squadron of the 90th Bomb Group. There were more planes available now, and the Allies were slowly gaining air superiority, but the Japanese still clung tenaciously to their foothold on the north shore of New Guinea. The battlefront had been shoved 50 miles up the coast, but it was still the same kind of fight that it had been when Jones and Larson had first come out to the sunny South Pacific in 1942. Briefing papers still said "New Guinea," and footlockers were still being shipped home without the airmen who had brought them.

Captain Charlie Jones died on March 16, 1943, during a mission in which he led the 19th Bomb Squadron's B-25s against the Japanese base at Lae. He was posthumously awarded a Purple Heart to go with his two Air Medals.

That same year, several more of the classmates of 1941 arrived to take their places in the Fifth Air Force. Bert Rosenbaum and Edgar Sliney came out with the 345th Bomb Group, and Robert Colleran and Jim Dienelt arrived with the 380th Bomb Group, a heavy bomber group flying consolidated B-24s.

While medium bombers were used against Japanese positions on New Guinea, the heavy bombers of the Fifth Air Force were typically assigned to missions against targets farther to the rear. Attacks on enemy shipping were generally unsuccessful because of the heavy antiaircraft fire from land and naval batteries. Shipping was, nonetheless, the most important target for the USAAF in the southwest Pacific. The vast Japanese naval base at Rabaul also received repeated pounding.

On June 2, 1943, Jim Dienelt was copilot on the crew of a B-24 sent on attack against three such targets. After a hazardous takeoff from a poorly lighted landing strip and a flight through rough and stormy weather, Dienelt's Liberator reached the target. Because of broken clouds with a ceiling of 6,500 feet, it was necessary that bombing be done from an altitude well within range of the antiaircraft batteries.

On the first bombing run, Dienelt's bomber was intercepted by six Japanese fighters, and he was forced to abandon the bomb run. The Japanese fighters returned to the attack on a second bombing run, and one enemy fighter was shot down. Although his bomber's hydraulic system, a portion of the bomb release mechanism, and one fuel tank were shot out, and the tail assembly was badly damaged, at least one of Dienelt's bombs hit the target. Throughout the mission, Dienelt displayed the courage and determination that won him an Air Medal. Most important, his plane reached home safely.

On June 11, 1943, Jim Dienelt's 531st Bomb Squadron of the 380th Bomb Group was sent to Kupang, a port at the western end of Timor Island in the Netherlands East Indies, 2,000 miles to the west of Rabaul, a mission for which the B-24s were uniquely suited. Dienelt's flight encountered heavy flak en route to the target. His own plane was severely damaged, and it was only his skillful flying and tenacious will that enabled him to drop his bombs on the target. On the way back from the target area, with a piece of phosphorous bomb burning a hole in a wing, his crippled ship was attacked by Japanese Zeros and forced to make a water landing. As he and his crew left the aircraft amid a shower of bullets from the Japanese fighters, the Liberator exploded and sunk, engulfing the crew in fire. No trace of any survivors was ever found. Jim Dienelt was posthumously awarded the Silver Star.

Few people were more devastated by Dienelt's death than his best buddy from Cadet Company E, Moon Mullins. Inseparable pals for four years, they had both gone to primary flying school at Tulsa, Oklahoma, where Mullins married Frances McLean and Jim married her sister, Kathleen. In March 1942, the two men graduated from flying school and received their wings. Though they both had wanted to go overseas, Moon was assigned to the 58th Fighter Group at Dale Mabry Field, while Jim was sent to the Pacific. On June 13, 1942, Moon's first son, Charles Love "Kayo" Mullins V, was born.

When Mullins learned Dienelt had been killed overseas, he was stunned. Dienelt's wife was shattered by the loss. Mullins, who had been the "class character" only two years before, became much more serious about everything, and tenaciously sought an overseas assignment.

In late 1943, Mullins was assigned to the command and general staff school course at Fort Leavenworth, Kansas, and by December was promoted to major. Mullins returned home for the Christmas holidays to see his new son, who had been born the month before, but he had to return to school by December 27. On December 26, the entire Midwest was socked in by fog, and all commercial flights had been canceled. Moon was determined to return to school on time, so he decided to fly himself.

He took off from Dale Mabry Field for Memphis, Tennessee, where he picked up a course classmate and continued on. Arriving at Kansas City, Kansas, Moon experienced radio trouble and could not contact the range station for instructions. Continuing on his flight plan, Moon was letting down on the radio range when his plane hit a tall tree at the entrance to the post at Fort Leavenworth. He and his passenger died instantly.

"No one could have landed in that fog with the equipment we had at that time," Mullins' classmate and friend Leon Berger said sadly.

By 1944, the fortunes of war had turned in the Allies' favor in the south Pacific. Air superiority had finally been achieved, and the Thirteenth Air Force was now active alongside the Fifth in the Far East Air Forces. Among the men from Black '41 then serving with the Thirteenth Air Force were Harry Harvey, who commanded the 42d bomb Group, and Charles Peirce, who was deputy commander of the 5th Bomb Group.

The Forgotten Front

In June 1941, men of Black '41 were scarcely aware that the United States was on a collision course with the Japanese in their quest for a great Asian empire. But in the ensuing year, the Japanese juggernaut moved swiftly, relentlessly, taking eastern China, Hong Kong, Singapore, Indochina, Thailand, Malaya and Singapore, and Burma.

The British viewed India—the crown jewel of their Empire—as the bastion from which there could be no retreat. In addition to its own intrinsic importance, India was also the terminus of the key route for supplies to Chiang Kai-shek's Nationalist forces in China. Matériel was delivered to India first, and then to China over the Himalayas, by air or by way of the Ledo Road to be tediously hacked out of the jungles and mountains of northern Burma. Tactically, the first priority was to stop the Japanese on the plains of Imphal to keep them out of India, although there was also concern that Field Marshal Rommel's Afrika Korps—if they seized Egypt, the Suez Canal, and Palestine— would head for the Kyber Pass, the historical invasion route into India that Alexander the Great and the Mongols had taken centuries before.

While the British worried about India, America undertook the task of supplying Chiang Kai-shek, the Chinese Nationalist leader, principally by air. The first step in this process was to establish air support bases at Karachi, Agra, Dumdum at Calcutta, and the valley stations in upper Burma. The air route from the United States was through Natal, in Brazil, across the Atlantic, through mid-Africa, on to India and over the lofty Himalayas—the "Hump"— to Kunming.

The American combat role was principally in the air. On the ground in China, the effort was mostly advisory. General Stillwell prodded Chiang, his "Peanut," to do more, and Stillwell led the initial efforts to retake Burma only to yield later to British Admiral Sir Louis Mountbatten in the final phases of the war.

The Class of 1941 was present. Pete Crow went over early and stayed late. His depot built the principal rear support base for air operations in Burma, over the Hump, and in China. P. J. O'Brien and Horace "Race" Foster came later with B-24s. They arrived as the Japanese were making their last intensive, fanatical efforts to hold Burma. Herbert Clendening came with the 823d Engineer Battalion to help build the Ledo Road. Tom Lawson arrived with the Quartermaster Corps. Hank Boswell showed up in Calcutta to help in moving the vast supplies arriving by sea. He sent them over the Ledo Road and to the valley bases for over the Hump runs, which by then had become the greatest airlift in history, presaging the Berlin airlift.

Major General Lewis Brereton, late of the retreating Far East Air Force in the Philippines, arrived in Ceylon in February 1942 with six planes and some currency wrapped in an Army blanket. His objective was to establish the Tenth Air Force as the USAAF operating command for the China/Burma/ India (CBI) theater. Established on February 12 in the least publicized theater of the war, the Tenth Air Force attacked Japanese positions in the jungles and rivers of Southeast Asia, frequently striking targets in and around Rangoon and Bangkok. As the Allied advance into Burma began in 1943, the Tenth was overhead providing cover.

Across the Hump in China, American pilots had been fighting the Japanese since before Pearl Harbor as members of the American Volunteer Group under Brigadier General Claire Chennault. The AVG, better known as the "Flying Tigers" because of their shark's-mouth–decorated P-40s, had been the first real challenge to the Japanese since they had invaded China in the 1930s. Though overwhelmingly outnumbered, the Flying Tigers exacted a severe toll in both men and planes from the frustrated Japanese by flying hit-and-run, guerrilla-style missions, catching their opponents off guard.

Because they were initially isolated from the outside world, the AVG continued to operate independently of the USAAF for over six months after the United States entered the war. On July 4, 1942, however, the AVG was incorporated into the slightly larger China Air Task Force (CATF), part of the Tenth Air Force. With American markings instead of Chinese, and with a more regular flow of supplies, Chennault's Flying Tigers continued to pick away at the Japanese.

Finally, on March 10, 1943, the CATF was sufficiently supplied and equipped to be upgraded to numbered status. It was redesignated the Fourteenth Air Force. For the duration, when other commands were achieving numerical superiority in their theaters, the Fourteenth fought as an underdog. While the Japanese were fighting the bulk of American forces in the Pacific, the Fourteenth persistently harassed them in China. Often launching their attacks from improvised bases spread across a rugged 5,000-mile front, the Fourteenth constantly disrupted rail and boat traffic along the coasts of Japanese-occupied China. In the skies over the vast Asian mainland, the Fourteenth shot down nearly eight enemy aircraft for every one it lost.

The China/Burma portion of the CBI theater was probably the most arduous in the war because the enemy consistently held the upper hand on the ground and air superiority, which the Allies had achieved elsewhere by 1943 or 1944, was always in doubt. Because the Japanese controlled all of China's seaports and had closed the Burma Road—the only overland route— the only way to bring supplies into China was to fly them in over the Himalayas from India. Flying over the Hump was a grueling ordeal fraught with fierce winds, blinding snowstorms, frigid cold, and Japanese interceptors. It was, in itself, a major air battle, which every man and every aircraft that served with the Fourteenth fought before facing the enemy in combat.

By the middle of 1943, the Fourteenth Air Force—which had begun life as a fighter organization—had a number of active bomber groups, notably the 308th Bomb Group. For the bombers of the Fourteenth Air Force, the primary targets were the facilities that made it possible for the extensive Japanese armies to carry out their land war against Chinese and Allied armies.

These included railways and river traffic in Thailand, Burma, and Indochina, as well as in China itself, and the seaports from which the matériel was initially unloaded.

One of the most important targets was Hankow, the vital Yangtze River port and supply center, which was earmarked for several major raids at the end of August 1943. On August 21, the 308th Bomb Group sent 14 B-24 heavy bombers and seven B-25 medium bombers against Hankow. Among the heavies was a Liberator piloted by Race Foster of the Class of 1941. The B-24s and B-25s took off from bases in western China, but unfortunately, due to weather conditions at the forward fighter fields, their escort was unable to take off, so it became necessary for the bombers to proceed to heavily defended Hankow alone.

As they approached the target area, the Liberators were attacked by approximately 80 to 100 enemy aircraft. A fierce battle, which lasted for almost an hour, took place, and two of the B-24s were shot down, including the group leader. As Colonel James Maher later reported, "The Japanese always tried to knock out the group leader first, in order to disorganize the formation."

The 308th Bomb Group lost two aircraft, and many of the returning planes carried dead and wounded personnel. However, a total of 59 Japanese fighters were destroyed, with an additional 13 probables.

General Chennault, after hearing the details of the mission, decided that another raid on Hankow's airfields would reduce the enormous fighter strength the enemy held there, so it was decided that all available B-24s, B-25s, and P-40s would be used on the morning of August 24. Race Foster, who was group operations officer, pleaded with the group commander to be allowed to lead the B-24s, knowing full well that the lead airplane would be the first one attacked. His duties as group operations officer were of great importance, but, throughout his combat career, he often asked to lead group formations. Group Commander Colonel Eugene Beebe at first refused, but finally, under Foster's urging, permitted him to go.

The B-24s took off early that morning, but due to weather in the valley, only seven planes proceeded to Hankow. At the rendezvous point, the six B-25s and the 22 P-40s and P-38s were picked up, and the group entered the target area with the B-25s leading and the P-40s flying top cover. During the target run the B-25s and the P-40s left the initial target, but Race Foster pressed home his attack in spite of the fact that seven B-24s were now without fighter cover.

The Japanese hit the American planes with approximately 60 to 70 fighters and pressed their attack heavily on the unescorted B-24s. The ensuing air clash was ferocious, and the outnumbered Americans were set upon by several dozen enemy interceptors. Again, the Japanese zeroed in on the heavy bombers, knowing full well the kind of damage of which they were capable.

Tracers crisscrossed the sky, and the air reverberated with the thunder of exploding fuel tanks. One after another of the big B-24 Liberators were cut out and shot down.

Four of the heavy bombers—more than half the force on the mission—were shot down. The Japanese, however, lost 24 fighters as well. Race Foster assembled the surviving B-24s that had fought their way through, and they started toward home. Just as the squadron thought they were home free, a single Zero appeared from the clouds directly in front of Foster's aircraft. Maher watched from his own B-24 as the Zero fired his machine guns directly into Foster's cockpit.

"In spite of the fact he was nearly knocked unconscious, and, where any ordinary person would have died immediately, Major Foster continued to try to fly the aircraft until he was forcibly pulled from his seat," recalled Maher.

The copilot took over the controls and landed the plane at the nearest American air base. Race Foster was pulled from the B-24, unconscious but still alive. Within moments he was in the hands of a flight surgeon, who treated him as best he could. To the surgeon's amazement, Foster fought for life from approximately 1:25 in the afternoon until about 10:00 that night, when he died.

"This was the most amazing, and normally impossible, display of raw courage and indomitable will to live that I have *ever* seen or heard of," the flight surgeon said later.

Race Foster's body was flown to Kunming, China, the next day, and on August 27 he was buried in a vault at the American cemetery located at the north end of the runway.

Immediately after the funeral, Colonel Beebe wrote to Foster's wife, Kathryn: "Race was wounded in the head by a Japanese bullet at about 1:45 P.M. on August 24, while he was leading a raid here in China. He never regained consciousness and died as a result of the wound during the night. The airplane was successfully brought back by the copilot and medical attention obtained immediately on landing. There was nothing left undone to save him. You will receive letters from Race's friends. They will help you to know how much we loved Race and how much we miss him. Be consoled by the thought that his job was to fight, that he was doing it, and that he died the way a soldier on active duty expects to die."

Pete Crow embarked from Charleston, South Carolina, with the Third Air Depot Group in March 1942, having been briefed that the group was headed to the Philippines. This destination was soon changed to Java, and thereafter to Australia. After 57 days at sea, however, they landed at Karachi, India, where they learned via Signal Corps radio that President Roosevelt had just categorically denied the presence of any American troops in India.

When the ship arrived, the Third Air Depot Group was assigned to build a huge overhaul base at Agra. Not all of Crow's assignments were work, however. After meeting the Maharajah of Bharatpur while investigating a plane crash, he was invited to the Maharajah's annual duck shoot for the Viceroy Lord Linlithgow. In a blind with a case of ammunition, plenty of Scotch, elephants to stir the ducks, and a half dozen runners for pick up, Crow shot 84 birds—more than had any of the British officers present. From then on, he became a fixture at Bharatpur, with tiger hunts and all.

The Third Air Depot Group helped to keep a lot of airplanes flying. They were the hub for over-the-Hump operations. Being at the end of the longest supply line in the world, their forte was, as it had to be, improvisation. They set up overhaul lines that turned out aircraft engines that delivered more hours of flight time than those shipped from the States.

In 1944–1945, Crow was the commander of the 84th Air Depot Group, and for a time acting commander of the 84th Air Depot at Bangalore. It was at Bangalore that William Pawley—along with the government of India—had set up Hindustan Aircraft Limited to support the AVG in China. The USAAF later took it over as a support base for the CBI theater. By 1945, it had 15,000 employees; overhauled aircraft, engines, and instruments; and served as the primary support facility in the theater for C-47s, B-24s, and B-25s. The key employees were American and Indian experts in their respective fields, and Bangladore's overhauled engines were good for more hours than stateside overhauls. The secret was meticulous tear-down and rebuilding by Indian employees who took great pride in their work. Bangalore later became a manufacturing center for an Indian version of the C-46 and is today a major site factor for Indian aircraft overhaul and manufacturing.

Dr. Baba, the "father of nuclear research in India," was often a visitor at Pete Crow's office at Bangalore. Evenings were spent in bull sessions on physics, mathematics, and metaphysics. Crow's job involved a lot of "make-do."

"Innovation was our salvation," he said later. "The farm boys were the ones we looked for. They knew how to make things run with baling wire and pliers. When our crew came back from Cawnpore with the Carew's gin, one of the fellows said, 'I hope I wasn't lying when I told the plant manager it was de-icing fluid for the airplanes!' "

After three and a half years, Pete asked Hank Boswell, who was then an intelligence officer in Calcutta, to put him on a boat at Trincomalea, Ceylon, which was headed back through the Suez to go home. Pete was not so sure that this had been a good idea, for the first morning out general quarters was sounded when a Japanese torpedo was sighted astern. Once stateside, he headed to West Point, where Tulah Dance had suffered the war years with half a dozen of his classmates in pursuit. Pete and Tulah—who was now 20 years old—were married in the Cadet Chapel, with classmates crossing swords.

Walking down the aisle, Crow caught the eye of a former Batt Board member, who, four years earlier, had joined in sentencing him to one hundred punishment tours on the concrete area of central barracks for a now-forgotten infraction.

9

Crusade in Europe

The Italian Campaign

The Allied Trident Conference opened in Washington on May 12, 1943, the same day that all organized resistance by Axis forces in North Africa finally collapsed. The Anglo-American Allies were now free to proceed with their next step, which was the defeat of Fascist Italy.

Operation Husky, the invasion of Sicily, began on the night of July 9, 1943, with an airborne assault, followed the next morning by amphibious landings by the American Seventh and British Eighth Armies. For the first time in the war, Allied forces had actually secured a foothold on the soil of one of the three Axis powers.

On July 19, the same force of American bombers that was earmarked for the great Ploesti raid on August 1 struck Rome. This, the first serious air attack on the Italian capital, combined with the initial Allied successes in Sicily, threw the Fascist government into convulsions. King Victor Emmanuel III fired Benito Mussolini on July 24 and replaced him with Marshal Pietro Badoglio, and secret negotiations between the Italian government and Allied leaders were begun.

The Allies, following the doctrine adopted at the Casablanca Conference, demanded unconditional surrender, and Badoglio's government pondered this as the Allies hammered at the German and Italian defenders of Sicily. These forces surrendered on August 17, and Badoglio capitulated ten days later, although an official announcement was delayed until September 3, as the British Eighth Army crossed the straits of Messina to land on the mainland of Italy. German troops, which were in place throughout Italy, immediately seized command of their former ally. Thus Italy went from being an Axis power to just another of Germany's occupied territories.

As Joe Gurfein said, "Until about 1935, Mussolini was given credit for upgrading Italy. Then we felt sorry for him as Hitler took over and made Mussolini look like a clown. Since Italy was on our side in World War I, we had hoped it would split with Hitler. It seemed to me in 1943 and 1944 that the Italian people did not want to fight Americans, and they gladly helped us and spoke of their families in the United States. Had Britain stood up to Hitler on Austria and Czechoslovakia, Italy might not have supported him."

One Sunday shortly before the Allied invasion of Italy, all the lights went out on the air base at Oujda, Morocco, where Joe Gurfein was stationed. The engineer responsible for the power contacted the local power company, which told him, "We're short of fuel and are forced to close down the generators on Sundays."

Since airplanes were landing every day and night, the air base had to have lights. The next morning, Gurfein went down to the power company to speak to the manager about the problem.

"Is there something the Americans can do to get the extra fuel for you?"

He said, "I don't need much, about 20 gallons of gasoline each week."

This was such a small amount that Gurfein told him that he'd be delighted to personally bring the manager the 20 gallons every week in his jeep. Gurfein then asked if he could see the power plant. The manager proudly showed him around the generators and the large coal-fired boilers the plant had. Upon seeing the coal, Gurfein asked why he needed the gas.

With a straight face, he shrugged. "My wife and I like to go driving on Sundays."

In July 1943, Joe Gurfein and his commanding officer were invited to dinner at the home of a local sheik who had been a chief of police for the Allied forces, but had been fired because of certain undesirable activities. The meal was served in a huge 15-room house with a garden. The sheik and his young son sat on one side of the table, with the colonel and Gurfein on the other. The women, who would bring the food and then disappear, served an excellent lamb stew with rice and mint tea.

About halfway through the meal, the sheik said something to his son in Arabic. The son then translated to Gurfein in French, who in turn passed it on to the colonel in English. The sheik wanted to know if he could have his old job back. The colonel said, "No, sorry."

This said, the sheik proceeded to remove all the food from the table, folded his arms, and said, "Good-bye." The Americans were then unceremoniously escorted to the door.

Not every reception was so lacking in warmth. Nearly a year later, while in Trapani, Sicily, Joe Gurfein and another officer were invited to a private home for Sunday dinner. Since they knew that the people had no food as a result of the war, the Americans had eaten a normal lunch before they arrived at their host's home. The house, which was bombed out and had only two usable rooms—one being the kitchen—was boarded up on one side, and there was a small wood-burning stove in the corner.

They sat down at the table and were served an eight-course meal, each course bigger than the previous, starting with pasta and ending with a T-bone steak. Being polite, they stuffed the food down, dying in agony with every bite. The food was excellent, and Gurfein attempted to tell the host's wife in broken Italian, "You are a good cook." She looked around the burned-out room and smiled, "Of course." It was not until later that he discovered that he had actually told her, "You are a beautiful kitchen."

On September 9, 1943, the American Fifth Army came ashore at Salerno, just 45 miles from the great port of Naples. The Germans launched a fierce counterattack two days later, but by September 15 the beachhead was secure, and on September 17 the Fifth Army linked up with the British Eighth Army driving up from the south. Naples fell on October 1, and the Germans retreated first to the Veturno River and then to their "winter line" along the Garigliano and Sangro Rivers.

Although the first six weeks that the Allies were on the Italian mainland were marked by relatively rapid progress, the period from mid-October to mid-January 1944 was characterized by rigorous combat in which a terrible human price was paid for every mile gained. The rocky terrain of the Italian peninsula greatly favored the Germans, who had dug into the formidable and heavily fortified Gustav Line.

A brutal stalemate ensued during the last quarter of 1943. The Allied advance was nearly brought to a halt at Cassino, the pivotal point in the Gustav Line. Prime Minister Churchill proposed a flanking action, and General Eisenhower, although he had serious reservations about the tactical viability of such a maneuver, finally agreed.

On January 22, 1944, the Allied VI Corps, composed of the American 3d and the British 1st Divisions, landed at Anzio, 50 miles behind the German lines and just 33 miles south of Rome. German Field Marshal Albert Kesselring met the Anzio incursion by rushing armor and infantry from his strategic reserve in northern Italy rather than from the Gustav Line. The invaders held fast against Kesselring's forces, but the net result was that instead of breaking the original stalemate with Anzio landing, the Allies now found themselves with *two* stalemates!

Not all close calls at Anzio were initiated by the enemy. Two days after the Anzio landings, Bradish Smith and General "Wild" Bill Donovan of the Office of Strategic Services (OSS) were offshore in the Mediterranean aboard a PT boat observing the action. Meanwhile, the boat carrying General Mark Clark, Commander of the Fifth Army, came astern the British commander's PT boat. Suddenly, a US Navy destroyer got trigger happy and started dropping shells all around the PT boats. Smith and Donovan immediately veered off and continued on in another direction without getting hit, but they later found out that during the hot and heavy shelling, General Clark's boat was hit and three men were killed, including the executive officer of the PT boat, and two others were wounded. It was a case of a simple snafu turned deadly, and one which, had Clark or Donovan been killed, could have changed the course of history.

In June 1943, Bradish Smith had been assigned to the OSS, the forerunner to the postwar Central Intelligence Agency (CIA). He spent a short time training in Washington, then went to Africa with Sergio Valensky, the famous Russian émigré who ran the Plaza Hotel in New York for years. They stayed in Algiers for a few weeks, until Valensky was sent on a secret mission. Smith was subsequently sent through Sicily and arrived at Caserta in October, not long after the troops had taken the city. Located east of Naples on the Volturno River, Caserta became the Allied Headquarters in Italy, and Bradish Smith was assigned to be in charge of the OSS office.

Later in 1943, Smith found himself billeted at the Achille Lauro's villa in Naples, which overlooked the edge of the Bay of Naples and commanded a beautiful view of Mt. Vesuvius, which very obligingly erupted for Smith and the OSS while they were there. "We had a ringside seat," he said. He also had a ringside seat for several German air raids.

The march north into the Italian peninsula was usually a brutal affair, virtually unencumbered by swift tank-cavalry charges, for there was seldom a large enough flat area to make such a maneuver feasible. After the Italians had surrendered, the Germans were determined to make the Americans and the British pay dearly for every foot of Italian soil. They had all the advantage. They had time, good engineers, mountainous terrain to defend, and four years of intense experience in modern warfare. They chose successive defense lines where the mountains on their side gave them dominant observation with penetrating south-to-north valleys that narrowed to easily mined and defended defiles.

On *both* sides, however, the Italian campaign was run on a shoestring. The Allies were putting most of their effort into preparing for their main invasion across the English Channel into France (where there was more room for large armies to maneuver), and the Germans were plagued by their extended Eastern Front in in the Soviet Union, which was becoming for them more terrible and bloody by the month.

When the American or the British commanders decided to make an end run up the coast, such as they did at Anzio, they had to struggle to get Eisenhower to even temporarily release any landing craft from his invasion stores in England, and they never received as many as they asked for. The Germans, on the other hand, were short of ammunition and other supplies. They did, however, have the advantage of "interior lines" and could transfer men and supplies from Yugoslavia or from their rear areas in France or the Eastern Front.

In the fall of 1943, the Allies found themselves in the Liri Valley. They were halted by a series of defense lines anchored on low hills and in stone villages nestled into the slopes. A small sugarloaf, Mount Porchia, controlled much of the valley floor. The 6th Armored Infantry were dismounted from their armored half-tracks, overloaded with extra ammunition and a second canteen of water, and all 3,000 men were formed for attack at dusk. By midnight, 40 infantrymen had reached the peak, and their regimental commander had radioed for 1,000 replacements. Within the next half hour, about 40 Germans counterattacked. They were veterans and knew that their best chance to retake the peak was before the attackers could reorganize. In the ensuing melee the American artillery forward observer, bent over his radio to call in defensive fire, was shot when he did not hear an American challenge. The German counterattack failed, and its remnants drifted north before daylight revealed the German's position on the valley floor.

The whole campaign was like this. As Bizz Moore, who was with the 1st Armored Division, recalled, "At critical times, after six hours of brain-stunning shell bursts and ripping small arms crossfire, the whole burden of success or failure came to rest on the shoulders of just 40 riflemen, isolated in the darkness on an unknown hilltop, not knowing if any help was at hand, or even could be at hand, in time to aid them. But their stubbornness and faith in each other, though unseen in the black gloom, prevailed against the bravery of the counterattacking German remnants. Veterans of five years of warfare, probably sick of the constant strain, the Germans were so disciplined that they came charging back as their last, best chance of retaking their mountain. It happened so often that the special discipline—the will—and that confidence in each other of just a few men was the decisive factor in a battle. As the British Marines proved in the Falkland Islands in 1982, it pays to have *elite* troops with that overwhelming *confidence* which will dare anything!"

The main advantage of having the beachhead at Anzio was that for the first time in the Italian campaign the Allies were the ones who were dug in, with the protection of carefully plotted defensive fire. The Germans pulled ten of their best assault infantry regiments in from Yugoslavia, France, and Russia to break the beachhead, and in the end, the Germans' best infantry was shattered, casualties were very heavy, and the remnants were taken prisoner, shell-shocked and staggering.

Ernie Durr of Black '41 left for North Africa in March 1943 as adjutant of the 38th Engineer Battalion, sailing on the *Gripsholm*, but soon after the landings in Sicily he succeeded in transferring to the 34th Division, which shipped out for Italy.

On February 1, 1944, Durr's 34th Division, along with the U.S. 36th Division, launched a frontal assault on the Gustav Line at Cassino. However, the German stronghold at the ancient Benedictine Monastery on Monte Cassino held out during February and March against one of the more ferocious air and artillery bombardments of the entire war.

Meanwhile, Horace Brown had left the United States for North Africa with the 88th Infantry Division Artillery on December 7, 1943, arriving in Casablanca eight days later. After nearly a week in a staging area, the division was loaded onto 48 rail cars for an exceptionally cold two-day trip to Magenta, Algeria, where they were in training until January 31, 1944. The division embarked for Italy the following day from Oran.

The deceptively short hop across the Mediterranean took about 10 days—longer than the trip across the Atlantic. The first night out, the Luftwaffe hit the convoy, and toward the end of the voyage they encountered a terrible storm, which scattered the convoy and forced it to reassemble off Sicily before proceeding into Naples on February 9, 1944. The 88th Division entered the Gustav Line on the night of March 1–2, relieving the British Fifth Infantry Division.

A Cavalryman Aloft

During a lull after the Anzio landing, Paul Skowronek happened to run into his classmate from Cadet Company G, Edwin "Bud" Harding. Skowronek was with the 2d Cavalry Division, and Harding was with the USAAF's 301st Bomb Group. He'd met Harding as part of a ground forces to air forces exchange for temporary familiarization duty. Skowronek was supposed to observe staff operations at the Air Corps headquarters in Foggia, Italy, and the commanding general to whom he reported noticed that Skowronek was a 1941 West Point graduate.

"That's a coincidence," he said. "One of my B-17 squadron commanders is also out of that class." It was Bud.

Skowronek attended a few staff meetings, and Harding took him on four bombing missions in his Flying Fortress, which he had named *Amazin' Maisie* after his wife. One of the longer missions against the Iron Gate lock system on the Danube River was considered a "double mission," so Harding nominated Skowronek for a five-mission Air Medal.

Having a cavalry officer as a waist-gunner on a Flying Fortress required some faith on the part of the crew and the crews of the other aircraft flying in the formation. Their lives depended on Skowronek firing the machine gun effectively if they were attacked by German fighters. Harding was an inspiring leader, well liked by his entire squadron and idolized by his crew. His confidence in Skowronek's ability was good enough for them. They lost some squadron aircraft to enemy ground fire, but none to fighters.

The U.S. forces had broken out of the Anzio bridgehead and were moving rapidly toward the capture of Rome, and as a result, Skowronek was called back to his unit. Harding flew Skowronek to Anzio, landing his B-17 on an improvised fighter strip there, and they never saw one another again, although both survived the war.

In 1944, when Skowronek was in Palermo, Sicily, waiting for transfer to a combat unit after the break up of the 2d Cavalry Division, by sheer chance he bumped into Joe Gurfein, who was off-duty from the nearby parachute school in Catania. Gurfein had been training French airborne troops for operations in France. Gurfein invited Skowronek to visit the school the next day, and soon after he arrived, Paul found himself accompanying Joe on a flight in a C-47, which was to drop about 20 students—after which he and Gurfein would jump on a second pass over the drop zone.

With only the most rudimentary instructions, the two classmates jumped together. Their chutes opened so close to one another that Gurfein was able to shout further instructions to his former classmate on the way down. For Skowronek, that first, thrilling jump over Sicily spawned a life-long interest in parachuting.

The March on Rome

Even as Eisenhower's great armada made its final preparations for Operation Overlord, the great cross-channel invasion of Normandy, the combined forces of the American Fifth and British Eighth Armies linked up with the troops from the Anzio beachhead, under the command of General Lucian Truscott, and set their sights on the Eternal City.

On May 11, the Allies finally launched the main offensive that was destined to take them to Rome. The artillery shelling began at 11 P.M., and the initial close-in objective was taken quickly. But the real fight began when the Allies tried to break the Gustav Line.

Along with Ernie Durr's 34th Division, Horace Brown's 88th Division was to play a principal part in the offensive. After three days and heavy losses by the 351st, the Gustav Line was breached and the 349th Infantry Regiment—to which Brown was liaison officer from the 337th Field Artillery Battalion—was compelled to fight through the rugged mountains on foot. Brown left his jeep and, with his radio operator, two Italian porters, and a mule, he joined the infantry on foot. They started in a column of battalions, with Brown and the regimental commander with the lead battalion.

No sooner had they gotten underway than a sniper opened fire on them. This single German kept the entire regiment pinned down until Brown called to a following company, which outflanked the sniper and eliminated him. In the melee, however, Brown's mule fell off the mountain, and all his supplies except the radio, which was being carried by the porters, were lost.

The infantry regiment kept moving, picking up supplies as they could. Finally, they had moved out of the 337th's artillery's range, and this time

Horace Brown had gotten a good mule pack artillery battalion to back up the 349th Regiment, which filled in as direct support since its fire was directed by Brown, his liaison officer, and forward observers.

Monte Cassino finally fell on May 18 as Brown's men fought their way through the mountains to the Proverno area, where the 337th Field Artillery finally caught up with them as soon as roads were available. The 349th Regiment went into reserve at Proverno since the breakout from the Anzio beachhead had just occurred ahead of them. The breakout had pivoted out—like a gate closing off in the front of the advancing troops. At Proverno Brown lost his first forward observer, who was killed in action.

The 88th Infantry was at Proverno only a short time because the American VI Corps, which had been involved in the breakout from the Anzio beachhead on May 23, soon ran into more than it could handle south of Rome, where it found itself fighting in two directions, north and east. So after May 25, the II Corps, composed of the 88th and 85th Infantry Division, was committed to the north.

One day, just as Horace Brown and his radio operator reached the crest of a ridge, they spotted a wounded German corporal. Fearing the man would be killed by his own artillery, Brown and his radio operator ran out and picked him up. He was severely wounded in the leg, but they managed to rig a makeshift stretcher. As they were carrying him down the back side of the hill to safety, the German looked up and said to Brown in good English, "Why do we have war?"

"*You* ought to tell *us*," Brown replied. "This one is all *yours*."

Field Marshal Kesselring made a brief attempt at forming a new defensive line in the Alban Hills, but by this time the Allies had gained momentum and Rome, which had already been declared an open city, was abandoned by German forces. The Allies subsequently took Velletri and Valmontone in the Alban Hills on June 2 and entered Rome unopposed two days later, as Kesselring retreated north to regroup.

The 88th Infantry Division was the first unit to enter Rome on the evening of June 4, 1944, after a very tough campaign over mostly mountainous terrain from the Gustav Line. Horace Brown, the regimental liaison officer from the 337th Field Artillery Battalion to the 349th Infantry Regiment, had walked with them.

"Now we were marching into Rome," he said to himself. "The crowds with beautiful women and good-looking men cheering and pouring wine were so dense we could hardly get through. We felt the war was over. When we marched past the Coliseum and other monuments that I had studied in my Latin classes, I knew I would never forget it. Southern Italy was a poor, bleak, mountainous place, but entering beautiful springtime Rome in combat formation with an infantry regiment and marching past the historical monuments was the most exciting and delightful of times."

On the north side of the city, however, reality returned in an instant when suddenly the German shells started coming in, demolishing a half-track and knocking out Brown's jeep. The cheering crowds evaporated within moments, and the 88th Division once again was moving north.

A wartime lesson was noted in Horace Brown's diary: "The peaks are never far removed from the valleys."

Normandy

The idea of a cross-channel invasion as the means to liberate France and defeat Germany had been agreed to in principle in April 1942, and it was at the Eureka Conference in November 1943 that Operation Overlord, the 1944 invasion, would be given priority over all other operations that the Anglo-American Allies would undertake in Italy or in the Pacific. On December 7, 1943, two years to the day after the United States entered the war, George Marshall informed Dwight Eisenhower that he was to be the Supreme Allied Commander for history's biggest military operation.

When Eisenhower arrived in England in mid-January 1944 to set up the Supreme Headquarters, Allied Expeditionary Force (SHAEF), American supplies had been flowing into the island nation for two years, but what had initially been a trickle now became a torrent. Nevertheless, a great deal of key matériel did not arrive until May.

To insure the success of Overlord, Eisenhower had amassed the largest force ever committed to a single operation in the history of warfare. The Allied strength assigned to the operation on land, sea, and air on June 6, 1944, was 2,876,439 officers and men. Ground forces included 20 US Army divisions, (with 41 more in the United States ready to follow a successful invasion), 17 British Empire divisions (including three Canadian divisions), one French division, and one Polish division.

Typical of the nearly three million soldiers scheduled to take part in Overlord, Bill Hoge left the 81st Field Artillery Battalion at Nome, Alaska, in August 1943 and went to Fort Sill to attend an advanced officers' course. While there, he met up with Horace Brown, who was in a class that was about to graduate. Brown left in September and returned to the 88th Infantry Division, which was alerted for deployment to Italy.

After the advanced officers' course, Hoge spent about three months at Camp Howze, Texas, with the 103d Infantry Division. He went to England in May 1944 as an officer replacement with the 7th Field Artillery Battalion of the 1st Infantry Division—"The Big Red One."

The men of Operation Overlord would make the trip across the English Channel aboard more than 5,000 ships and landing craft supported by six battleships, two monitors, 22 cruisers, 93 destroyers, 255 mine sweepers, and

159 smaller fighting craft, not including motor torpedo boats, PT boats, and mine layers.

Overhead, the British First Tactical Air Force and American Eighth and Ninth Air Forces committed 5,049 fighters; 3,467 heavy bombers; 1,645 medium, light, and torpedo bombers; 698 other combat aircraft; and 2,316 transports and 2,591 gliders. Eisenhower had ordered a massive air campaign against the highway and rail net of northern France in the week prior to the invasion to "isolate the battlefield." A major part of this effort was directed at the area around Calais—the point on the French coast closest to England—in an effort to convince the Germans that the invasion was actually going to take place there.

However, Overlord's *true* landing zone was a 50-mile section of the Normandy coastline between Cherbourg and Le Havre. This region was in turn divided into five zones or "beaches." The westernmost—Utah and Omaha—were assigned to General Omar Bradley's US First Army, while the others—Gold, Juno, and Sword—were assigned to the Brits and Canadians of General Miles Dempsey's British Second Army.

The invasion date—D-Day—was originally set for June 5 but was postponed one day because of bad weather. The first landings, by paratroopers of the U.S. 82d and 101st Airborne Divisions, were made at 2 A.M. on June 6. They were followed by a withering naval bombardment of the German fortifications along the intended beachheads, which began at sunrise. The leading waves of landing craft hit the beaches of Normandy at 6:30 A.M.

Even after this unprecedented shelling, the Germans were able to put up a fierce defense. Landing craft and tanks were hung up on obstacles, and fire from fortified machine gun emplacements chewed up soldiers attempting to make their way across the narrow, heavily mined strip of sand that separated the cold waters of the English Channel from the Norman cliffs.

Losses were extreme, but as D-Day wore on, the German defenses were eventually overcome. By the end of the day, the Allies' beachhead at Utah was ten miles wide and four miles deep, and the Gold/Juno/Sword beachhead, which was 20 miles long and five miles deep, reached the edge of the city of Caen. The stiffest German resistance had come on Omaha Beach, where towering cliffs gave the Germans a defensive advantage. Nevertheless, the tenacious Yanks of the V Corps held on and refused to be pushed back into the sea.

Bill Hoge landed on Omaha Beach with the survey section of the 1st Infantry Division Artillery. He did not have a specific job, having been assigned to division artillery as an "excess officer." Having excess officers gave the assault division the capability of replacing combat losses without delay. On the morning of D-Day plus one (June 7), he was sent to the 7th Field Artillery Battalion to replace a liaison officer who had been killed on D-Day.

The 1st Division Artillery Survey Section was scheduled to land on Omaha Beach at 10 A.M. on June 6, the unexpectedly fierce resistance resulted in the entire timetable being revised. They finally got ashore about 7 P.M. and arrived several hundred yards to the west of where they expected to be, in the 29th Division sector. Because they landed on an incoming tide instead of an outgoing tide, all of their equipment in the soft sand was quickly flooded. Hoge and the survey officers set off for dry ground carrying the transit, a stadia rod, and a few other essential items and found that the beach just above the high-tide line was covered with soldiers of various units, completely lost and disorganized. They were simply huddled there, ducking whenever they heard an incoming round. Several senior officers were exhorting them to get on their feet and start moving, but they weren't responding. Some munitions trailers a little farther up the beach from the pounding surf were burning, and soon they exploded and added to the general confusion.

Hoge and one of his lieutenants kept the survey section together, and as soon as they got more or less organized, they led the section off the beach via the nearest exit—which turned out to be a 29th Division exit—and found themselves accompanying an infantry battalion of the 29th Division, which had recently landed. There was some intermittent sniper fire from the hedgerows on both sides of the ravine through which the road ran, but it was not serious, and soon the group came to the village of Saint-Laurent-sur-Mer, passed through, and found the 1st Division Artillery command post. They dug foxholes and spent a cold, wet night in them.

Back in England, Ham Avery, assistant operations officer at Ninth Air Force Headquarters, maintained the huge 10-foot by 40-foot status board depicting all of the air and support units that were assigned to Overlord. The board served as a timetable of commitment and readiness status for all units, but Avery also spent several days and several sleepless nights actually visiting the units themselves as liaison contact to insure first-hand that "the board" reflected accurate information.

On the Normandy beachhead, Bill Gardner of Cadet Company F at West Point had come ashore with the 116th Regiment of the 26th Infantry Division. Cut down by machine gun fire as he dashed across the beach, he was the first man from Black '41 to die in the battle to liberate northern Europe.

By daybreak on June 7, Overlord had been deemed a complete—albeit costly—success, and a quarter million Allied troops were now in France. Thousands of troops came ashore during the night and moved into assembly areas. "By morning, the whole situation looked a lot better," Hoge recalled.

Although the Germans had 60 divisions in France, only ten were committed to the vain attempt to counterattack. Hitler insisted that the Normandy landings were merely a diversionary maneuver and that the Allies' main invasion would come at Calais. Thus, he refused to release the forces which, if they had been deployed immediately, might well have been able to push Eisenhower's troops back into the English Channel.

As it was, Field Marshal Erwin Rommel—still Germany's top tactical field commander—had to make do with what he had and fight a desperate battle of containment. By July 18, Rommel had gathered 27 divisions in Normandy, but it was too little, too late. Two days later, he was a participant in the failed attempt by a group of dissident generals to assassinate Adolf Hitler. When the conspirators were arrested and tortured to death, only Rommel was permitted to commit suicide.

Within a week, the Allies had consolidated the beaches into a single beachhead, installed artificial harbors called "Mulberries," and continued to pour personnel and supplies ashore at an astounding rate. Although they were unable to successfully counterattack, the Germans compelled the Allies to pay dearly for every yard of ground they gained. The hedgerows—actually solid walls of entwined shrubs—that existed throughout the province of Normandy provided effective defensive positions and thus helped to prolong the battle of Normandy for well over two months.

Harry Blanchard of Black '41 was transferred to England in 1943 in anticipation of Overlord, and it was here that he met Jean Russell, a young English woman whose husband had been killed while serving with the Royal Navy in the Pacific. On Easter Monday, April 10, 1944, Henry and Jean were married in a twelfth century parish church near Winchester. After a quick honeymoon at Bournemouth, Blanchard rejoined the 9th Signal Company bound for Utah Beach. Just 11 days later, on June 17, the 9th Infantry Division closed the Cherbourg peninsula and reached Barnesville. However, four signal corps jeeps were caught in German machine gun fire while on reconnaissance near Bricquebec. It was here that Henry Blanchard "stood his last retreat" and joined the Long Gray Line. He was posthumously awarded the Purple Heart and the Bronze Star.

The son he never saw, Henry Nathan Blanchard III, was born in Winchester, England, on February 2, 1945. He was raised in the English countryside by his mother and stepfather, Kenneth Goodall, a farmer near Andover. He later visited his father's classmates at their 45th class reunion.

By June 18, the day after Henry Blanchard's death, the German troops at the port of Cherbourg were surrounded. The city finally fell nine days later, although the defenders had destroyed the port facilities, which would not be repaired for more than a month. The British captured Caen a month later, and on the first of August, General George Patton's newly arrived Third Army spearheaded a drive through Avranches to the Atlantic Ocean at Saint-Nazaire.

The Breakout

When the Americans achieved a breakout from their positions below Saint-Lô on July 25, 1944, morale was extremely high. After two months of dreadful, static warfare reminiscent of World War I, the Allied forces were moving at 35 mph—the top speed of an M4 Sherman tank—on the smooth highways of northern France. Men were smiling for the first time in weeks.

The success was not, however, without cost. Tom Reagan had gone overseas with the 28th Infantry Division and had become regimental adjutant for the 110th Infantry Regiment. The 28th, known as "the Bloody Buckets," was pressed into combat shortly after the Normandy Invasion and took part in the heavy fighting after the breakthrough at Saint-Lô. It was on August 1, during this fighting, while establishing an advance command post for the 110th, that Tom Reagan was killed.

Paul Duke, too, was killed while making a road reconnaissance miles away from the nearest friendly troops. A German sniper shot the young major.

Paul Duke's battalion commander, Lieutenant Colonel E. M. Fry, Jr., found himself with the one job a commander dreads most. "I remember the day he was wounded back in July," Fry wrote to Duke's wife. "He came in with a big bandage on his head, and he was given a Purple Heart decoration. The same day his promotion to major came, that afternoon he got three letters from you. He was so elated he could hardly talk, but he didn't miss a minute of duty."

A few weeks after the invasion, the commander of the Seventh Field Artillery Battalion made Bill Hoge the headquarters battery commander. The officer who had been commanding the battery was an electrical engineer who was highly qualified in communications work but not at all interested in the rest of the administration of running the headquarters, so Hoge was instructed to tighten up discipline, motor maintenance, food service, and local security. As Hoge had begun to discover, many of the practices that were being used in the 1st Division bore little resemblance to what was taught at Fort Sill. For example, Hoge learned at Fort Sill that a battalion would normally have four radio channels. In Normandy, however, a battalion was allowed only two because there were so many units crowded together.

On the night of August 1, Omar Bradley's plan was for VII Corps to swing around to the west behind a German corps that was confronting the British. The First Army's VII Corps, including the 1st Infantry Division, 3d Armored Division, and 104th Infantry Division, was in the forefront of the breakout from the beachhead area at Avranches, and Bill Hoge accompanied the 7th Field Artillery Battalion reconnaissance party, which had left the 1st Division assembly area at nightfall. The team continued until midnight, encountering no resistance. Meanwhile, the 16th Infantry had stopped for the night just short of Gavray, and Hoge's team pulled into an adjacent orchard and prepared to bivouac.

They had just begun to unroll their bedrolls when a squadron of German attack bombers came overhead and dropped flares. The Yanks remained motionless, but the Ju-88s came back and dropped what seemed to be a limitless reservoir of antipersonnel bombs. At one point, Hoge thought, "God, they ought to be empty by now."

But they weren't—the bombing lasted through much of the night.

Nearly everyone in Hoge's team got a Purple Heart that night. There had been 11 officers in the party when they bedded down. One was killed, seven were seriously wounded, and only three were still available for duty, among them, Bill Hoge.

As Hoge would recall, the hero of the evening was Sergeant Silverberg, a pharmacist from the Bronx. "I had never liked him before. He was notoriously uncooperative when you tried to get help from the medics for any reason, but that night he was cool as cool could be—deft and skillful. I think it was largely his work that saved many of the wounded."

Another soldier who did a wonderful job of gathering up the wounded and getting them loaded aboard trucks for the trip to the first aid station was a sergeant named LeYangie. When it was all over, Hoge complimented him on the job and said, "Did you once serve a hitch in the medics?" LeYangie said no. Hoge asked, "Where did you learn that trick of rolling up blankets and using them as litters?"

"In the Boy Scouts," the sergeant replied.

Meanwhile, Mike Greene finally arrived overseas. Larry Greene had gone over with the 1st Armored Division in 1942, but Mike—assigned to the 11th Armored—did not reach England until September 1944. Late in 1943, Mike Greene, an operations officer, and the 11th Armored Division were sent to the desert training center in California. Shortly after their arrival, the division's executive officer was killed during a demolition training exercise when one of the instructors dropped a detonator on the table of explosives. In April 1944, Mike became squadron commander of the 41st Cavalry Reconnaissance Squadron when the commander fell ill. Since he was holding the only vacancy for a lieutenant colonel, Mike became executive officer when a lieutenant colonel came into the division to take over as squadron commander and remained in this capacity when the squadron was sent overseas.

On April 3, 1943, all of the Black '41 classmates in the 11th Armored Division—among them Hill Blalock and George Pickett—were promoted to major on the same day. They were all between 24 and 26 years old, and when the post newspaper came out, the headline read, "11th Armored Promotes Minors to Major."

Virgil Bell, the regimental commander who called Blalock and Pickett the "boy wonders," told the new majors, "I was a captain for 17 years and you were captains for six months. There's no justice in the army."

To many men in the Class of 1941, Bell was remembered as one of the men who literally *made* them into combat soldiers. As Pickett would say later, "Virgil Bell knew what he was doing and he just took a couple of young braves that had a year of bouncing around as lieutenants and gave us very responsible jobs. Blalock's job called for a major, but he was a first lieutenant when he was assigned to it. My job called for a major too, but he assigned me to it as a first lieutenant. Then the old man promoted us to captain."

When Mike Greene and the 11th Armored Division arrived in England in September 1944—diverted from a landing in France—they immediately went to the Salisbury Plain, where armored units were being trained for the assault on enemy enclaves along the coast of France prior to being shipped to Europe.

By August 6, the Americans had seized Avranches. The First Army had turned east toward Mortain and Paris, while Patton's Third Army drove south. Suddenly, on August 7, Field Marshal Günther von Kluge, who succeeded Rommel, launched a massive counterattack from Mortain with the objective of splitting the First and Third American Armies and throwing the now highly successful breakout into confusion. When the Germans attacked fanatically at Mortain, the 39th Division temporarily fought them to a standstill, while one battalion of the 30th became surrounded on high ground—a place called Hill 317—just east of Mortain.

It soon became clear that the only unit in a position to blunt the attack was the 35th Infantry Division. The division advanced slowly, not only stopping the German attack on its front, but also pushing the crack SS troops back to the Mortain-Barenton Highway. This accomplished, the now exhausted 35th had the job of undertaking a frontal assault on Hill 317. To achieve this objective, Major General Paul Baade, the division commander, had only one remaining fresh battalion, the 1st Battalion of the 320th Infantry Regiment, commanded by one of the youngest battalion commanders in the United States Army, Black 41's Major Bill Gillis.

The captain of Army's football team in 1940, Gillis had translated his aptitude for leadership into an extraordinary military career. In his first battle as a battalion commander, he was faced with a situation that would have been most perplexing to a veteran. But failure never once occurred to him. Gillis calmly analyzed his mission and looked over his tools. Little did he and the men of the 1st Battalion realize as they assaulted the heights of Hill 317 that they were fighting one of the great battles of the war.

Shells from the 35th Division Artillery filled the air and slammed into the German positions as P-47s bombed and strafed. The 1st Battalion fought its way forward, and nightfall of the first day found them at the foot of Hill 317. Throughout the night and most of the next day, the situation remained fluid. For hours it was questionable as to whether Gillis was gaining ground or whether he was himself surrounded.

However, at dawn on the third day, the 1st Battalion broke through the German lines and relieved the beleaguered battalion of the 30th Division. German documents later revealed that members of the German General Staff decided that the war was lost when the counterattack at Mortain failed. General Eisenhower later described the battle as one of the turning points of the war: "Had the German tanks and infantry succeeded in breaking through at Mortain, the predicament of all troops beyond that point would have been serious, in spite of our ability to partially supply them by airplane."

The 1st Battalion of the 320th Infantry received the coveted Distinguished Unit Citation. Bill Gillis received a Silver Star, the first of many decorations. The citation read, in part: "Throughout this action, Major Gillis, although himself wounded in the hand, accompanied leading elements of his battalion and inspired the troops under his command by his skillful leadership, tenacity of purpose, courage, coolness, positive action, and utter disregard for personal safety."

Von Kluge's counterattack had almost succeeded, but Patton's Third Army beat it back, and by August 13, the Yanks controlled the entire Loire River line from Saint-Nazaire to Angers. In the first two weeks of August, the Allies were able to capture five times as much ground as they had during all of June and July.

The Allied Blitzkrieg

In August 1944, five years after they had added the word "blitzkrieg" to the world's lexicon, the Germans found themselves on the run from one.

The Allied pursuers had doubled the size of the force they landed with in June. The British Second Army now contained four corps and had been joined with the 21st Army Group, along with the Canadian First Army. On the American flank, Bradley now commanded the 12th Army Group, which contained Patton's Third Army and the First Army that was now under the command of General Courtney Hodges. By August 25, the 12th Army Group surrounded Paris—which had already been abandoned by the Germans—and the French 2d Armored Division, which was attached to Hodges' First Army, led the victorious Allies into the City of Lights. By the time Paris was liberated, the Allied blitzkrieg was skimming across the north European plain like a pat of butter across a hot skillet.

On September 11, Patch's Seventh Army linked up with Patton's Third Army near Dijon. In the six weeks since the Allied breakout from Normandy and the four weeks since Anvil, virtually all of France had been liberated, and the Allied combat forces were literally moving as fast as the movement of provisions permitted. Brussels was emancipated on September 3, and Hodges' First Army secured Amiens, Sedan, and Liège by September 11.

Simultaneously, the Soviet forces on the Eastern Front were tying up 60 percent of Germany's armor in their great summer offensive that forced the Germans west of Russia's prewar border for the first time since 1941. Predictions that the end of the war would come before Christmas now did not seem so far-fetched.

Meanwhile on August 15, the United States Seventh Army, which was composed of both American and free French units under the command of General Alexander Patch, successfully landed in southern France. This maneuver was spearheaded by the American VI Corps under Lucian Truscott, who had led the breakout from Anzio three months earlier. In contrast to the Normandy invasion, however, Operation Anvil went ashore near charming Saint-Tropez in perfect weather against minimal German resistance. Except for its defense of the ports of Toulon and Marseilles, the German Nineteenth Army basically broke and ran.

Truscott had been handed a fabulous opportunity to chase down and destroy the Nineteenth Army, thus routing the entire German defense of southern France. However, with memories of Anzio and Normandy still fresh, the Seventh Army was configured to move slowly against moderate to heavy resistance. When they found it possible to move quickly, they didn't have a supply network set up to support a blitzkrieg. Patch sent two French divisions against Marseilles and Toulon, while agreeing to let Truscott's VI Corps chase the Nineteenth Army as far as the supply system would permit. Truscott moved two spearheads north through the Rhone River Valley, catching the Germans in a pincer on August 21 at Montelimar, 250 miles north of the invasion beaches.

The ensuing six-day battle saw a mismatch of disorganized Germans facing Americans at the end of a strained and circuitous supply line. Ultimately the Allies prevailed. The entire Rhone Valley was now in Allied hands, and the Nineteenth Army was destroyed as an operational tactical unit.

In early August, shortly after the capture of Rome, Joe Gurfein was involved in planning for the invasion of southern France. One morning a young lieutenant came in as one of a group of replacements. It was Gurfein's job to see that they had training in demolition, so he said to the lieutenant, "Take 25 men along with explosives and equipment, whatever you need. Wander throughout the countryside and look for blown-out tanks to practice on or a building that is partially collapsed."

About five o'clock that evening, the lieutenant returned and reported that they had found a tall, steel tower and had demolished it. Shortly after he left, Armed Forces Radio News reported that saboteurs had blown up the last remaining radio tower for Radio Free Europe.

Shortly after the Allied troops from southern France linked up with the troops from Normandy, Joe Gurfein was sent to Paris. While there, he had a free afternoon and decided to go to a nearby cafe on the Avenue des Champs Elysée to watch the Parisians pass by while he enjoyed the sunny day.

As he was sitting there, a charming young lady of about 20 approached him, and he asked her if she would like a drink. After a dance or two together, she said, "You are the first American I have met and you are just like I pictured an American soldier would be. I have fallen desperately in love with you!"

She then said she had a hotel room just around the corner and asked him if he would like to go there with her. Gurfein told her he was sorry but he had very little money.

Seemingly embarrassed, she said, "I do not want any money, I just want to make love with you." They had talked for another 15 or 20 minutes, when she finally said, "Well, how much money *do* you have?"

He replied, "Practically nothing—just a few dollars."

At this point, the fair young lady stood up and said, "I see another friend of mine over there, so *adieu,*" and off she went.

As Gurfein and two other men were driving south through central France, returning from Paris, they were stopped by a black unmarked Mercedes Benz limousine, which pulled up next to them. One of the men in the limousine came over to their truck and asked if they had any gas they could give him for his car. Gurfein told him they had only enough to return to Nice, and drove on.

About an hour later, the limo overtook Gurfein's truck again, blocked the road, and three men got out with pistols drawn, saying they *really* needed the gas. Gurfein's truck had a driver in front, and he and the sergeant were in back with a gas can and other items they had picked up in Paris. They also had a machine gun mounted in the rear. Gurfein signaled the sergeant to load his machine gun, so when the three men went to the rear of the truck, they came face-to-face with a .50 caliber Browning. They retreated quickly.

The Yanks drove on, but about 20 miles down the road they were again stopped, this time by a roadblock manned by the French underground, who asked them if they had seen a big, black limousine with three men in it. It turned out that the men so desperate for gas were actually three German generals attempting to escape from western France to Switzerland.

On his return to Nice, Gurfein was given the job of studying the mountain crossings into Italy through the Alps for a possible attack to the east by Allied troops. There was very little available in the way of detailed maps, so he went to the local library to see if they had anything on geography. To his delight, he found a report by Napoleon's engineers on crossing the Alps. Gurfein translated it, updated it—the roads were the same, just better maintained—and submitted it to headquarters. He received kudos from headquarters for his brilliant study.

Northern Italy

After the capture of Rome, II Corps of General Mark Clark's Fifth Army pushed north against active resistance until it reached the Lake Bracciano area, where the 1st Armored Division took on the pursuit action. On June 13, Horace Brown and the 88th Division moved south of Rome to the Alban Hills for refurbishing and replacements, and nine days later, as the Fifth Army advance continued, they moved north of Rome to the Tarquinia area for more training. On July 4, they were alerted that the 1st Armored Division's advance had ground to a halt, and the next day they moved north again to relieve a combat command of the 1st Armored just south of Volterra.

Horace Brown crawled on his stomach to the top of a hill to see Volterra, the division's initial objective. The view was spectacular in the clear, golden light of dawn. Volterra is a heavily strengthened medieval fortress on a good-sized hill on a plain. Brown's heart fell as he realized that their job was to attack it.

As soon as the 349th Regiment was in position, the 337th Field Artillery prepared to attack. On July 8, the 105mm howitzers opened up and the infantry attack began with a holding action in front, while the main effort came from the east, where the hill slope was gentler.

Observing the battle from the south side, Brown saw the finest artillery shooting he'd seen from the 337th. The infantry was held up by machine guns and 20mm fire, and his forward observer with the attack battalion called for fire and successfully knocked out all the guns, which were only two hundred yards ahead of the infantry. After that, the infantry was able to take Volterra and the Germans were pushed into retreat.

Action by action, the 88th Division moved north until, by the end of July, they had reached the Arno River in the vicinity of San Miniato, just west of Florence. The Arno River was now between them and the Germans.

The division was moved to the Villa Maena area to train, refurbish, and prepare for the Arno River crossing. At this point, the battalion's commanding officer was transferred to Headquarters Division Artillery and the operations officer took over the command, so it was decided that Brown would come back to the battalion as intelligence officer and also work with the operations and fire direction people.

On the night of September 25, the 337th moved north with the 442d, forded the Arno, entered Florence with little opposition, and went into position on the western edge of the city. The Allied drive was now moving north opposite the German positions on the Gothic line.

Hugh Foster had gone into Italy as a signal company commander assigned to the 53d Signal Battalion, which was designated to provide all internal communications services for the II Corps main headquarters rear command post and—if one existed at any given moment—the II Corps advance command post.

By this time, that which passed for the "front lines" was often an indistinct, thinly held, and ambiguous no-man's land that existed as a "line" only on maps, so the commanding officer of the II Corps had selected a tentative site for a new advance command post while flying over the front line at 10,000 feet. Hugh Foster was a member of the reconnaissance party sent to check out the site, which turned out to be behind the German lines. The recon party wound up roaming enemy territory, while Foster posted signs marking where the message center, crypto, and telephone switchboard were to go.

Setting up communications for command posts—which were constantly being relocated—was a major part of Hugh Foster's job. Basically, his mission was to verify road access, defensibility, communications access, protective cover, concealment, ground conditions, adequacy of usable space, absence or existence of mines and booby traps, and probably some other items of concern to the engineers. The construction company reps were primarily concerned about wire and cable routes into the site, but Foster's concerns were broader. He had to select dispersed locations for the telephone switchboard van, teletype van, message center tent, crypto van, radio sets, and telephone carrier equipment trucks. The radio sets had be to located some distance from the command post proper to reduce the likelihood of enemy radio direction-finding equipment locating the command post. That necessitated finding a separate, defensible location with reasonable access, from which wires and cables could be run. At the same time, the message center had to be close to an entrance to the area so that messengers would not have to drive around the whole command post to deliver or pick up messages, and it also had to have space to park a couple of vehicles nearby. The crypto van required special security and a position near the message center tent.

The center of the communications area usually needed to be in a convenient location with respect to the rest of the command post layout to simplify and minimize the amount of telephone wire to be installed. Therefore, Foster usually "claimed" a general area at the outset, based on his expectation of where the other elements of the command post would settle. For instance, it was not advisable to place the message center near the commanding general's area; he would not appreciate the noise of motor messengers coming and going all night. So Foster stayed near the II Corps Chief of Staff or Headquarters Commandant while they prowled around and made their choices. Later, he returned to the area they had selected, made adjustments if necessary, and posted signs marking the specific locations.

At this point Foster had to find a bivouac area for his signal company. They could not bivouac *in* the command post area—and would not wish to—but Foster was required to locate his company within two miles of the command post. At times the two-mile restriction could prove to be onerous. On one occasion the command post was set up in a school house in a "terrain bowl"—an oval, flat area with mountains all around. Foster had to place the company in that bowl, so he got as far from the command post and as far up the mountain sides as possible. One night, the Luftwaffe bombed the command post and hit the command post building seven times. All the bombs were duds.

Occasionally, Foster went through this drill only to be told the following day that the situation had changed and he was going to have to look for another command post location. In the latter days of the war in Italy, when the Fifth Army was racing across the Po Valley, the II Corps command post was moved three times in one day. By this time, Hugh Foster was spending all his time in a jeep on command post reconnaissance or looking for a bivouac area for his signal company. He used drivers in shifts, while he slept in the jeep en route.

In most cases the command post locations were reasonably safe and secure, so the process of reconnaissance was generally fairly routine. But that wasn't always true. In Florence, the commanding general of II Corps selected a large city park along the banks of the Arno River. There were numerous paths and roads, and Hugh Foster drove and walked around there for more than two hours posting signs for his elements and checking the road accesses. The next day he was told that the Headquarters of Fifth Army was going to use that area and II Corps would have to look elsewhere.

Foster went back and gathered his signs, and that afternoon, a Fifth Army signal battalion went in to make their wire installations. In two hours they lost 15 men due to booby traps.

For Horace Brown and Hugh Foster, being in Florence was a pleasant interlude. However, it was quickly followed by Brown's worst experiences of the war. On September 20, 1944, the 88th Infantry Division joined the Gothic line offensive for the push across the North Apennines to break into the Po Valley. It was a grinding, day-by-day campaign in very mountainous, desolate country, with the weather worsening almost daily.

It was a mountain-by-mountain battle through the medieval cities and hill towns of La Fine, Cappello, Battaglia. Defense against the violent German counterattacks was like a modern rendition of the Alamo. Casualties were heavy. "I'll never forget the forests in the Firenzuola area," Horace Brown remembered, "where the trees were chopped off head high from the artillery fire."

Fifth Army's advance finally ground to a halt in early November. "It was one of the most frustrating experiences of my life," Brown recalled, "when I crawled to our observation post and could look down into the Po River Valley about 9,000 yards away but knew that we could not get there—at least not then."

Horace Brown was now going into his second "winter line" in Italy. The artillery would now remain in position; the mud and roads would restrict any significant movement until the spring of 1945. The infantry occupied sectors and then was relieved for rest and training, only to later relieve a unit in another sector. The winter line was a time for retraining, refurbishing, and for integrating new replacements into the units.

In early December, Brown was transferred to the headquarters of the 88th Infantry Division Artillery to become S-1 and S-4 (Personnel and Supply), and three months later he was promoted to major. At almost the same time as Brown reported to division artillery, Brigadier General Thomas E. Lewis arrived as the new division artillery commander. He and Brown got along well. Lewis enjoyed traveling to the front line units, and because Brown, unlike many other staff officers, had spent all his time with the infantry, Lewis asked Brown to go with him.

One day, General Lewis called him in and told him that Brigadier General Niblo, the Fifth Army ordnance officer, was flying in for a visit to investigate complaints about the condition of the equipment after nearly a year in combat. He added that he wanted Brown to get Niblo a command car in which to ride around the area, and Brown said he'd get the best command car he could find.

"You will not," Lewis barked. "You will get him the *worst* command car you can find."

Brown then had each battalion send up its worst command car. They were all terrible. He picked the most dilapidated one—doors banging ajar and top sagging—then tacked the general's star on it and sent it off to pick up Niblo. After a while, Niblo arrived at the Command Post, hanging onto the car to keep from falling out, and mad as hell. However, he had gotten the message, and his help in procuring essential equipment markedly increased.

Airpower over Europe

A month after the United States entered the war, the USAAF began to establish units in Europe for the conduct of the war against Germany. The Eighth Air Force was established in England on January 28, 1942, where it joined with the RAF Bomber Command to form the Allied Strategic Air Forces that carried on the strategic air offensive against Germany and occupied Europe.

The Eighth Air Force was destined to become the largest of the USAAF numbered air forces, with thousands of Boeing B-17 and Consolidated B-24 heavy bombers, as well as many hundreds of fighters, not to mention the men who flew them. George Brown and the other men of the 93d Bomb Group, the heroes of the Ploesti raid, had a large number of classmates flying bombers in the Eighth Air Force, including Cliff Cole, 95th Bomb Group; Wharton "Mike" Cochran, 290th Bomb Group; Bill Brier, 323d Bomb Group; Ralph Freese, 381st Bomb Group; Tom Corbin, commander, 386th Bomb Group; Edwin Brown, deputy commander, 401st Bomb Group; Jack Bentley, commander, 429th Bomb Squadron (who was shot down and captured in 1943); Harold Norton, 486th Bomb Group; Dave Kunkel, Commander, 534th Bomb Squadron; Eric de Jonckheere, 613th Bomb Squadron; Lew Elder, commander, 710th Bomb Squadron; and Clint Ball, executive officer, 351st Bomb Group (who took home *three* Distinguished Flying Crosses).

On July 4, 1942, three weeks after Halpro hit Ploesti, the Eighth Air

Force made its first attack on occupied Western Europe—a mission against four German airfields in the Netherlands. It was largely a symbolic Independence Day gesture. The first heavy bomber mission by the Eighth Air Force took place on August 17, more than two months after Halpro. Although this was another scaled-down operation, it gave the Germans a taste of the kind of firepower they would soon be facing at the hands of what would become known as "the Mighty Eighth."

In January 1943 at the Casablanca Conference, the mandate of the Eighth Air Force was confirmed in the now-famous directive that stated that the primary objective of the strategic bombing offensive was "the progressive destruction and dislocation of the German military, industrial, and economic system, and the undermining of the morale of the German people to a point where their capacity for armed resistance is fatally weakened."

Contrary to much of what would be written four decades later, the US Strategic Bombing Survey report, published immediately after the war, showed that this objective had been fulfilled. Meeting this objective, however, was costly in terms of both men and planes. In addition to Joe Tate, three other men from Black '41 gave their lives to the Eighth Air Force's strategic bombing offensive against Hitler's Reich.

One of these was Tommy Cramer, who was assigned to the 68th Bomb Squadron of the 44th Bomb Group. Cramer had been commissioned into the Coast Artillery Corps, but he went directly to the Air Corps' Randolph Field, Texas, from which he graduated in the first wartime basic flying class on January 9, only a month after Pearl Harbor.

The 44th arrived in England on October 9, 1942, and Tommy Cramer was promoted to captain the next day. He participated in several diversionary missions over the European coast during November and led his squadron on its first actual bombing raid—against Abbeville Airfield in France—on December 6, 1942. He completed five missions before February 1, 1943, and won his first Air Medal. Two weeks later, he was on his tenth mission, returning from Dunkirk, when his B-24 Liberator was heavily damaged by antiaircraft fire and forced out of formation. Eight German fighters jumped the crippled aircraft, and three of the four Liberator engines were knocked out. Three of his crew jumped to their deaths over the Channel, but their actions served to lighten the airplane enough that Cramer was able to crash-land in the waves on an English beach, saving the surviving crew to fly again. For this action, Tommy Cramer received the Distinguished Flying Cross on February 15, 1943.

By April 6, 1943, he had received his first Oak Leaf Cluster to his Air Medal then led the 68th Bomb Squadron in the famous May 1943 attack on Kiel, in which he lost five out of 20 B-24s in his unit. On June 6, 1943, he won his second Oak Leaf Cluster.

In June 1943, he wrote: "Out of the first ten pilots I came over with, only two are flying regularly now, but things are looking up since the last two missions. La Pallice and Bordeaux were easy, with no losses, and I don't think we'll be going to places like Kiel anymore; furthermore, the unit is receiving reinforcements finally, and I think we're going to get along okay."

By the end of June, the 44th Bomb Group Eight Balls, along with the Traveling Circus of the 93d Bomb Group, had been loaned to the Ninth Air Force at Benghazi, Libya, in anticipation of the great Ploesti raid. In his last letter, Cramer wrote: "The hot weather has really set in here in the Middle East and I got to swim in the Mediterranean yesterday—you've never seen such blue water. If it weren't for a little barbed wire and a sign announcing the Eighth Army School of Mine Removal, I'd think it was the beach at Fort Monroe. The only females within field glass range are camels and jackasses— I must try to get a more even tan."

Cramer was a major by now, and despite a desire to return home, he assumed the responsibility, as both squadron commander and operations officer, for breaking in one of the new crews on its first combat mission. It had become customary for one of the "older" hands to do this. Cramer chose to do it himself rather than ask someone else to do it. On July 2, he was trying to help a new crew to successfully accomplish their first combat mission when the plane was shot down over Lecce, Italy, with all hands lost.

It took time for the strategic bombing offensive to become effective, and the Eighth Air Force faced serious obstacles during 1942 and 1943, including a lack of aircraft and supplies—the same problems that American forces faced everywhere. But by the beginning of 1944, the Eighth Air Force was capable of putting a thousand heavy bombers over German targets. Operation Argument, undertaken during "Big Week" (February 20–25, 1944), dropped eight million pounds of bombs on Germany, damaging or destroying 90 percent of the German aircraft industry.

In November 1943, the striking power of the Eighth Air Force's heavy bombers was augmented by the medium bombers of the Ninth Air Force, designated for the *tactical* air offensive against German-occupied Europe. While the Eighth was earmarked to shell strategic targets deep within enemy territory, the Ninth Air Force was assigned targets more closely associated with ongoing land operations on the Continent. The Ninth Air Force was to play a key role in the Normandy invasion in June 1944 and in maintaining air supremacy over the battlefields of western Europe until the end of the war.

Another problem that had faced both the Eighth and Ninth Air Forces, but was eliminated by early 1944, was the lack of fighters to escort the bombers to their targets. The bomber squadrons had suffered at the hands of Luftwaffe interceptors. But when the two air forces finally had adequate coverage by North American P-51D Mustangs, the Luftwaffe came in for a rude shock. The scrappy Mustangs not only had the speed and reliability to take on the

Messerschmitt Bf-109 and Focke Wulf Fw-190 fighters, but also had the range to accompany the heavy bombers to targets deep within the Reich.

Two of the pilots in the Eighth Air Force's VIII Fighter Command were Andy Evans, deputy commander, and later commander, of the 357th Fighter Group, and Fox Rhynard, who was originally with the Ninth Air Force's 86th Fighter Squadron and later commanded the 356th and 35th Fighter Squadrons.

Fox Rhynard had flown 84 combat missions in the Mediterranean and rotated home only to sign up for a second turn. Rhynard's second combat tour was as commander of the 359th Fighter Squadron of the 356th Fighter Group, flying P-51D Mustangs, based at Martlesham Heath, east of Ipswich in East Anglia, England. The 359th's missions were mostly as fighter escorts to B-17 and B-24 heavy bombers over Germany, but on occasion they'd be released for a fighter sweep over Germany, a deep patrol searching for enemy fighters—or any other aircraft—to engage and destroy. Often, on return from escort missions, Rhynard would split off half of his squadron to strafe trains, truck convoys, and enemy airfields.

His classmate in Black '41, Andy Evans, had been with the 33d Pursuit Squadron in Iceland in 1942 and 1943. The following year he was transferred to the Eighth Air Force in England, where he served until 1946.

Evans had gotten his wings in March 1942, and in July he went to Iceland, where he was assigned to the 33d. The primary mission of the 33d Pursuit Squadron's P-40 Warhawks was not only to defend the Icelandic ground forces in the event there was an attack, but also to survey the sea lanes that passed near Iceland. Basically, just by being there, they were fulfilling their job. During his 13 months in Iceland, Evans flew 44 of what were termed "operational missions." He saw a couple of enemy aircraft but never actually engaged one. However, one of his fellow pilots, Joe Schaeffer, was the first American pilot flying from an American unit to shoot down a German aircraft in the European Theater of Operations.

In August 1943, Evans came back to the States. Assigned to a fighter squadron at Dover, Delaware, he couldn't wait to get out of the assignment and to Europe. Finally, he went to England in September 1944 and was assigned to the 357th Fighter Group at Leiston, whose primary mission was to escort Eighth Air Force B-17s and B-24s in their bombing runs over Germany. Andy Evans completed 41 combat missions and 220 combat hours. Not every mission, as he points out, was in combat.

"Frequently, our very presence kept the Germans from attacking the bombers, and sometimes, when weather didn't permit the bombers to take off, we would fly a sortie mission over Germany at ground level—so to speak—attacking air bases and whatever other military targets there might be that were appropriate to a low-level attack by fighters."

While at Dover, Evans flew the Republic P-47 Thunderbolt. The venerable "Jug," as it was known, was a highly regarded fighter aircraft, but the P-51 Mustang—notably the P-51D—was considered to be nothing short of sublime, and Evans talked his unit commander into letting him fly a P-51. "I have always been grateful," he said later. "I consider it the best fighter aircraft ever in history."

On January 14, 1945, the Eighth Air Force sent 650 heavy bombers against Berlin. They were escorted by 15 USAAF fighter groups that engaged 250 Luftwaffe interceptors, killing over 100. The 357th Fighter Group itself destroyed 57 German Bf-109s and Fw-190s over Berlin—the largest number of aircraft destroyed in a single air battle over Europe. Andy Evans personally destroyed four enemy aircraft that day. Added to a pair of earlier victories, the four kills on January 14 made him a fighter ace. Two aircraft were also confirmed as having been destroyed by him on the ground, so he was credited with a total of eight aircraft. All of these were either Messerschmitt Bf-109s or Focke Wulf Fw-190s. Evans also encountered the Messerschmitt Me-262, which was the first jet aircraft ever to be deployed in combat, but as he said, "Frankly, I never got into a position to shoot at one, let alone shoot him down."

With the strategic bombing campaign underway from England involving the RAF Bomber Command and the USAAF Eighth Air Force, a new strategic air force, the Fifteenth, had been established in Italy on November 1, 1943. Its role was to conduct a campaign against targets in Italy, southern Germany, southern France, Austria, and other targets, including Ploesti.

Except for those assigned to the Eighth, there were probably more men from the Class of 1941 in the Fifteenth Air Force than any other, and certainly more of them were squadron or group commanders: Edwin "Bud" Harding, commander, 301st Bomb Group; Willis "Bill" Sawyer, commander, 343d Bomb Squadron; Rod O'Conner, commander, 429th Bomb Squadron; Samuel Parks, commander, 456th Bomb Group; Clarence John "Jack" Lokker, 465th Bomb Squadron; Bob Horn, commander, 721st Bomb Squadron; Bruce Cator, commander, 776th and 352d Bomb Squadrons; Fred Ascani, 816th Bomb Squadron; and John Earl Atkinson, commander, 831st Bomb Squadron.

The 465th Bomb Squadron was stationed at Foggia, Italy, in 1944. After he had been in Italy a few months, Jack Lokker assumed command of the 465th and was promoted to lieutenant colonel. He had completed about 40 combat missions and received the Distinguished Flying Cross with three clusters, the Air Medal, and the Purple Heart for missions that included targets in Italy, France, Austria, Romania, Hungary, and Yugoslavia, as well as in Germany itself. On the morning of November 20, 1944, he led a formation of B-24s against an oil refinery at Blechhammer in Silesia, that corner of eastern Germany that is now part of Poland. The weather turned four squadrons back, but Jack maneuvered his 150 bombers through. Meanwhile, radar had alerted the German defenders to expect a smaller than usual force. "Ammunition need not be conserved," was the order.

Through the clouds, target identification was difficult, so Lokker made a last-ditch decision to do a 360-degree turn at the initial point in order to circle the target and get a fix. However, two times around allowed more time for the antiaircraft gunners to track, and the enemy's opening salvos made a direct hit on the midsection of the lead ship. Fighting for control long enough for five of his crew to bail out, Lokker miraculously jumped clear from the plane.

Jack Lokker and Captain Duckworth, his copilot, bailed out together and landed safely, but they were captured by a German farmer, who took them to his farm and left his wife to guard them while he looked for more airmen. About 1 P.M. the farmer's wife permitted Lokker and Duckworth to escape. They headed toward the Oder River, hoping to reach Poland and receive help from the Polish underground. They hiked for about five hours, but at dusk they ran into a German patrol that immediately pursued them with rifle fire.

Duckworth was finally captured, but he was far behind Lokker. The last time Duckworth saw Lokker alive, he was running into a thick clump of underbrush with two Germans chasing him. He was declared missing in action and was not among the men liberated from German prison camps when the war ended. From reports confiscated after the war, it was discovered that his body was buried in a small cemetery in Langsleben in Silesia.

Casualties in the Black '41 class were high among fighter pilots of the Fifteenth Air Force. Among those who died were Marsh Carney and Jim Walker. Elk Franklin met his death during the third week of April 1944, as fighters of the Fifteenth and Twelfth Air Forces were in the midst of air-to-ground attacks in Tuscany while supporting the Fifth Army's slow advance up the Italian peninsula. Targets for that week had included bridges, railyards, and roads from La Spezia to Arezzo, and it was to Arezzo that Franklin was assigned on April 20. He dived against the railyard there and released his bombs square on the target, but as he pulled out, his P-47 suffered a direct hit.

Classmate Mac Home had gone to North Africa as a fighter pilot, but early in the Italian Campaign he became liaison officer with the 64th Fighter Wing. As his own request, however, he returned to the more active combat flying role. Home talked incessantly about the future of aviation, his dreams for a career in the postwar Air Force, and his devotion to the advancement and development of the airplane. Flying fighter planes seemed the only answer to all of Mac Home's hopes and ambitions in the service. The Mustang was his "dream ship." On February 18, 1944, after 76 combat missions, Home volunteered to make an experimental flight to test a new dive-bombing technique. As he was bringing his plane into a vertical dive to permit close observation by high-ranking officers, a structural defect in the plane caused the right wing to collapse, and the plane crashed without his having any chance to bail out. He was posthumously awarded the Distinguished Flying Cross. It was one of many citations he had received, including the Air Medal with Oak Leaf Clusters.

Not So Fast

In September 1944, buoyed by six weeks and hundreds of miles of stunning successes, the cocky and self-confident Allied land armada that had exploded into Normandy and swept across northern France had arranged itself at Germany's doorstep. The 21st Army Group under British Field Marshal Bernard Montgomery, which also contained the First Canadian Army and the Second British Army, gathered in northern Belgium on the Netherlands border. Between Luxembourg and southern Belgium, opposite the German city of Aachen—the gateway to the Roer, the Reich's industrial heartland—was Omar Bradley's 12th Army Group, which contained Hodges' US First Army, Patton's US Third Army, and the newly formed US Ninth Army under General William H. Simpson. In the Alsace, opposite Strasbourg, was the 6th Army Group, which contained Patch's US Seventh Army and the French First Army.

These forces comprised an almost continuous line of steel from the North Sea to the Swiss border. Only in the rugged Ardennes highlands of southern Belgium—the area between the zones controlled in force by the First and Third Armies—were Allied troops spread relatively thin. The area was so rugged, the Allied staff believed, that the Germans were unlikely to counterattack there.

Back in Britain, the First Allied Airborne Army, under USAAF General Lewis Brereton, was ready for deployment. Opposing the recently victorious Allies were two formidable obstacles: the Siegfried line, or West Wall on the border of Germany, and the Rhine River, a natural barrier consisting not only of the wide river itself, but also the steep cliffs that form its gorge. Populating this area were ten million soldiers of the Wehrmacht. It was clear to General Eisenhower and his Army commanders that the relatively easy progress—the Allied blitzkrieg—of August and early September was now over.

Patton's Third Army reached Nancy on September 5, and within a week the 35th Infantry Division faced the German stand on the Rhine-Marne Canal and the Sanon River at Dombasie below Nancy. Major Bill Gillis was a battalion commander with only 45 days experience, yet he was remembered for his great success at Hill 317 and his commanders were starting to see him as the model for the kind of dash, vigor, aggressiveness, and leadership that would be the hallmark of the men who would be needed to lead the US Army after the war.

On September 15, Gillis took his 1st Battalion of the 320th Infantry Regiment on a crossing of the Rhine-Marne Canal and the Sanon River. The Germans stubbornly opposed the crossing with mortar and machine gun fire from commanding positions on the hills. The direct assault over unimproved bridging constructed under intense direct enemy fire was reminiscent of the memorable offensive conducted—also under the command of Bill Gillis—by the Army football team against Notre Dame four years earlier.

As he had been in France a month before, Gillis was in the thick of the battle, amid the tracers and the smoke, leading elements of his troops and moving among them to direct the attack. He waded and swam across the river and canal several times under heavy enemy fire. In the words of his Distinguished Service Cross citation, "His courageous leadership and exemplary conduct under fire so inspired his men that they were able to force the crossing successfully against heavy odds."

As soon as Gillis helped to spearhead the Third Army drive across the waterways south of Nancy, the Germans evacuated the city rather than be surrounded. This gave the 35th Division the momentum it needed to wheel east to establish a front on the edge of the Foret de Gremecy. On the morning of September 30, however, the Germans launched a heavy counterattack. Gillis, being the kind of soldier that he was, immediately went forward to check his battalion's lead positions. "They'll *expect* to see me," he said casually.

The Germans used heavy mortars as well as 88mm artillery to soften the Yank positions in a scenario reminiscent of the situation two weeks before on the Sanon. In the ensuing chaos of exploding shells was Bill Gillis. A mortar round exploded practically on top of him. By the time the medics reached him he was barely alive, but somehow he clung tenaciously to life before finally losing his last battle the following day.

His final combat citation for the Bronze Star Medal read in part: "For heroic service . . . in the vicinity of the Foret De Gremecy, France, on September 27, 28, and 29, 1944 . . . For a period of two days, until he was killed by enemy mortar fire while in the area of one of his front line companies, Major Gillis led the attack of his battalion with tireless energy, inspiring his troops by his constant presence at the front, and displaying sound tactical judgment which resulted in repulsing numerous German counterattacks . . ."

In addition to his Distinguished Service Cross, Silver Star, Bronze Star Medal, and Purple Heart, Gillis was awarded the Distinguished Service Order by the British government, the croix de guerre with silver gilt star and the croix de guerre (with a vermilion star) by a grateful France.

"I read the citation for his Distinguished Service Cross awarded the night Bill Gillis died," said Bill Hoge, with whom Gillis used to sneak out of the

West Point barracks for nocturnal visits to New York City. "He waded back and forth across that river seven times, encouraging his battalion and getting them over the river. That sounds like Billy."

At the age of nine, Bill Gillis acquired his dog Jimmie, and the two became inseparable pals. Nothing was too good for Jimmie, and even on his birthdays Bill would throw a party for Jimmie, complete with candles and treats. The last party they celebrated was when Bill was a senior in high school. The cake had nine candles. While Bill was away at Schreiner Institute and West Point, Jimmie was cared for by Becky, the nurse who had helped raise Bill from the day that he was born. As if in a last gesture of loyalty to his master, Jimmie died from natural causes on the same day that Bill was killed in action.

By mid-September 1944, the major challenge facing the Allies was one of supply. With over 50 divisions on the ground, the need for a continuous river of food, fuel, and ammunition was acute. Seizing the port of Antwerp, one of continental Europe's largest, was of the utmost importance, as was clearing out the German units that controlled the Scheldt Estuary north of Antwerp, that port's access to the North Sea.

To capture this area, and southern Holland as well, Field Marshal Montgomery developed the strategy for Operation Market-Garden, which was to be the most massive airborne operation in history. His plan was to drop the British 1st Airborne, the American 82d—Bob Rosen's unit, and the 101st Airborne Division behind enemy lines around Arnhem in the Netherlands. They were then to retake key bridges in the area as the four divisions in the XXX Corps of the British Second Army hit the adjacent front.

Writing about Operation Market-Garden later, Eisenhower said that it "unquestionably would have been successful except for the intervention of bad weather. This prevented the adequate reinforcement of the northern spearhead, and resulted finally in the decimation of the British airborne division and only a partial success in the entire operation. We did not get our bridgehead [across the Rhine at Arnhem], but our lines had been carried well out to defend the Antwerp base."

In fact, Market-Garden was very nearly a calamity. The three airborne divisions and a Polish brigade were dropped behind enemy lines near Nijmegen, Holland, on September 17, with the British dropping into the the middle of a major German troop concentration west of Arnhem. The inclement weather that Eisenhower blamed arrived on September 18, which prevented air resupply and air support for the hapless 1st Airborne. They fought on courageously, praying for the arrival of the XXX Corps. The corps did arrive on September 22, a day after the paratroopers holding the Arnhem bridge had been pounded into submission. Antwerp itself was taken on September 30, but it was not until November 28 that Allied ships were able to make full use of its port.

Following graduation from command and general staff school in January 1944, Bob Rosen was sent overseas to England to serve in a staff job, but in July he finally received an assignment for which he had long been lobbying. Detailed to the jump school of the 82d Airborne Division, he won his coveted paratroopers wings on July 29. From the letters he wrote home during this period, it is evident that he had, at last, found his place. He was prouder and more satisfied with his job as company commander in the 82d than with anything he had ever done.

The joke that used to go around about Bob was that "in Brooklyn they are saying 'Wait until next year.' Do they mean the Dodgers are going to win the pennant or Bob will graduate?" The *Howitzer* recalled that his classmates didn't know about the Dodgers, but "the borough could count on Bob." So could the 82d Airborne.

Rosen jumped with the 82d into the cauldron of Nijmegen. On September 29, he led a portion of his company in a charge into enemy positions. He paused in the middle of a bullet-swept street to direct his men and exposed himself to continuous sniper and machine gun fire as he moved back and forth through the lines with snipers less than 75 yards away. As he crossed an open street to obtain tank support, he was wounded by snipers but refused to be evacuated until the tanks were in a position to bring effective fire on the enemy. In so doing, Rosen set in motion the events that permitted the Yanks to push the enemy back and to launch a successful counterattack. For Rosen however, it was his last battle—he died of his wound the next day. He was posthumously awarded the Silver Star.

By October 13, the First Army had cracked the Siegfried line at Aachen. They gained control of the city nine days later. Again, they were chronically short of supplies and their movement forward was painfully slow as they headed through the Hertgen Forest and turned toward Cologne.

The 87th Infantry Division went overseas early in October of 1944, with Bill Kromer leading Company A of the 345th Infantry Regiment. In the short period before combat that the 87th had spent in England, Bill conditioned his men physically by long hikes over the hilly countryside. Football games with Companies B and C and company parties on Saturday night were Bill's ideas to keep his company from getting stale prior to their first great test, and it was at one of these parties that two of Bill's men introduced the song and skit entitled "That Mean Old Captain Kromer," the last two lines of which were, "The best damn CO of them all, that mean old Captain Kromer."

After a quick trip across the Channel and an even quicker trip across France, Company A, 345th Infantry engaged the enemy at Metz on December 6. It is recalled as typical of Company A's loyalty and discipline that the first man wounded would not permit himself to be evacuated until he had gotten permission from Bill Kromer. Kromer guided Company A through its initial baptism of fire at Metz, its first attack at Rimling in the Saar Valley, and across the German border into Medelsheim. In three weeks, a green company had become a battle-tested combat unit.

On November 22, the Third Army took Metz and pushed into the Saar Valley. On the same day, the 6th Army Group, storming through the Alsace, succeeded in capturing Belfort. They reached Strasbourg on the Rhine the following day.

The 69th Infantry Division was picked to be part of the Seventh Army for the invasion of southern France, and Ralph Hetherington, who had done such a memorable job at Anzio, was again commanding the 69th Armored Field Artillery Battalion. After the invasion, Hetherington had been decorated and promoted to lieutenant colonel. The colonel who took part in the ceremony wrote to his wife, Elizabeth: "I had the pleasure and honor of pinning Ralph's silver leaf on. I believe he was the first of his class to be promoted to lieutenant colonel."

After the Seventh Army's incredible dash through southern France, they were closing in on the Rhine at Strasbourg. On December 1, 1944, after the capture of the city, Hetherington was completing a reconnaissance mission when the jeep in which he was riding struck an antitank mine. He was killed instantly. Another officer wrote of him, "I don't know of anything that hurt us as much as losing Ralph. He was the finest example of a battalion commanding officer. He was extremely aggressive, and loved by all who knew him."

On the Roer

In the Ninth Army sector north of Aachen, the Germans opened the floodgates of the dams on the Roer River, flooding the region below and making progress virtually impossible. Among those affected was Major George Pickett. A year before, while Pickett was attending the command general staff college at Fort Leavenworth, the 11th Armored Division was reorganized from regiments to battalions and separate combat commands. When Pickett got out of Leavenworth just after Thanksgiving 1943, he was made the operations officer of Combat Command B of the 11th Armored Division.

Combat commands were created in November 1943 to increase the range of control and combat effectiveness of each division and to put maneuverable battalions in position to be easily transferable between combat commands. No combat command had any administrative power. Combat commands were strictly that; they were not set organizations. The combat command headquarters' staff could be used to control maneuver battalions as assigned by the division. From a tactical standpoint, this system enabled a commander to better form his task forces: He could shift his battalions back and forth from one combat command to another. He could place the bulk of his tanks under one colonel, the bulk of his infantry under another, and mix them up any way he wanted to. It was a very effective system, and it ultimately led to the US Army adopting the brigade system, which it has had since the Vietnam war.

By the time that the 11th Armored Division arrived in England, George Pickett was the operations officer for Combat Command B, commanded by Colonel Wesley Yale. Pickett was put on special duty with the the 2d Armored Division, then fighting in Holland. During the first week of November, Combat Command B was shipped to Holland, and George Pickett relieved General I. D. White's operations officer. Pickett had been sent to the 2d Armored for two reasons. First White wanted to give some of his staff a rest; second, the divisional brass wanted to—as was the expression in those days—"bloody" the new arrivals, to give them some experience in the line before the full division was moved into the line. In other words, they wanted some of the senior commanders of the 11th Armored to have had *combat* experience before they put the new division into action.

George Pickett had been with the 2d Armored for nine days when it was ordered into the line. When Pickett arrived in Holland, Combat Command B was bivouacked near the large slag mine at Alsdorf on the north side of the Hertgen Forest. At this time, the Roer River was basically the front line. The reconnaissance squadron of the 2d Armored was at Baumen in an old castle on the Rhine, and the command post was located in a bombed-out school building. The building had originally had three floors and a basement, but the first two floors had been shelled and bombed so completely that the top floor no longer existed. With a little work and ingenuity, however, it was usable as a command post. The Germans used dams as part of their defense scheme. By opening the sluice gates on the dam, they could flood the area with water, and the Allies could hardly maneuver in the mud. The third morning, Pickett woke up in his bottom bunk and turned his head to find the water was about one inch from reaching the bottom of the bed. Everything he owned, including the clothes he was supposed to wear that day, was under water. He carried out his duties the rest of the day wearing a blanket and a pair of boots that someone had loaned him.

When he got back to the 11th Armored Division under orders, Pickett found out that he was executive officer of the 42d Tank Battalion, the unit that he would be with almost to the end of the war. The commander of the 42d Tank was a man named Joe Ahee, who had come to the 11th Armored from the 7th Cavalry. He was a graduate of the University of Arizona and a good battalion commander.

10

The Battle of the Bulge

A Watch on the Rhine

To the Germans, it was an offensive that was known as Operation *Wacht am Rhein* (Watch on the Rhine). To the Americans who fought to turn the Germans back, it was known simply as the Ardennes Offensive. However, once American reporters saw what it looked like on the map of the European battle front, posterity would never know it as anything but the Battle of the Bulge.

Wacht am Rhein was a desperate, yet bold, gamble, a last-ditch effort to force the Anglo-American Allies to drop their demand for an unconditional surrender and accept a negotiated truce. Tactically, the German plan was to concentrate three heavily reinforced armies—the Fifth Panzer Army under General Hasso von Manteuffel, the Sixth Panzer (later Sixth SS Panzer) Army under SS General Josef Deitrich, and the Seventh Army under General Erich Brandenberger—for a massive strike against the weakest link in the Allied line. This was, of course, the mountains and forests of the Ardennes, an area only lightly held by the VIII Corps of Hodges' First US Army, and, due to its rough terrain, theoretically the least likely place for a German counterattack.

During November 1944, a major battle was being conducted in Hertgen Forest as the 2d Armored Division was making plans to attack along the Roer River. The Allies knew that the Fifth and Sixth German Panzer Armies were carried as reserve units to counterattack in the event that the American First and Ninth Armies attacked into Germany. Allied intelligence people insisted that the Fifth and Sixth Panzer Armies were going to be the main German counterattack force. As it turned out, the Fifth and Sixth Panzer Armies were actually the units that the Germans were assembling to break through in the Ardennes.

The German strategy called for surprising and overwhelming the Allies in the Ardennes, then driving to the huge Allied supply dumps along the Meuse. Once these were under their control, the Germans would turn north to capture Liege and Antwerp, thus driving a wedge between Montgomery's 21st Army Group and Bradley's 12th Army Group. If the tactic worked, it would be a masterstroke.

Wacht am Rhein had originated with Adolf Hitler himself and was a clear example of how he imposed himself upon his generals as a great military strategist. He was a ravenous reader of military history and easily imagined himself as a soldier-statesman in the mold of Napoleon or Frederick the Great. His Ardennes plan dated to the end of July 1944, when the Allies broke out of Normandy and began their push eastward across France. It had been only a week since a failed attempt on his life, and the führer was badly shaken, both physically and emotionally. He wanted action, and he insisted on a great counteroffensive on the Western Front.

Within the German high command, Hitler's plan fell in line with the thinking of Field Marshal Wilhelm Keitel and General Alfred Jodl, who felt that Germany's decisive moment would lie in a successful operation on the Western Front. It ran counter to the ideas of General Heinz Guderian, the German commander on the Eastern Front, who naturally believed that the war would be won or lost in *his* sector.

As history now shows, it was a pointless turf squabble. The war was lost on *both* fronts, and Germany's rapidly diminishing reserves would prove to be no match for America's inexhaustible industrial capacity or the Soviet Union's limitless manpower pool. American cannons and Russian cannon fodder: it was only a matter of time. However, Hitler—and by their compliance, his generals—were willing to bet against the odds.

Hitler saw *Wacht am Rhein* as a chance to recreate and relive the brilliant 1940 offensive in which his troops, in a dash through the Ardennes, had outflanked the Maginot line and pushed the British into the sea at Dunkirk. However, while Hitler's geographical situation recalled 1940, the tactical situation was starkly different. In 1940, Germany was *the* undisputed superpower in Europe. Four years later, the Reich was encircled by the largest amalgam of military might in history. Still, Hitler had 24 divisions with which to smash only four American divisions scattered through the Ardennes, and he had the element of surprise on his side.

He arranged for the operational chain of command to run directly from himself to Field Marshal Walter Model to the three armies, bypassing Field Marshal Gerd von Runstedt, Commander in Chief for the Western Front. Hitler wanted no interference with *Wacht am Rhein*, no possibility that any of his hand-picked forces would be diluted or diverted from this important undertaking.

December 16–18

Wacht am Rhein was launched at 5:30 on the morning of December 16, 1944, two weeks behind the original schedule. Nevertheless, it achieved complete surprise along a 40-mile wide front. Dietrich's Sixth Panzer Army led the attack, having been assigned to the north wing with orders to cross the Meuse River at Liege and strike out to form a line between Maastricht and Antwerp. Theoretically, Manteuffel's Fifth Army would then advance to the Meuse, while Brandenberger's Seventh Army would push south to Luxembourg City and cover the south flank.

The first contact occurred between the I SS Panzer Korps and the inexperienced United States 99th Infantry Division in the area of Hofen. Much to everyone's surprise, the 99th confounded the superior SS force by holding its positions. In the early hours of December 17, however, a breakthrough spearheaded by a 1st SS Panzer Division task force under the command of SS Obersturmbannführer Joachim Peiper occurred at Honsfeld.

At Malmedy, Peiper's task force captured a number of Americans and turned them over to SS guard units, who marched them into a field and executed them. The Malmedy Massacre was only the first of several atrocities committed by SS troops over the next three days, which left 350 American POWs and 100 unarmed Belgian civilians dead. According to the subsequent investigation, Hitler had ordered that the lead units in the *Wacht am Rhein* attack should create a "wave of terror and fright and . . . no human inhibitions should be shown."

George Pickett had been with the 2d Armored temporarily until the morning of December 15, when he was ordered out to rejoin his assigned unit, the 11th Armored Division. Though the front line was in Holland and Belgium, pockets of Germans still occupied some of the French harbors, and the 11th Armored was being brought across the Channel to knock out the German defensive enclaves.

On morning of December 16, Pickett and eight other officers were returning to the 11th Armored in a two-and-a-half-ton truck. They were supposed to head south and report to Bradley's headquarters in Luxembourg City. Without knowing it, they drove through Malmedy just after the German spearheads had passed through. If they had been there 90 minutes earlier or later, they would have all been scooped up and shot.

For the first few hours after Peiper's panzers had gone through, there was still a lot of American traffic drifting down that road, oblivious to the offensive. The outfit behind Pickett's stopped to eat and was captured by German troops. Most of them were among those massacred at Malmedy. By noon, however, the news of the Malmedy Massacre had spread, engendering a degree of paranoia, especially since there were some Germans in American uniforms driving captured jeeps.

By the end of the day on December 17, the German panzers had made considerable progress against the 99th Infantry Division and the adjacent 2d Infantry Division. But the northern shoulder of the assault at Hofen was rebuffed by the Americans, and the plan for the Sixth Panzer Army to pivot to the north was frustrated.

December 19

In what seemed to be an ironic symbol of the disasters of World War I, Eisenhower met with Generals Bradley, Devers, and Patton at Verdun—site of the costly Allied victory in 1917—on December 19 to discuss strategy. In both Washington and London, newspapers were now describing "The Battle of the Bulge" as a major disaster for the Allies, but Eisenhower later recalled that around the table were "only cheerful faces."

These men viewed the unsightly "Bulge" on the map quite differently. They saw the German penetration not as a serious threat to the overall Allied front, but as a desperate effort backed by minimal resources that seriously exposed major elements of German military power to vastly superior Allied forces. Everyone at the table knew that it would have been a much tougher situation for the Allies if they had met these same German units after they'd dug themselves in to a defensive line in Germany. "Hell, let's have the guts to let the sons of bitches go all the way to Paris," Patton suggested wryly. "Then we'll *really* cut 'em off and chew 'em up!"

Eisenhower told him the Germans would never get past the Meuse, but Patton's remark highlighted the fact that the farther the Germans went, the more exposed and open to annihilation they were. It also underscored the fact that many commentators at home were suggesting that Paris really *was* in danger. The Ardennes offensive presented more of an opportunity to the Allies than to the Germans, but there was a great deal of damage control to be done before that opportunity could be realized.

The Allied plan called for a temporary army group command realignment, under which Bradley would command all forces south of the Bulge, while Field Marshal Montgomery would command forces to the north, including the US First Army. The overall strategy was to continue a holding action on the north flank, complemented by a three-division counterattack from Arlon in the south, which would be launched by Patton on December 22 or 23. Patton would then punch his way through to Bastogne, which had been reinforced on December 18 by the 101st Airborne Division. From Bastogne, Patton would extend his own penetration toward Houffalize, with the objective of linking up with elements of the First Army.

Bastogne was the central pivot of the Ardennes offensive for both sides, because it was the hub for the seven major highways in the region. Out of an ironically inopportune sense of expedience, Manteuffel's Fifth Army had bypassed Bastogne on December 18, and by the end of the following day— when the 101st Airborne Division, under Brigadier General Anthony C. McAuliffe, arrived to reinforce the city—it became an American island in the German-controlled Bulge.

December 20–22

On December 20, the main thrust of *Wacht am Rhein* passed to Manteuffel's Fifth Panzer Army, which had just taken Wiltz on the road to Bastogne.

As Patton and Bradley prepared for their counterattack, the Germans continued to roll westward. The weather, meanwhile, went from bad to worse. Rain mixed with snow turned to heavy snow. Muddy roads turned into icy slush. With their tanks painted white, winter uniforms for their troops, and sleds to carry supplies, the Germans were much better prepared than were the Americans.

With the First and Third Army counterstrikes still several days away, the only thing that the beleaguered Yanks defending the Bulge could count on was Allied air superiority, but because of the weather, the planes were grounded. These were the darkest days of the campaign for the outnumbered protectors of Bastogne and for those facing the Fifth Panzer Army on the west and the restless Sixth Panzer on the north.

Up north, Peiper's SS task force came close to seizing several bridges over the Meuse before being turned back on December 20. The Americans at St. Vith came under attack at the same time and were forced to withdraw three days later, having taken heavy losses.

Bastogne remained isolated. Supplies dwindled, but morale remained high, with the biggest boost coming from an enemy "ultimatum." About noon on December 22, four Germans entered the lines of the 101st Airborne under a white flag. The message from General Heinrich von Luettwitz was: "The honorable surrender of the encircled town" should be accomplished in two hours on threat of "annihilation" by German artillery.

General McAuliffe disdainfully answered "Nuts!" and Colonel Harper, commander of the 326th Airborne Regiment, hard-pressed to translate the idiom, compromised on "Go to hell!"

Lieutenant Colonel Paul Danahy, intelligence officer of the 101st, saw to it that the story was circulated—and appropriately embellished—in the daily periodic report: "The Commanding General's answer was, with a sarcastic air of humorous tolerance, emphatically negative." Nonetheless, the 101st expected that the coming day, December 23, would be difficult.

Patton ordered the Third Army to move out. "Drive like hell!" he told his tank commanders.

As George Pickett later pointed out, Patton's theory prior to the Bulge was that "a tank-versus-tank fight was like a pissing contest between two skunks." Patton had said that "you didn't fight tanks with infantry, you destroy infantry with tanks and let your tank destroyers destroy the enemy tanks."

"But the first damn thing Patton did when the Germans attacked in the Ardennes," Pickett said wryly, "was to shove every tank he had at them. So there went that theory."

December 23–24

The 11th Armored was just getting ready to make the channel crossing when the Battle of the Bulge erupted. Mike Greene's 41st Cavalry Reconnaissance Squadron went over on the first ship.

When the 41st landed in Cherbourg, they moved out at once, not, as previously planned, to the French coast, but to the Bulge. Greene had been directed to "keep moving to the front as soon as you possibly can, and to deploy into action as you arrive there." The 41st Cavalry drove off the Cherbourg docks in third gear and didn't downshift until they hit the Belgian border. Even as the squadron was arriving at Neufchauteau in the vicinity of Bastogne, the rest of the 11th Armored Division was still disembarking in Cherbourg.

It was Christmas Eve when Mike Greene drove into Neufchauteau. Once on the front, it was clear to him that the lines had collapsed. There were all sorts of wild stories about the Malmedy Massacre and Peiper's SS troops infiltrating Allied lines disguised in American uniforms. "You can't tell the Americans from the Germans," everyone seemed to whisper. On the 41st Cavalry's first night at the front, there was a Luftwaffe strafing attack over the town.

"We didn't know where we were in relation to anybody else," Greene recalled. "We're just trying to get the squadron into position to be prepared to do something when it got to be daylight."

Despite the chaos and confusion, the fortunes of the Allies changed dramatically on December 23 with a sudden break in the weather. Air controllers from London to the Ardennes reported "visibility unlimited." Under brilliant cobalt blue winter skies, Ninth Air Force fighter bombers roared into action, blasting German supply and communication lines and bombarding the panzers on the leading edge of the battle. Meanwhile, just before noon, 241 USAAF C-47 transport aircraft dropped over 140 tons of ammunition, food, and badly needed medical supplies to the still-surrounded 101st Airborne defending Bastogne, which continued to anticipate the promised arrival of the 4th Armored Division, part of Patton's Third Army spearhead.

The day before Christmas also dawned crystal clear, and another 160 transports completed their missions over Bastogne. This time, their cargo included an urgently needed team of doctors who arrived at the encircled city by glider. The accompanying fighter bombers concentrated a great deal of their bombing runs on the Fifth Panzer Army forces that surrounded Bastogne.

As the weather turned turbulent again on Christmas Eve, the Germans regrouped for another major offensive against Bastogne, which was planned for Christmas Day. Orders had arrived from Model's headquarters insisting that the Fifth "lance the boil" that existed within the Bulge.

To the north, at the Sixth Panzer Army front, the 82d Airborne Division, under General James "Jumpin' Jim" Gavin—the man who had given the men of the Class of 1941 so much practical instruction a few years before—held a key segment of the front on the Salm River, a position that was exposed to the Germans on three sides. If the 101st was on an island at Bastogne, then the 82d was on a precarious peninsula on the Salm. Armored divisions and the 84th Infantry were maintaining the line from the Salm to the Meuse at Dinant, while the 82d Airborne faced several divisions of the II SS Panzer Korps. This Korps included the Führer Begleit Brigade, an elite unit that had originally been formed as Hitler's personal SS escort battalion. Its inclusion in the order of battle was indicative of the importance that the Führer placed on *Wacht am Rhein*.

December 25–27

The German Christmas Day offensive began with a nighttime Luftwaffe air strike and a major thrust by the 15th Panzer Grenadier Division into the city of Bastogne from the west at dawn. The German tanks ran into a wall of resistance from the determined Americans, and the Grenadier Division was horribly mauled and forced to retreat.

George Pickett spent Christmas Day on the Meuse River, opposite Dinant, Belgium, with the first elements of the 11th Armored Division to arrive after Mike Greene's cavalry. At that time, everyone still assumed Peiper's SS task force was in full force, but it was determined later that he had literally run out of gas. The 11th Armored was to hold the Meuse River, but it turned out that the Germans never got that far. Later, the Americans were told to go to Bastogne.

On Christmas Day, there was to be no Christmas celebration for George Pickett's 42d Tank Battalion, though they did cook Christmas dinner. The tanks were deployed along the river bank; the command post, as well as the battalion headquarters' kitchen, was located in a bombed-out house. Joe Ahee, Pickett, and two staff officers ate in a room in the house without a roof, under the cold winter sky.

Christmas dinners were available all up and down the Meuse River. The men had gotten an issue of turkey as well as sweet potatoes, peas, fruit cake, and coffee. Everything but the turkey came out of cans. Everybody remembers the sweet potatoes because they came in big number 20 cans, as did the English peas. Typical of a front line, a lot of the men ate off the back of tanks and jeeps.

On December 26, the Germans made another attempt to overwhelm the Bastogne defenders, but a combination of spirited soldiers and American fighter bombers overhead ended their hope of winning this vital crossroads. As the sun rose on December 26, the advanced echelons of the 4th Armored Division of Patton's Third Army rolled into Bastogne from the south, after four rigorous days of battling through the German Seventh and Fifth Panzer armies.

Meanwhile, back on December 16, the 334th Infantry Regiment of the 84th Division had been moving out of the line back to Palenberg, Holland, for some much-needed rest and recuperation. Black '41 classmate Charlie Murrah was riding in his jeep, followed by five kitchen trucks and enough borrowed personnel carriers to move all the foot soldiers to "some relatively good living." However, before they'd gone ten miles, they were stopped by the 84th Division supply officer, Colonel Channon. "Follow the MPs," he said to Murrah. "You're headed for Marche, Belgium."

Charlie Murrah got out of his jeep, put the Battalion personnel officer into it, and told him, "Go get transports with the machine gunners and other riders, and meet us in Marche."

He then got into the cab of the first kitchen truck and followed the MPs, who were standing alone at each crossroads and fork to guide them. By some miracle, both parts of the battalion had settled into a schoolhouse in Marche by midnight. On the way, Murrah's group went through the little town of Hottern, Belgium, and when the borrowed transport got back there after dropping off the 3d Battalion, Hottern had been captured by the Germans.

Normally, a battalion will cover 1,000 to 1,200 yards with a full 800 troops, but the next morning the battalion was deployed to cover 5,000 yards east of Marche with only 520 men. Verdenne, Belgium, was the center of their position, and Murrah put his command post about 1,500 yards to the rear on *top* of a hill to make certain that the radio would reach all the companies.

On December 19, the 116th Panzer Division slammed into Charlie Murrah's 3d Battalion, 334th Infantry at Verdenne. Before the onslaught was contained, five battalions would be jammed into Murrah's original 5,000 yards.

The weather was terrible, so they were not able to mass all the corps artillery on single targets. However, thanks to the artillery's proximity fuses—being used for the first time—the 334th did not lose the position. These proximity fuses exploded the shells 30 feet above the ground and could kill soldiers in foxholes. Murrah's men took 468 prisoners out of two cellars in Verdenne where they had run to escape the shelling.

When it was over on Christmas Day, Charlie Murrah said, "Thank God my retreat plan didn't have to be carried out. It would have, had the sun not come up on Christmas, allowing the Air Force to fly. I'll never forget those planes. You could literally walk from wing to wing, there were so many."

December 28–31

On December 28, Brigadier General Frank C. Collins' 87th Infantry Division and General Charles S. Kilburn's 11th Armored Division were assigned to Patton's Third Army to lead the march from Bastogne to Houffalize that was to be launched on December 30. This offensive would also include the 101st Airborne and Combat Command A of the 9th Armored Division, as well as ten battalions of VIII Corps artillery.

During the attack, Kilburn divided his 11th Armored into Combat Command A, under Brigadier General Willard H. "Hunk" Holbrook, and Combat Command B, under Colonel Wesley W. Yale. This way, the 11th Armored would be able to provide more adequate tank support for the 87th Infantry.

Meanwhile, on December 29, Manteuffel was fleshing out a plan for a German strike, code-named *Nordwind* (North Wind), against the jugular of the Third Army's lifeline into Bastogne. This maneuver, spearheaded by the 3d Panzer Grenadier Division and the Führer Begleit Brigade, was also scheduled for December 30.

The two opposing forces moved south at daybreak. Combat Command A of the 11th Armored got into a shoot-out with the German panzers, and seven M4 Sherman tanks were knocked out at Chenogne. On the western flank, Combat Command B, along with the 41st Cavalry Reconnaissance Squadron and assault units of the 87th Division, met no enemy resistance until "all hell broke loose" near Remagne just after noon.

On December 23, the 87th Division had received orders to move to Reims in France—a more centrally located position for movement against the point of the German bulge. In the 87th Division, there were two men from the Class of 1941. Bill Kromer was Commanding Officer of Company A, 345th Infantry, and Jack Murray was Division intelligence officer.

Bill Kromer's 345th Infantry Regiment had spent Christmas Day on the road and the next three in a field bivouac near Reims. The man known as "that mean old Captain Kromer" used every minute of the time available to drill his men and correct the battlefield mistakes he had observed in Company A. He was constantly striving for improvement, and the men would soon fully recognize how valuable this training had been.

At Reims, Kromer also made no secret of the fact that he wanted the

war to end quickly. He missed his wife and twin baby daughters, Pamela and
Marcia, born just a month after he went overseas. There was something about
spending Christmas thousands of miles from kids he had never seen to make
him want to get it all over with, and *soon.*

After a long, cold ride in open trucks to the vicinity of Libramont, Kromer
and the other company commanders quickly drew up a plan. The enemy was
believed to be preparing an advance to the south, with American cavalry
holding a thin screen. The plan was simple. The 87th Division would attack,
pass through the cavalry, and drive the enemy back.

At that point, the regimental commander looked at Kromer. "Company
A will lead out. Advance guard formation."

Kromer slept little that night. There was too much to do. Company A
had to be ready.

At dawn on December 30, Company A jumped off, but at the outskirts
of Moircy, Kromer's lead platoon walked into a fire fight with the Germans.
Kromer quickly deployed the rest of the company and informed the battalion
commander of what he could make of the situation. Kromer then decided
that he needed a better handle on what was going on, so with Lieutenant
Guy Allee, his forward observer, he went in search of a vantage point to
control the attack.

From a concealed operating position, the two officers watched as Company A's heavy weapons platoon began to take part in the action. Just as
they were starting to make some progress, Kromer got a report that the first
platoon had run into some German tanks farther into town. Unable to see
the action taking place among the buildings of Moircy, Kromer decided that
he'd better move forward to investigate personally. As he stepped from his
covered position, a German machine gun cut him down.

Bill Kromer died within a few seconds, as a first aid man feverishly
attempted to stop the flow of blood. He was awarded the Silver Star posthumously for his gallant leadership in this action. He would never see Pamela
or Marcia, nor would they ever know him. His hopes for future Christmases
would be unfulfilled.

As would be written later in his obituary, Bill Kromer's story did not
end on that cold December day. "He trained us right," said his first sergeant.
"There were times when we thought it was tough, but combat showed us
how right he was and how necessary his training methods were. Captain
Kromer was the finest soldier I have ever known. Bill Kromer may not have
gotten up from the ground that day, but his spirit marched at the head of
Company A, 345th Infantry, throughout the war. The men he trained, fought
with, and died for made him their model. 'Mean Old Captain Kromer' became a legend—a symbol of all that was soldierly in an officer and a leader
whose presence could always be felt."

George Pickett didn't even hear that Kromer was killed until the summer of 1945. "I knew Bill very well when we were cadets," Pickett said, "but not in the war. You're fighting your own little fight and you don't know what's going on a mile away."

The 87th Division's clash at Moircy had been a nightmare, but they had fought amazingly well for an untried unit rushed into battle only 24 hours before. They overwhelmed the Germans that had killed Kromer and the others at Moircy, and by New Year's Eve had captured Remagne as well.

December 30 ended with heavy casualties on both sides, but the Germans had lost their initiative. The wind had been knocked out of *Nordwind* by the Americans, who now had the momentum for a follow-up attack on the first day of the new year. General Kilburn regrouped his 11th Armored during the night of December 30 in order to consolidate the entire division for a drive into the icy roads and knee-deep snow of the Rechrival Valley on New Year's Day. Despite serious traffic jams encountered as the 11th and 6th Armored Divisions attempted to reposition their growing force of tanks during the early hours of December 31, the assorted units under Third Army control managed to assemble a formidable force with which to awaken the crumbling panzer division at the dawn of 1945.

New Year's Eve in the field was miserable and far from festive. At that time of year in Europe it gets dark by 4:30 P.M., so when George Pickett's 42d Tank Battalion got into Rechrival it was already night. The houses in the town, which was a small agricultural village, were laid out in clusters. There were a lot of haystacks within the town, and some of the hay had been brought in to put in the mangers, which were attached to the houses so people could go out their kitchen doors and attend to their animals.

The Germans had left an artillery observer behind in Rechrival, hiding in an attic of one of the buildings. When the 42d Tank Battalion came into town, he continued to monitor its movements. George Pickett stopped his tank, climbed down from the turret, and walked into the very building where the German artillery observer was hiding. He addressed the Belgians who were there in French. "Where are the Germans?"

"They have all gone."

Pickett asked them again, "Are there *any* Germans left?"

"No," they assured him.

Believing them, he continued to the house next door and set up his command post without conducting a search. The German observer stayed in the attic another 30 minutes—long enough to send messages to his own artillery—then he sneaked out and got away.

Later that night, the Germans started shelling Rechrival. Fortunately, the only damage they inflicted was to set the haystacks on fire. Although the Germans were able to adjust their fire by using the burning haystacks, the Americans suffered only five wounded in the task force during the entire night.

January 1–2

The war's last year began in freezing cold under a leaden, overcast sky, as sleet and snow pelted the troops unmercifully throughout the Ardennes.

General Kilburn had decided that the main New Year's thrust by the 11th Armored Division would be made by Combat Command B. After a slow start on the morning of January 1, Kilburn's offensive ran into fierce opposition from the tanks of the 3d Panzer Grenadier Division and the infantry of the Führer Begleit Brigade.

The weather was very bad, and the Germans were wearing white camouflage with white helmet covers, so that in the snow they were virtually invisible. The Yanks didn't have any winter camouflage, so they confiscated all the tablecloths, bedsheets, and pillowcases they could find in Rechrival and started using them for snow capes.

The Germans had also painted their tanks white. By contrast, Pickett's 42d Tank Battalion, with their GI-colored tanks, looked like like bulls-eyes in the snow. The panzers didn't have any trouble seeing the M4s, but the Yanks had a hard time spotting the Germans.

Black '41 classmate Hill Blalock and Combat Command B were the first to go, with an 8:30 A.M. charge against Chenogne. The infantry task force attacking from the south was supported by massed rank and artillery fire, and by noon the town was securely in American hands. A tank task force followed up from Houmont, smashing the German opposition in the woods northwest of Chenogne and rejoining the infantry task force. Reorganized, the command then launched its full power northeast toward the German stronghold in the Des Valets woods. The tank task force led the attack, followed closely by the infantry. The reserve tanks remained in position near Chenogne as Combat Command B entered the Des Valets woods, eliminated all remnants of enemy opposition, then dug in for the night in a position overlooking Mande Saint-Etienne from the north edge of the woods.

Meanwhile, Combat Command A held off until noon, when they were sure that Combat Command B had succeeded at Chenogne. Combat Command A then began by attacking into the heavily defended Hubermont-Millomont-Rechimont triangle. The leading tank force made some progress but was slowed by a heavy German infantry and armor counterattack from the northwest tip of the Des Valets woods. It was to be the worst day of the war for Pickett's 42d Tank Battalion.

That morning, General Holbrook had come up to confer with Joe Ahee, and they decided to assign George Pickett the job of taking the first two companies in as the assault force. Pickett went over the ridge at Rechrival, with the light tank company to protect his right flank. Ahee's task force had

15 tanks in the company when they jumped off that afternoon, and that night there were two.

Pickett himself braved heavy antitank fire to lead a flanking attack over a dangerously exposed ridge above Hubermont. This proved to be the straw that broke the camel's back. The fire from Pickett's tanks, along with artillery and air support that his spotters brought to bear, succeeded in smashing German resistance.

The bloody battle raged through the afternoon, with incredible losses on both sides. Pickett's advance tanks of Combat Command A reached Hubermont at twilight, but because the armored infantry had not kept up with the Shermans and because of the fear of a German counterattack, the tankers pulled back to set up their defensive lines for the night at Rechrival. The 6th Armored Division also gained a great deal of ground during the first two days of the new year, despite heavy resistance from the panzers and the Luftwaffe.

The successes of the 11th Armored had come at a dreadful price. Two hundred and twenty men had either been killed or were missing, 441 had been wounded, and 42 M4 Shermans and 12 light tanks had been lost. The events of the day had earned George Pickett a bronze star. On the other side, it was the beginning of the end as Model and Manteuffel realized that the Ardennes offensive was over and that it was time to assume a defensive posture in order to hold the ground that had been gained.

Combat Command A was supposed to take three towns, and by dark they had taken two of them. The next day, they received word that they were to turn the front over to the 17th Airborne Division. Combat Command A pulled back and their tank units became the back-up force for the airborne troops.

January 3–14

As late as January 3—even as both the First and Third Armies were on the offensive—Hitler held on to the illusion of a potential victory in the Ardennes and so compelled Model to plan for another "lance the boil" attack on Bastogne for the next day. As morning crept over the snowy woods on January 4, the battlefield was fogbound, effectively grounding all of the Allies' tactical airpower. For the next fortnight, the situation would remain the same: heavy overcast with heavy snowfall every day but one. It was destined to be a tank and infantry showdown.

For the next week, panzers and Shermans plunged through the snow, jockeying for position in the hamlets and on the back roads of the hills between Bastogne and Houffalize. On January 3, the Germans made modest gains against the Americans, causing no less a figure than George Patton to comment on the following day that "we can still lose this war."

By nightfall on January 5, Model tacitly admitted the show was over when he pulled Manteuffel back from Bastogne. Nevertheless, the Americans perceived little evidence of a withdrawal until January 11. By then, German commanders were getting edgy. They could see that it was only a matter of days before the Allies closed off the Bulge, trapping troops that would subsequently be needed for the defense of the Rhineland.

The 11th Armored Division regrouped for a major attack along the Longchamps-Bertogne highway northwest of Bastogne. Combat Command A led the drive that began at 10 A.M. on January 13. An hour later, an enemy counterattack of approximately 20 tanks developed on the east flank, but massed artillery fire, adjusted by liaison planes, smashed the German effort. Six enemy tanks were hit and set afire, and the rest withdrew to the northeast.

George Pickett's 42d Tank Battalion went sailing through the mortars without any trouble. The first two tanks in the woods were Joe Ahee's and George Pickett's. They came to a little patch of woods about halfway between Longchamps and Bertogne, where they decided to halt and wait for the rest of the battalion to catch up.

As Pickett said later, it was getting "a little bit lonesome with just the commander and the executive officer waiting for the lead company to catch up." The others arrived four minutes later, and they immediately became involved in a fight with some German tanks that were pulling down the road in front of them.

As they pulled away from the woods, General Holbrook arrived in an M5 light tank with a great big general's star plate on it. All of this was being watched by the German mortar observers, who, with a good pair of binoculars, could see that it was a brigadier general's tank.

Pickett remembers that Holbrook then proceeded to "exhort everybody to greater glory, shook hands and told us we were doing a great job. Then, just as suddenly as he arrived, he turned around and left! The whole time he was there, the Germans never fired a round, but the minute that his tank got over the next crest, about 20 mortar rounds hit all at once in the area where his tank had been. They had used it as a center point at which to fire."

When the Germans ascertained that the 42d Tank Battalion was using that particular patch of woods, they shelled it again and killed practically the whole recon platoon, including the lieutenant commanding it.

It was about two o'clock that afternoon when leading tank elements of Pickett's 42d Tank Battalion cut the crossroads southwest of Bertogne and formed an arc on the high ground to the south and east of the town. Pickett's tanks blasted Bertogne during the remainder of the afternoon, and enemy forces were whittled down to about two companies of infantry with 11 tanks in support, four of which were destroyed and the rest immobilized. When nightfall came, Combat Command A consolidated its position astride the enemy's main line of resistance between Gives and Compogne and remained in a dominating position over Bertogne.

Under cover of darkness during the night of January 13–14, the Germans occupying Bertogne withdrew in an attempt to form a defensive position farther east at Compogne.

"They kept pressuring us to move forward," Pickett recalled, "and I think that the division commander was afraid that General Patton didn't think he was moving fast enough. So things got a little hectic, and you could tell that people were getting nervous."

On January 14 at 10 A.M., Combat Command A resumed its attack to clear the Nom de Falize woods. Meanwhile, the 41st Cavalry attacked in the same direction, along the east flank, to clear the Les Assins woods. Concentrated enemy small arms, machine gun and mortar fire, together with some artillery, slowed the advance, but by 3 P.M. the infantry task force had fought its way to the north edge of the Nom de Falize, and was joined about two hours later by the 41st Cavalry along the Longchamps-Compogne road. Heavy fire across open ground came from the Pied du Mont woods and Compogne, where they ran into heavy German fire and were forced to stop.

After dark that night, representatives of the 11th Armored Division met with members of the 101st Airborne Division at Bastogne to coordinate plans for an attack they hoped to launch the following day. This was also the night that George Pickett had the command post in Compogne, and the Germans counterattacked. Pickett was trying to control the attack from a half-track near a farmhouse, while the Germans were up on a ridge. Shells were hitting the roof of the farmhouse over his head while he was trying to direct the attack and to figure out what was going on.

All of a sudden Pickett looked up and discovered that "there was nobody there but Krauts!" He and the headquarters people started shooting .50-caliber machine guns at them.

"I don't think the Germans were supposed to *retake* the place," he said in retrospect. "I think they had just shoved them in there to hold us long enough so that they could pull out, because if they had pushed their attack, they would have scooped up the whole battalion headquarters. They had flak wagons, but our tanks were all in other places, and they had hit a soft spot."

Sherman versus Tiger

When the 42d Tank Battalion had gone into action on January 12, they had been issued one company of the new M4A3 Sherman tanks with 76mm guns. The 76mm gun on the M4A3 was a far better gun than the 75mm gun on the basic M4 Sherman, but it could still be out-gunned by the Germans. The 76mm gun on the German Mk V Panther tank was like that of the M4A3 Sherman, but it was a long-barreled gun, which had been originally designed as an antitank gun. George Pickett found that his tanks could do better with a 76mm than a 75mm, yet neither would knock out a Mk VI Tiger tank, except if it could be used to shoot it in its back or side. On January 1, the Americans had discovered—to their chagrin—that a 75mm shell from the M4 Sherman would bounce off the armor of the Tiger tank, while the German tank gunners' rounds would go right through the Sherman tank. "This got to be a little disconcerting!" Pickett commented the following day. While the Mark V Panther had a 76mm gun, the gun on the Mk VI Tiger was an 88mm. In the field, the American tank crews ran into far more Mk Vs with 75mm guns than Tiger tanks with their 88mm guns. However, the "88" had an almost mystical reputation, so when a tank crew *did* run into a Mk VI, it never forgot the experience. Thus, the term "88" got to be the standard explanation of everything. "We ran into an '88' " became the vernacular for almost any encounter with German tanks or antitank guns.

Most of the tanks fielded on the Western Front by the Wehrmacht in 1945 were Mk V Panthers. But it was not unusual for the 11th Armored Division to come across a Mk VI.

In the fight at Rechrival on January 1, the 42d Tank Battalion had encountered a great many VIB *King* Tigers, as well as both Tigers and Panthers. The King Tigers were identifiable by their extra long 88mm gun and by their distinctive silhouette, although the Germans had painted them white, which made them hard to see against the snowy landscape.

The Mk VIB King Tiger was the largest tank on the Western Front. It had an 88mm gun and six inches of steel armor plate, as well as about six to eight inches of reinforced concrete on the front slope plate. The Mk VIB moved very slowly, which would theoretically make it more vulnerable, but, as Pickett described it, "If you're going to take on the largest tiger in the jungle, you usually are very careful. When it moved in slow, you were very careful about how he approached you and you approached him."

Some of the later models of the Mark VIB did not have the reinforced concrete, but American tankers were still coping with six inches of steel. Even without the concrete, an American 75mm tank gun would not shoot through the frontplate of the King Tiger tank. The only way the Allies could cope with these tanks was to maneuver to get side shots at them or to set them on fire with white phosphorus, which was done only occasionally. The Americans tried to lob white phosphorus into the German tank, where it might get in the eyes of the German tank commander or blind his crew long enough for the Americans to get under cover.

George Pickett himself ran into two Mk VI Tigers on either side of the road at Houffalize. However, as he pointed out later, "I don't think the 42d Tank Battalion got into an actual gunfight with more than seven or eight Mk VIs. If people started to tell how many Tigers they ran into, you'd want to take it with a grain of salt because the majority of German tanks we saw, both Panthers and Tigers, had been burned up by air strikes. They had never even gotten to the front. In fact, we also found some that had been parked just off the road in perfect condition. They were out of gas, and the crews had just gone off and left them."

However, the German vehicle that was the deadliest of all was not the Panther or the Tiger, but rather the Panzerjaeger (tank hunter). Instead of a turret, it had a 75mm gun mounted directly on the chassis, so that its silhouette was three to four feet lower than that of a tank. This meant that the Germans could shove in and around haystacks and behind stone walls, making it extremely difficult to find and harder to hit. Some of them had 88mm guns, but the bulk of them had 75mm guns. "The damn thing was, they were about half height of a tank, and when they'd paint it white, you just couldn't *see* the rascal in the snow!" Pickett recalled. "They were blowing hell out of us, and we were having trouble even figuring out where they were firing from."

As Pickett soon discovered, when advancing, the GIs became the *target*. "The enemy sees you coming before you see him because he's concealed. Consequently, he usually gets off the first few shots. That's like getting the first couple of punches in a prize fight. Sometimes, that's all it takes. This is where the Panzerjaeger was most deadly. As we went forward, he opened up. So as we were advancing, the low silhouette and high velocity gun of the Panzerjaeger became a tanker's nightmare. That little mobile full-track, with a high-velocity gun and armor protection, was our nemesis because that little baby could *move*."

In general, German infantry weapons were seen by many on the Allied side as superior to those that the Americans had developed. For example, the German Sturmgewehr 44 was the original assault rifle, the weapon from which the term 'assault rifle' originated. "If you ran into a bunch of Krauts with those, you were in trouble, because we didn't have anything to compare with an assault rifle," George Pickett recalled. "It had both the full automatic and semi-automatic features. A lot of us felt that the US Army–issue Thompson M1928 submachine gun—the 'Tommy gun'—was worthless in competition with a gun like the Sturmgewehr 44 because the Tommy gun was useful only up to about 50 yards. A Sturmgewehr shot a 7.92mm Kurz round that was a lot like the 7.62mm round the Russians had designed into the AK-47. The case length on the round shot by the AK-47 was roughly comparable to the case length that the Germans used on the 7.92mm round, the '8mm short,' so they had a round that a sniper could use very effectively at around 300–350 yards. When a German got within 75 yards of his target, he just flipped the switch and went to full automatic, which gave him a far better advantage close in. An American with a M1 rifle could compete with a German with a Sturmgewehr 44 at 300 yards, but when the German got within 50 yards, there was no competition," Pickett also explained.

The assault rifle fit the German tactics perfectly. Americans had a tendency to form skirmish lines and move forward under rifle fire to gain superiority. What the Germans did in reaction to this was to set their machine guns and mortars up to the American position, and then their infantry would run forward—using every bit of cover and concealment—until they got within 50–60 yards of the American position. Then they would attack, throwing grenades and firing fully automatic weapons. The Americans found, to their horror, that such a volume of firepower at 50 yards was very effective.

The Night That Belonged to Mike Greene

In terms of morale, supplies, airpower, and sheer overwhelming numbers, the Allies had a clear superiority. On January 14, Field Marshal von Runstedt had personally appealed to Hitler to authorize a withdrawal from the Ardennes before it was too late. Neither the great Prussian field marshal nor the little Austrian corporal had ever met an American major named Mike Greene, but it didn't really matter. It was too late. The night of January 15–16 belonged to Mike Greene.

Picking up the pieces after two brutal weeks of fighting, the 11th Armored Division made slow, steady progress during January 15, and it appeared as though they were on the threshold of being able to "pull the string" on the garbage bag that was the Bulge. On January 16, Pickett's Combat Command A was to continue its attack to secure a bridgehead on the river east of Rau de Vaux, advancing rapidly on the Bertogne-Houffalize road to seize its assigned portion of the front. The 41st Cavalry Reconnaissance Squadron was to continue as far east as Houffalize.

In an effort to make contact with VII Corps, Combat Command B was to assist the attack of 101st Airborne Division units at Vaux as well as Rachamps, then attack and seize the southeast tip of the woods a mile north of Vaux and probe northeast along the Noville-Houffalize highway.

Combat Command A moved out at 8 A.M. on January 15. An infantry bridgehead over the Rau de Vaux allowed the quick construction of a treadway bridge by engineers, and the tank force then pushed rapidly northeast through Mabompre in pursuit action. The infantry task force followed.

The squadron command post was at Bertogne late on the afternoon of January 15 when General Holbrook came up with instructions from General Patton. Holbrook approached Greene and told him simply, "Take a task force and, no matter what, get around to the north and west and establish contact with the First Army near Houffalize."

Greene put together a small task force composed of Troop D and E (the assault gun troop) and the 2d Platoon of Troop A in Company F, the tank company. The 2d Platoon, commanded by Lieutenant Ellenson, and the tank company, commanded by Lieutenant Mullins, were up in the northeast of Bertogne in the village of Rastadt, where they had been in contact with the enemy, and action was still going on. They had been instructed to wait there and join the task force when Greene arrived.

With Troop D and the assault gun troop, Greene left Bertogne around 5:30 P.M., just as twilight faded into night, advancing northeast and east to Rastadt. The snow was quite deep, and by the time the last vehicle rumbled out of Bertogne, it was pitch black. Meanwhile, they could hear the sounds of gunfire as the battle continued to rage off to their right.

One platoon was sent ahead of the column as a point unit in order to conduct necessary route reconnaissance. As they were cutting through farm fields just north of the Pied du Mont woods, they encountered a minefield. Visible in the snow were recent vehicle tracks, so in order to avoid delay, Greene ordered his men to follow these tracks across the minefield. They had just started to advance when the left front wheel of his half-track hit a mine, and Greene was thrown to the ground. Uninjured, he continued to lead his men across the minefield and through a patch of woods to rejoin the trail to Rastadt.

When they reached Rastadt, the tank company and the Troop A platoon were assembled as they had been directed. Troops of Combat Command A task force, driven from positions east of Velleroux, were falling back into Rastadt, which was still blazing with fires set during fighting earlier that day. The town's inhabitants were in a state of complete confusion and excitement. Light from the burning buildings was the only illumination in otherwise total darkness. Consequently, the noncommissioned officers were having a difficult time reorganizing their units. Greene called a meeting of all group commanders to discuss strategy. When they moved out from Rastadt, Mike's group consisted of four reconnaissance platoons, three assault gun platoons, and a tank company.

"When I departed Rastadt that night," Greene recalled, "I did so very reluctantly. I had no conception of the potentialities of the next few hours. Indication of enemy locations was confused, as no daylight reconnaissance of the route ahead had been made, and I had visions of our being another heroic, but uselessly lost, battalion. However, I felt that I had a compact, effective force and a sound, workable plan of operation."

In the region through which they were moving, the only road of any consequence was a single-lane dirt road from Rastadt to Bonnerue, thence north to the Ourthe River at Grinvet; all other routes were logging trails through the forest. The woods were quite dense, due the government's reforestation program. Since it was dark, Greene used a flashlight to read a trail map as he and two other commanders led the advancing column on foot, walking ahead of the vehicles to guide them.

After a grueling 12-hour journey, at around 6:30 the next morning, just as traces of dawn began to appear in the eastern sky, the column emerged from the woods and descended into the Ourthe River Valley. As they rounded a turn near an unfinished water mill at the foot of a hill, their goal came into view some 600 yards to the east. There, perched on the ridge ahead, stood Houffalize.

Although Greene believed that his task force had not been seen, they began drawing fire from their right flank. Since the city was on a hill and they were approaching it from the flats, Greene ordered the assault gun platoons into firing position. He sent Troop D to the right to clear a position on the high ground that would provide him with an overview of Houffalize so he could direct the fire of the assault guns.

At about the same time, Greene could see troops advancing toward his men in the snow from the north. At first he feared that they might be German reinforcements attempting to outflank him, but as they drew closer, Mike recognized the lead element of the First Army's 2d Armored Division. They came across the river and established contact with his task force just below Houffalize.

Mike Greene pulled off his glove and shook hands with the commander. The Bulge was closed. The German troops in it were now cut off from the rest of the German forces. For his part in sealing the Bulge, Mike Greene received a Silver Star.

Meanwhile, the main attack of the 11th Armored Division, which was being led by Combat Command A, rumbled up the main road. When they got to where Greene had positioned his Troop D, Greene pulled his men out, and the division took over the responsibility. Later, Greene's squadron was broken up to be used as reconnaissance units spread over the front, and it rarely operated as a single unit again.

In January 1990, the town of Houffalize invited the Americans back for a 45th anniversary celebration. "There were about six of us from the 11th Armored Division there—five from the 2d Armored Division and two from the 17th Airborne," Mike Greene recalled. "We enjoyed an all-day celebration. It was really quite interesting to go back 45 years later. By coincidence, one of the fellows of the 11th Armored Division who came to the anniversary celebration had been with me in the operation.

"What I can't get over is that, standing there in broad daylight, we could just barely see those little logging trails through the forest. How we could have seen anything back then in the pitch dark, I don't know!"

The Little Black Book

On the morning of January 16, George Pickett's 42d Tank Battalion was part of the 11th Armored Division's main thrust to exploit the tactical advances made possible by Mike Greene's hard night's work. However, shortly after they moved out, they found themselves stymied because of terrain and a German roadblock. They had to figure out a way to flank the opposition.

All at once, Pickett and Joe Ahee noticed two Belgian kids on skis.

That sight ultimately turned the whole situation around for the 42d Tank Battalion. A half hour later, Pickett and the battalion S-3 took 42 tanks down a ski slope. "If you've ever seen a tank on a ski slope, you have seen something!" Pickett said later.

The slope was too steep for the tanks to go straight down, so they had to be angled and turned. They "skidded and banged and boomed around" but were able to bypass and outflank the German position, and from then on it was smooth sailing—or skiing, as the case may be.

"There weren't any Germans shooting at us going down that hill, but I'll tell you, it was a thrill!" Pickett laughed. "We had tanks scattered all over hell, but the Lord was with us."

Between the paved road and the ski slope, there was another little hill, only about 50 feet high. When the tanks got to the bottom of the ski slope, they immediately started going uphill again, and as they did so, they lost momentum. When they hit the road—which was covered with ice and snow— one of the tanks overturned in the ditch.

Just as Pickett was trying to figure out how to upright the toppled Sherman, he heard sirens and horns and looked up to see a jeep with a four-star general in a gleaming helmet. It was General George Patton. Behind him was what looked like an army of newspaper people. "What happened?" Patton asked. "How'd that tank turn over? Who took those tanks down a ski slope?"

Ahee told him the whole story. Patton spoke sharply to Pickett for letting the tank turn over, then shook his head, smiled, and directed his motorcade to move on.

As it turned out, not a tank was lost. The only casualty was a broken leg suffered when the tank turned over. The 42d had tank recovery vehicles with cables, and they had the tank back in operation before nightfall.

Patton's own famous remark on the subject was, "If you could take tanks through Louisiana, you could take them through hell, and that's why the first armored units had trained in Louisiana in 1941."

Joe Ahee was decorated for the ski slope incident, but because Patton had caught him with a tipped over tank, Pickett thought it was the end of his career. However, it turned out that Patton had listed Pickett in his little black book where he kept the names of potential battalion commanders. Several months later, Pickett was promoted.

When the north-south and the south-north offenses of the 11th and 2d Armored divisions met in force in Houffalize that afternoon, the "boil" was effectively pinched out.

The Battle of the Bulge was finally over.

11

Victory in Europe

On to the Rhine

It had been only 30 days between the time Model launched *Wacht am Rhein* and the night that Mike Greene closed the Bulge. The Germans had sustained 200,000 casualties, the Americans half that number. By this time, the Allies had lost 30 days from their schedule, but the enormous quantities of men and matériel that Hitler had sacrificed in the Ardennes had greatly weakened his defense of the Rhine River Valley and, as such, probably *saved* the Allies at least 30 days. Furthermore, some of the best combat units in the Wehrmacht had been destroyed or painfully battered. The losses taken in the Ardennes were irreplaceable.

By January 20, 1945, Hodges' First Army and Patton's Third Army were ready to move forward again, and south of Strasbourg, Patch's Seventh Army had just smashed the German forces in the pocket around Colmar. All the Allied forces were now within striking distance of the Rhine, and Eisenhower was determined to allow the enemy no respite.

Between February 4 and 9, the Allied leaders—Roosevelt, Churchill, Stalin, and their staffs—gathered at Yalta in the Crimea for what would be their last meeting and the last Allied summit before the final collapse of Germany. Stalin, the host for the conference, ran the show. He planned the agenda, and his bottom line was empire-building. Roosevelt was haggard and in ill health. With only two months left to live, he was not the same forceful world leader that he had been when the "Big Three" had last met at Tehran in November 1943. Had he been, the shape of postwar Europe that finally evolved in the early 1990s would probably have come much sooner.

There were a number of issues on the table at Yalta, but the centerpiece of the conference was the fate of Germany and Eastern Europe. The Big Three divided the still-unbeaten Germany into five slices. There would be one for each of them, plus one for France and another to compensate Poland for territory that Stalin had seized in 1939 while he was still allied with Hitler and that he refused to give back.

The exhausted Roosevelt was willing to make concessions to the wily Uncle Joe, hoping that in the peaceful postwar world the Soviet dictator would mellow into a more magnanimous gentleman. Stalin agreed with a sly wink to his Allies' proposal for free elections in Poland, Czechoslovakia, and Hungary. He knew that the only electing in postwar Eastern Europe would be done by the man who controlled the Red Army, because the Red Army would control Eastern Europe.

Tactically, the Yalta Conference discussion centered on which armies should take the major capitals still in Axis hands. The Anglo-American Allies had taken Rome and Paris; the Red Army, Warsaw. Stalin felt that the Red Army should also take Vienna, Prague, and Berlin. Churchill strongly disagreed. He wanted the Anglo-American forces to take Berlin, the symbol of the German Reich and the city whose capture would signal the end of the war. Roosevelt, weakened by illness, was too tired to object, but the Prime Minister was determined. Berlin, as far as Churchill was concerned, would fall to whoever got there *first*.

The Red Army, meanwhile, had turned south to sweep Hungary and the Balkans after securing Warsaw; it had yet to initiate a final push into Germany. The British, French, and American forces, having beaten the Germans back at Colmar and in the Ardennes, were poised for the final drive to Berlin. The Siegfried line had been penetrated, and now all that stood between them and the capital were the fabled headwaters of Teutonic myth and legend—the Rhine.

Even before the Bulge was closed, the Germans withdrew most of the remnants of the Fifth and Sixth Panzer armies from the Ardennes. They were, in effect, pulling out their main tank strength. Consequently, when the 11th Armored Division moved out on the morning of January 21, it didn't encounter heavy resistance. The Germans had mined the area very carefully, however. George Pickett ran over one of their mines, which tossed him up and out of the jeep.

"I remember watching the world turn over twice, and I landed in a snowdrift."

As he groggily got up and tried to walk away, somebody yelled, "Keep still. You're in the middle of a minefield."

Pickett stumbled, turned over, hit the ground, and passed out. When he came to, he was in the aid station of the 55th Infantry. The explosion had covered him with smoke and dirt and charred his clothes, so he looked worse than he was. However, his leg *was* injured and he couldn't walk, so he was evacuated to a field hospital where he was kept for a week while his leg was wrapped and put in a cast.

Pickett went back to the 11th Armored Division in time to join the 42d Tank Battalion as it moved up to Manderfeld on the Siegfried line in early February. For this action, the 42d Tank Battalion was attached to the 87th Infantry Division, and the balance of the division—three infantry battalions—was at Gross Kampenburg to the south.

As soon as the 87th Infantry Division blasted a hole in the Siegfried line, the 42d Tank Battalion and its attached units—now known as Task Force Ahee after Joe Ahee—pulled out and went south toward the Kyll River. On February 8, the Allies initiated their own "Watch on the Rhine," a massive offensive that was to provide the opening bars to Hitler's *Götterdämmerung*.

Joe Gurfein was at Allied Airborne Headquarters outside of Paris, helping to plan the Rhine crossing. It was his job to study the best plan for dropping paratroopers on the east side of the Rhine. To his dismay, he discovered the land near the river was subject to flooding. The Allies were afraid the Germans would open the dams to make the area unusable. It became necessary for Gorfen to find out how far back the floodplain extended before dry ground appeared.

Remembering his discovery of Napoleon's map of the Alps in 1944, Gurfein decided to go to the library. Sure enough, he found an 1888 map there that showed the highest point of flooding on the Rhine. He could thus easily pick the areas of dry land that the Allied leaders needed.

By March 5, the 11th Armored's Combat Command A, with Task Force Ahee as a key element, had reached the Budesheim-Wallersheim are. The key to a rapid advance to the east now lay in getting across the Kyll River.

With the capture of Gerolstein on March 6, the 90th Infantry Division was able to erect a Bailey bridge across the Kyll at Lissingen, and at 6 A.M. on March 7, Combat Command A was ordered to cross the Lissingen bridge site and attack through the 90th Infantry Division bridgehead in the direction of Kelberg.

Task Force Ahee had begun crossing the bridge at 10:30 A.M., but in the meantime, rain mixed with snow had turned the countryside into a swamp, making movement more difficult for the task force as it closed in on Kelberg. When the task force finally arrived around nightfall, they discovered that the Germans had blocked the road leading out of the town, and that the town was zeroed in with a brigade armed with *nebelwerfers*—the multiple rocket launchers GIs had nicknamed "screaming mimmies."

About five miles behind them, the road went around a hillside that had a sheer drop on one side and sheer rise on the other. The Germans had moved in self-propelled guns to cut the road at that point, but this shelling didn't affect Task Force Ahee, which it was already *in* Kelberg.

Kelberg was a critical road center on the master ridge between the Kyll and Rhine valleys and was the last hope for an enemy stand. It was a typical old-world town where all the buildings were built of stone, and many had stone fences. As a result, some of the houses could be, and were, made into forts. When the Germans fired their artillery, every round hit the Americans. Task Force Ahee had taken the village, but the Germans had now blocked the exit, so when the Yanks got into the town, they were forced to stop. The Germans almost certainly had an artillery observer in the village itself, because as soon as the American tanks stopped, a "screaming mimmies" barrage began. Since the Germans had the range down cold, rockets continued "banging and booming" the tanks.

Although they did not have the command post set up in town at that time, Joe Ahee and George Pickett came into town with a jeep at 5:45 P.M., just as the command tanks appeared. Ahee's operations officer had also come in and was walking down the road when a rocket hit nearby. He was carved up with so many shrapnel fragments that he had to be evacuated for two months.

In the next salvo, Joe Ahee got a round that lodged itself between his testicles. Pickett was then faced with the most embarrassing thing that an executive officer could have to do for his commander: pull the fragment out.

The fragments from a *nebelwerfer* are very crude. Usually, the metal on a hand grenade is striated in such a way as to yield what's called a "pattern fragment" when it explodes. In other words, the metal will break along the corrugations. On a *nebelwerfer* the fragments were more like ordinary shell fragments, except that they were machined from a much coarser grade of metal.

Pickett yanked on the fragment. It was stuck! He could see the end of metal through Ahee's pants. Ahee, who was in a great deal of pain, said hoarsely, "Pull that damned thing out!"

Pickett said, "I can see it, but it's stuck!"

"Well, jerk it out!"

"Joe, if I jerk it the wrong way, it will ruin you!"

"Just do it. Now!"

Gripping the fragment tightly, Pickett jerked it out. Ahee just stood there, with blood dripping out the bottom of his pants, and the two of them stared at the fragment just extracted. It turned out to be a lot bigger than either of them had imagined.

Next Pickett had to get both Ahee and his operations officer back to the medics. He found that the first aid station was back up the road, so he left the two wounded senior officers under cover in one of the buildings. In the meantime, the leading tank company had shot up the German barricade; Pickett directed them and an infantry company on the other side of Kelberg toward Andernach to set up an outpost. He then contacted the Combat Command A headquarters, and it was quickly brought forward. Additional in-

Left: Alexander "Sandy" Nininger. He was awarded Black '41's only Congressional Medal of Honor. (Courtesy of the U.S. Army)

Below: At 12:30 P.M. on April 9, 1942, silence descended over the Bataan Peninsula as the defenders, among them Black '41's Hector Polla, surrendered unconditionally. (Courtesy of the U.S. Army)

Left: 1st Lieutenant Horace Brown and his wife Chick at a Garden Party in Muskogee, Oklahoma, in June 1942. (Courtesy of Horace Brown)

Right: Mike Greene as a lieutenant in the 42nd Armored Regiment, 11th Armored Division at Camp Polk, Louisiana, in 1942. (Courtesy of Michael J. L. Greene)

Below: Captain Hamilton Avery (upper right) and part of the crew of the *Daisy Mae* at Benina Air Base in Benghazi, Libya, in 1943. (Courtesy of Hamilton Avery)

Paul Skowronek receives the Air Medal from General W. D. Crittenberger for combat missions over Italy and Rumania during World War II. (Courtesy of Paul Skowronek)

Andy Evans, commander of the 357th Fighter Group in the 8th Air Force, in England in 1944 after he had shot down eight German aircraft. (Courtesy of Andrew Evans)

Bob Rosen was a paratrooper with the three Allied airborne divisions being dropped behind enemy lines near Nijmegen, Holland, on September 17, 1944. (Courtesy of the U.S. Army)

Several Black '41 classmates took part in the Battle of the Bulge, countering the last great German offensive of World War II. (Courtesy of the U.S. Army)

George Pickett (far right) with the Duke of Saxe-Coburg-Gotha, shortly after American forces captured Coburg, Germany, in April 1945. (Courtesy of George Pickett)

Above: Within a month of the closing of the "Bulge," American troops were streaming en masse into Germany itself. (Courtesy of the U.S. Army)

Right: Colonel George "Blizz" Moore competing in the 400-meter cross-country run of the pentathlon in the 1948 Olympics. He earned a silver medal in the event. (Courtesy of George Moore)

Joe Gurfein (center) and Linton Boatwright (right) receive the Croix de Guerre from French Lieutenant General Montclar after the battle of Heartbreak Ridge in Korea, October 1951. (Courtesy of Joe Gurfein)

George Pickett went to Korea as an armor officer only to find that there were no armored infantry units or armored divisions in Korea, only recon tanks and armored artillery scattered among the various divisions. The last big IX Corps tank operation in Korea was on September 26, 1952. (Courtesy of the U.S. Army)

Right: Mike Green served as battalion commander of the 510th Tank Battalion in Germany, in 1955 and 1956. (Courtesy of Michael J. L. Greene)

Below Left: Paul Skowronek and the Soviet attache in Potsdam, 1965. Skowronek served as Chief of the U.S. Military Liaison Mission from June 1963 to May 1967. (Courtesy of Paul Skowronek)

Below Right: General George Pickett, commanding the 2nd Infantry Division in Korea, 1966. (Courtesy of George Pickett)

Left: Lieutenant General Pete Crow served as the Comptroller of the U.S. Air Force from 1968 to 1973. (Courtesy of Duward Crow)

Right: General George Scratchley Brown served as the Chief of Staff of the USAF from 1973 to 1974. From 1974 to 1978, he served as the Chairman of the Joint Chiefs of Staff, making him the highest ranking member of the class of 1941. (Courtesy of the U.S. Air Force)

fantry came in a combined tank–infantry attack, supported by artillery. By 6:20 P.M., the town was under Allied control, and a complete breakthrough was completed.

All through the night Doc Richert, the battalion surgeon, patched up wounded on the side of the road by the light of burning tanks. He even amputated one man's leg by the light of a burning vehicle. The night was pitch black, so Combat Command A was directed to suspend the attack for the night and in the meantime bring up all trailing elements. Temporary defensive positions were posted around the town, and Company B of the 63d Armored Infantry was sent forward a mile to the east of Hunnerbach to take control of a stream crossing and establish a line of departure for the next day's operations.

Meanwhile, the first units of the First and Third Armies had reached the Rhine on March 6. Eisenhower was fully expecting that getting a bridgehead across the Rhine would take weeks because the retreating Germans were blowing up all existing structures after using them to make their crossing. Even though the American troops had been told that securing an intact bridge was of the highest priority—Eisenhower had even authorized a case of champagne for the unit who found such a bridge—no one really expected it to happen. With their customary thoroughness, the fleeing Germans had wired every single bridge on the Rhine gorge, and one by one they were demolished—all but one.

At 1 P.M. on March 7, a patrol from the 9th Armored Division of the First Army—the unit to which Sears Coker of the Class of 1941 was assigned—emerged from the woods on a high bluff overlooking the Rhine about ten miles south of Bonn. Below them in the hazy light of the gorge, Coker saw the Ludendorff railroad bridge at Remagen. It was still standing.

In the chaos of retreat, the bridge's defenders had faded away, and the Germans had been reluctant to blow it up because their artillery was scheduled to use it as an escape route. As the Americans of the 9th Armored approached the bridge, the Germans attempted to detonate the explosive charges on the span, but nothing happened. The charges failed to explode.

Infantrymen, supported by M26 Pershing tanks, swarmed onto the Ludendorff. By 4:30, the Allies poured across the 1,069-foot bridge facing only weak, sporadic fire.

As the First Army secured the Remagen bridgehead and took Bonn and Cologne, Patton's Third Army turned south, moving parallel to the Rhine to link up with Patch's Seventh Army. Patton had advanced quickly to seize Trier on March 2; so quickly, in fact, that before he could tell Bradley of the victory, a message had come down from the 12th Army Group Commander that Patton should not risk attacking the city. He wryly told his boss that he had already liberated the city and asked the sheepish Bradley whether he should give it back!

Bradley let the Third Army keep Trier, and Patton scampered off to swallow Coblenz, situated at the confluence of the Rhine and Moselle Rivers, on March 7.

Despite the rain, snow, and wind, Patton was passing orders down from Third Army Headquarters that the 11th Armored would have to advance, at all costs, on March 8. Combat Command A moved out at daylight on March 8 with the 41st Cavalry leading. At 1:25 P.M. leading elements reached Mayen, and shortly afterward troops entered and secured the city against light resistance. At 6:15 P.M., contact was made with the 4th Armored Division five miles to the east.

When the Americans reached the Rhine at Andernach that morning, they discovered that the Germans had erected a bridge and were trying to retreat across it. The 63d Armored Infantry attempted to capture the pontoon bridge intact because no one in the task force yet knew about the Remagen bridge about 15 miles to the north. The standing orders from the Third Army were simply that any time anyone found an intact bridge over the Rhine, they had to get that bridge no matter what the cost.

The 63d went in quickly, but the natural line of withdrawal into Andernach was such that by the time the 63d got into position to attack, it was outnumbered. The Germans held the bridge until they got most of their equipment across, and then they blew it up.

When an enemy is concentrated and in a formalized defense position, the attacking unit is typically at the disadvantage, especially if it is outnumbered. In an armored unit, the advantage comes, as George Pickett describes it, when "you break them up and keep them running."

George Pickett never did get to see what the 63d was doing because his orders were to send three small forces out to try and block the roads on which the Germans were still coming into Andernach. Although the task force had broken through, the enemy was still behind them. "Lord knows how many Germans were behind us," Pickett recalled. "We had just broken through and kept going through. The Germans got orders to pull out and they started coming in behind and were drifting across the roads we had just driven down!"

On this particular morning, while the Germans were fighting on the east side of the Rhine, Pickett moved north along the west side to block several of the places where the Germans were trying to set up other pontoon bridges. He captured a great many prisoners, but there was not much fighting. Most of the enemy action involved the Germans trying to hold their bridgehead so they could get as many people as possible across the Rhine.

The Colmar Pocket

While the Battle of the Bulge captured the headlines and most of the attention of Allied planners, the pocket of German resistance that still remained at Colmar proved to be an even more long-lasting problem. Colmar, located 35 miles south of Strasbourg and 150 miles south of Bastogne, was situated in the Alsace region, that slice of land adjoining the Rhine that was part of either Germany or France, depending who won the previous war. From 1871 to 1914 it was in Germany. After 1918 it was in France, and after 1940 it was back in Germany, but as 1945 opened, it was in the path of the 6th Army Group.

The First French Army had reached the Rhine near Mulhouse to the south, and the Seventh US Army had taken Strasbourg to the north. General Devers had assumed that the former could take out the Berman Nineteenth Army. But weather, combined with the unexpected tenacity of the Nineteenth Army, frustrated the French. After January 20, with the Bulge pinched off, Eisenhower gave Devers the resources he needed to be able to throw four American divisions against the Colmar pocket.

On January 25, however, it was the Seventh Army's 42d Division—the famed "Rainbow" Division that had distinguished itself in France in 1918—that was in the forefront.

Fran Troy of the Class of 1941 had arrived in Europe in November 1944 with the 42d Infantry Division. Just over two months later, he took over the command of a company in the 242d Infantry. On the first day, as his regiment began its bloody task of eliminating the Colmar pocket, he was killed by a sniper bullet.

A Guest of the Reich

Tuck Brown had gone overseas in March 1943 as headquarters battery commander of the 132d Field Artillery Battalion of the 36th Infantry Division and had taken part in the invasion of Sicily. At Salerno on September 12, 1943, Brown and two other officers found themselves pinned down by small arms fire for over two hours. They fought back, but one of the men was killed, and soon after, a German infantryman got close enough to throw a grenade, killing the other and knocking Brown unconscious with a concussion. When he awoke, he was a guest of the Reich.

Once in custody, Brown and the other prisoners were stripped of their equipment and loaded on a train bound for Luckenwalde in Germany, a POW cage presided over by a sadistic officer called Hauptman Williams.

After they arrived at the prison, the Germans took everything but Brown's shirt, trousers, socks, and shoes. With much shouting and shoving, they threw him into a solitary cell, where he remained for the next 12 days. It was a dank, cold cell, and he soon became ill. Having no handkerchief, he blew his nose on his socks and put his bare feet in his shoes.

As he was throwing Brown into his cell, Hauptman Williams calmly reminded him, "We have not yet reported you as captured, so we can kill you any time if you do not cooperate!"

There were no windows in Brown's cell, but an electric bulb burned day and night. The door was solid wood with a covered peephole at eye level on the outside that only the guards could push aside. The guards never spoke to any of the prisoners and only pointed when Brown pulled a cord indicating that he had to use the latrine. Once a day a one-armed Russian placed a bowl of soup with a spoon in it on the floor near the door to Brown's cell.

Because he had disposed of the branch insignia from his uniform before being knocked unconscious, Brown's interrogation was a strange experience, and he was treated as though he were an infantry officer. On his fourth day in solitary, a single sheet of paper was slid under his door. It listed his correct name and rank—although his branch and unit were wrong—and contained a list of questions for him to answer. However, it was not until the eighth day that the Germans slipped a pencil under the door. On the eleventh day, he was finally interrogated by Hauptman Williams.

"It makes no difference that you have not written answers to my questions," Williams said. He went on to boast about how smart German intelligence was and showed Tuck the roster of an infantry outfit commanded by a Captain Brown.

"The Americans will lose the war because we are . . ." Williams said, letting the words hang for a moment, ". . . simply smarter fighters."

On the twelfth day, Brown was released from solitary and sent to OFLAG 64 near Suzubin in Poland. There the group he was in was passed in an inspection by American security and then deloused and assigned to a barracks. Then the "Bird" began to sing. The "Bird" was a POW who read the latest news from BBC. The radio was well hidden from the Goons (German Officers Or Noncoms), the worst of which was Hauptman Zimmerman, a dyed-in-the-wool Nazi. The inmates always had watchers keeping track of Zimmerman and his two security officers, whom they had nicknamed "Ferret" and "Weasel."

After becoming an experienced "watcher," Brown was assigned the task of "Officer in Charge of Dirt Disposal" for the inevitable tunnel project. OFLAG 64 was built on black loam, but below the surface was yellow clay. The prisoners were sure that Ferret and Weasel would not miss a clue like yellow clay, so Brown's dirt disposal command had a vital function.

Captain Don Stewart (Class of 1940) and Brown, using Stewart's engineering manual, devised a plan to put the dirt from the tunnel in cardboard boxes and place the boxes in the attic of the barracks. The tunnel was to start in the center of Barracks 38, under a cast iron kettle that sat on a concrete floor surrounded by ashes from past fires. Don and Tuck calculated what each beam would hold and found that the attic could take sufficient dirt to complete the tunnel.

Tunnel work had started in November 1943 and continued through

June 1944. Lieutenant Colonel Yardley was in charge of the entire project, under the general supervision of the "Escape Committee." Diggers stored dirt in the tunnel and underground changing room until nightfall, when a chain of men would bring the boxes of dirt to the outside door, fire-brigade style. There was no inside entrance to the attic, so the prisoners had made a small door just above the main barracks door. Brown's crew of eight would go into the attic, and Yardley would sit on top of the barracks door and hand up boxes of dirt to Brown, who passed them on, again fire-brigade style.

By June 1944 there were only about two hours of real darkness each night, so the inmates of OFLAG 64 had to work quickly, and the process was not without its moments of near disaster. One night, Sid "The Mouse" Waldman slipped off of a beam and broke through the ceiling, crashing to the barracks floor below. After making sure that Waldman had survived in one piece, everyone promptly went back to work. By 5:30 the next morning, the barracks ceiling had been repaired and the GOONS never noticed.

Another heart-stopping moment came a few nights later when the Weasel managed to sneak past the Allied gate watchers to within 50 feet of the tunnelers before he turned on his flashlight. Yardley and Brown were the first to see him. He was evidently blinded by his own flashlight, as he never saw the team as they started climbing down from the attic. Brown, who was only about 20 feet away, went last. Yardley quietly closed the upper door and the men all ran to their bunks. The Weasel never found out.

By June the tunnel project had passed under the electrified fence and the barbed wire. The mass escape was only a week away when a coded message was received in a BBC broadcast that indicated the United States government was ordering a stop to all mass escapes because the Germans had just murdered 55 British officers who had escaped from Stalag Luft II at Sagan.

After the tunnel work stopped, Tuck Brown was made "Shower Officer." He was responsible for supervising the men when they brought their bedding and clothes to the ovens to exterminate the bedbugs and while they took showers themselves. All twenty shower heads could be controlled from a single faucet, and everyone had one minute to get wet and three minutes to soap up. Brown turned the water on again for three minutes so they could wash the soap off. "The good part of the job was that it was warm in the shower room," Brown recalled. "However, the bad part was looking at so many bodies which were no more than skin and bones, and realizing that I looked just as bad."

The months dragged on as the erstwhile tunnelers—now roughly 1,500 US Army officers—patiently shuffled through the camp routine at OFLAG 64, listening to the Allied progress on the BBC and wondering how long they would have to remain there. Christmas, and its attendant hopes and dreams, came and went with only a few packages from home, courtesy of the Red Cross.

At noon on January 20, 1945, the Germans notified the inmates that they would leave OFLAG 64 the next morning. There was no indication of where they were going, but there was jubilation among the men over the abandonment of the hated camp. That evening the prisoners burned chairs and tables, and the barracks were warm for a change. Tuck Brown and his buddy "Treetop" Truett (Brown never learned his real first name) made a small sled and got ready to leave.

The next morning, after delaying the departure as long as possible to let the Red Army get closer, the Allied prisoners straggled out of camp behind Oberst Schneider, the German commandant, who rode in a surrey drawn by two mules. A bizarre procession headed west toward Germany and away from the advancing Russians. There was no east-bound traffic.

The prisoners walked in the left lane, the German refugees in the right lane. It was a surreal scene—a flashback to the 19th century. There were covered wagons pulled by oxen, sleighs pulled by horses, and a few troikas with strange, high harnesses over the horses. The OFLAG 64 guards walked down the middle of the road. At the end of the bitter cold day, the prisoners were grateful to sleep in a barn.

January 22 was still quite frigid, and now the prisoners could hear Soviet artillery behind them. They came upon a sprawling manor house that stood at the end of a lane and had small shacks on either side for the Polish workers. That night they slept in the plantation's huge cow barn, which held over 100 cattle and had a concrete floor and separate stalls for each animal. Truett and Brown put their blankets behind bulls and milked cows for food. The city boys slept behind cows, and in the middle of the night Brown heard screams, then swearing, as the uninitiated got soaked.

The next day, the group began their march in a fine mist that froze as it hit their faces. Again, there was an endless wagon train of refugees. Beside the road they could see a locomotive with a large plow attached to it tearing up the railroad ties.

The road they were on followed the tracks, and after dark the prisoners came to the railroad station. Just as they were bedding down, they heard a woman crying out in pain. Running to investigate, Tuck Brown and a couple of other officers found her deep in the throes of labor. A US Army Medical Corps officer offered to help, but she would not allow an American doctor near her, screaming that she would rather die first. That night one of the guards reported that it had reached 40 degrees below zero, but the Yanks never learned whether the German woman or her baby had survived until morning.

On January 24, there was no letup from the penetrating cold, but at least there was hot water available, and it tasted good with coffee and the black bread that some of the men had saved.

As they marched through a deserted Polish village whose name they never learned, Truett and Brown slipped away from the column and hid in a small house. They were trying to figure out what to do next when some German soldiers located them about two hours later and put them back on the road at the end of the column. Behind them was a four-day march in sub-zero temperatures across war-torn Poland. Ahead of them there was only the same road and the same frozen, snow-covered countryside.

Many of those who couldn't take care of their feet lost toes, but Brown's wife had sent him two pairs of woolen socks and a pair of fleece-lined slippers for Christmas. He kept one pair of socks pinned inside his coat and alternated socks at each rest stop. He put the slippers on at any long halt.

The next morning, the prisoners awoke to find the German guards gone. Colonel Paul Goode (Class of 1917), the senior American officer, took charge and decided to send patrols to try to contact the Russians. "The rest of you just sit tight," he directed.

As the sun rose in the brittle, gray sky, the Germans returned as mysteriously as they had left. As it turned out, the German officer of the guard had just told his men that the war was lost and they should go home. When Schneider discovered what he had done, he had the officer arrested and requested new guards. The new guards, who arrived at 2 P.M. that day, were Latvians. One of them told Tuck Brown, "We hate the Russians more than anyone in the world. The Germans we hate the second most." The Latvians had even refused to learn to speak German, despite the fact that they fought against the Russians while wearing German uniforms. They said that the country they liked best in the world was the United States. As Brown observed, politics sometimes puts people in impossible positions.

On January 27, the prisoners were each issued a quarter loaf of bread and a dab of margarine. Around 9 P.M., the Germans reported: "The United States and Britain have accepted the unconditional surrender of Germany. United States troops will be arriving soon in Danzig. If the Russians do not stop, the United States and Germans would fight them." Later that night, the prisoners found out that this information was just German wishful thinking.

During the night, one of the Russian prisoners died in the pig pen which they had been using as a latrine. However, things went on as usual the next morning. It was snowing heavily, and the guards decided to wait until the snow stopped before the pathetic column moved out. There was a cup of potato soup for each man at midday, but the snow continued, so the prisoners spent the entire day in the barn.

On January 29, the snow stopped and Schneider ordered the forced

march to resume. Walking was difficult because of the deepness of the newly
fallen snow. In fact, one of the guards dropped dead that evening just as
they came upon the barn where they were going to stay that night.

Truett confided to Tuck Brown that he had gone about as far as he
could go. When the Germans said they had freight cars for those who could
not keep up, both Brown and Truett opted for the freight cars. After much
shouting and pushing, the guards managed to cram 65 Yanks into a cattle
car that was big enough for 8 cows or 40 men—if most of them stood. The
German guard—the 66th person in the car—was an old man who gave the
Americans no trouble.

As his eyes became accustomed to the dark, Tuck Brown discovered
that his classmate James "Diamond Jim" Forsyth was also in the car. Forsyth had been a company commander in the 320th Infantry Regiment of
the 35th Division when he had been captured in 1944. Brown had seen him
in passing at OFLAG 64, but they had been in different barracks. Now the
two were reunited in a fetid cattle car in windswept northern Poland.

It was soon determined that Tuck Brown was the senior officer in the
car, so he was in command. He assigned Forsyth to one end and John Shirk—
an all-American football player from the University of Oklahoma—to be in
charge of the other end. There was not enough room for everyone to sit
down simultaneously, so they further divided into small groups of two or
three. In each group, one would stand, one would sit, and one would try
to lie down. Rotation times varied by group. Hammocks were slung in the
four corners of the car for those who were in the worst shape. The car had
a small stove with a chimney out the roof, but though it was cold outside,
the men did not light it. In such a crowded space, body heat alone was
sufficient to keep them warm.

Late that night, another car with American POWs was attached to
Brown's, and the train pulled out. However, around noon the following day
the two cars were disconnected and parked on a siding, and that evening
they were connected to a refugee train. Early the next morning, the train
stopped and the Yanks watched as men, women, and children got out of
the refugee train to relieve themselves. They took out a number of their
dead—men, woman, and children—throwing their bodies down beside the
tracks.

On January 31, the two carloads of GIs were unhooked from the refugee
train and reattached to a freight train. They had stopped at a village, and
several of the Americans went into a local shop and traded cigarettes for
bread—"*Brot für Cigaretten, bitte?*" Brown was able to get two huge loaves,
and when the traders divided the bread up among all present, each man
ended up with about two slices.

On the first day of February, the train crossed the German border and
arrived in Stettin (which is today part of Poland). Tuck Brown approached

a group of Red Cross people there, but they would give the Americans nothing. The following day, nearly two weeks after they had left OFLAG 64, the Americans found themselves in the sprawling Tempelhof marshaling yard in Berlin. They went from one railyard to another, asking for food, before they finally found some Germans who would issue a single cup of barley soup to each man. It was the first food that they had gotten from the Germans during the six-day train trip.

The German capital was in horrendous shape. The destruction from the Allied bombings was incredible. Only the shells of buildings were standing for as far as the eye could see from the railroad yard. The Americans spent their first night in Berlin sleeping in the same cattle car, under nominal German guard. The next day, they heard the air raid sirens. The guards and train men scattered to shelters, leaving the Yanks locked in the freight car. About six feet off the floor of the car was an opening about six inches high and 30 inches long, for ventilation. Several men jerked this open, and they were able to see hundreds of Eighth Air Force B-17s against the blue sky. The din of the 88mm antiaircraft guns and the thunder of the 500-pound bombs was fearsome. The Americans watched as several B-17s were shot down and the crews in their white parachutes, drifted down over the city.

Suddenly, the bombs started to hit the ground all around them. The freight car shook, and all the men were enveloped in dust. The raid, which lasted over an hour, was actually the second largest raid of the war on Berlin, and the Tempelhof marshaling yard was the primary target for Edwin "Ted" Brown's 401st Bomb Group. Brown didn't know that he had two classmates in the middle of his target area that day. A year later, when they were both at the University of Illinois getting master's degrees, Ted Brown showed Tuck Brown his aerial photographs of the raid, with the railyard completely obscured by dust and smoke.

After the raid the railroad workers, who were mostly big Russian women, brought in ties and rails, and within an hour they had repaired the damaged track in the railyard. Before the day ended, the Americans were shuttled through two other Berlin railyards en route to Stalag III-A at Luckenwalde.

The barracks at the camp were cramped, with bunks stacked five high. For the entire group, there was only one small table and four chairs. With nowhere else to go, the men spent most of their time in their bunks in the crowded barracks.

From February 5 through April 18, Tuck Brown and Jim Forsyth shared their compound at Luckenwalde with several hundred American and British prisoners, about a hundred Poles, all the Norwegian resistance fighters captured by the Germans, and five Italian doctors. The Italian doctors were assigned to Brown's personal care by Colonel Walter Oakes, the senior Allied officer, because they spoke no English and Brown could communicate with them in halting French. They had been turned over to the Americans for safekeeping because the Germans said they might be killed if the other Italian prisoners got to them.

The reason the doctors were so despised was because of what they did while they had been assigned to an Italian POW camp near Stettin. Italian enlisted men were given 350 calories of food a day and worked until they could work no more, at which time one of the doctors would paint a cross on the man's head. Then one of the German guards—or so one of the doctors told Tuck Brown—would hit the man on the head and kill him.

One day, one of the Italian doctors put his finger on Brown's face. "The impression I made did not spring back," he said. This means you have edema—or dropsy—but I think you might recover with proper food."

Brown had also been concerned, as he thought the excess fluid might be nephritis, which he had had when he was in high school. A few days later, the Germans took the Italian doctors away, and no one ever saw them again.

Luckenwalde had many compounds with fences between them. The Russian and French compounds were the largest. The Americans were unhappy with the French because they each received an American Red Cross parcel every week, and when the Americans asked that they share the parcels, the French refused. The Americans did not get a single parcel in two-and-one-half months of confinement at Luckenwalde.

Late in February, some RAF officers were brought into the same compound where Brown was. The British were a different lot, and they gave each of the Yanks half a loaf of bread, which could be stretched out over several days and augmented the cow turnips. These turnips were as large as a person's head, and eating them was like chewing on wood. Soup made with cow turnips was horrible. The only meals that were worse were those made of grass soup, which was so harsh that many of the men couldn't keep it down.

In late March 1945, the German doctor who sat in on all the German staff meetings let it be known to the American officers that the commandant had received orders to kill all the POWs. The commandant had stated that this must be done in an orderly manner, so after a staff study lasting into early April, the Germans devised a plan: They would shoot the RAF officers first, then the American officers.

In the meantime, the Allies had concocted a plan of their own to confuse things for their German captors. They exchanged uniforms. The first time RAF officers lined up to be marched out of the camp for execution, the Germans found that about one-third of those in the formation were actually Americans. In their thorough way, they had checked IDs and then run the Americans out of the ranks, with much shouting and waving of arms.

Another staff study was undertaken so that the Germans could rework their grisly plan. This second study was never completed, however, because most of the German officers really didn't want to kill the prisoners. They just wanted to stall and save their own lives, too.

Verlautenheide Ridge

On February 22, 1945, the Ninth Army was preparing a major assault—called Operation Grenade—to cross the Roer River, with an objective that included seizing the city of München-Gladbach north of Cologne. The First Army's VII Corps was detailed to protect the Ninth Army's right flank. Within VII Corps was a microcosm of the 1st Division's 7th Field Artillery Battalion and a battery commander named Bill Hoge.

Verlautenheide Ridge was a very unusual piece of terrain that ran generally in an east-west direction. At its east end was the village of Eilendorf, and at the west end the village of Verlautenheide. Eilendorf, which is close to the city of Stollberg, was in Allied hands, while the Germans still held Verlautenheide and the beech woods to the east of that town. The eastern end of the ridge was higher, and the ridge flattened out in a plateau as it approached Verlautenheide. On the south side, which was in American hands, the ridge sloped up gradually through a beautiful, lush meadow. On the north side—then controlled by units of the Fifth Panzer Army—the ridge fell away precipitously. The slope was almost vertical, although it was covered with small saplings and underbrush. There was a stone quarry on the northeast side of the ridge; because of the excavations, the slope there was also vertical.

Verlautenheide Ridge is the highest ground in the vicinity, and from its summit the view is unobstructed. In clear weather, you can see for miles across the north German plain to München-Gladbach. With high-powered binoculars, Bill Hoge could actually see tanks being unloaded in München-Gladbach.

"This area was an artilleryman's dream," Hoge said years later, comparing the valley below him that day to a film he had seen about the migration of wild game across the African veldt. "There were all sorts of troops out there—infantry on foot, horse-drawn artillery, trucks, Volkswagen jeeps, and tanks—about everything you can imagine. Mostly, they were all headed in our direction."

At dawn on February 23, as Operation Grenade jumped off, Hoge immediately began adjusting fire on the targets that looked most lucrative, but many of the targets were out of range of his 105mm howitzers. In the meantime, VII Corps had attached the 186th Field Artillery Battalion, with its 155mm howitzers, to the 7th Field Artillery under Hoge's command, and these weapons had a much better range.

Company E of the 16th Infantry was dug in along the crest of the ridge. The riflemen were delighted to have fire support. As soon as Hoge completed adjusting fire on one target, several riflemen would holler, "Hey, captain. Shoot *my* target next." Hoge would then determine as quickly as possible who had located the most likely target, and train his guns on it. This assistance from the infantrymen saved a lot of time.

Hoge's position got a little mortar fire late in the afternoon of February 23, and he himself took a couple of small mortar fragments, one in his right shoulder and one in the third finger of his right hand.

At daybreak the next morning, the Germans launched a very determined attack on the ridge. "We could hear them as they were climbing up the slope," Hoge recalled. "It was quite dangerous to look down over the slope because you were silhouetted against the sky and always drew fire, but we could toss grenades over the sides, and we inflicted a great number of casualties this way. They kept coming though, and damn near got us. At one point, Company E's commander asked me to call for fire on our own position, but I refused. The ground on the top of the ridge was very rocky, and our own men were not well entrenched. I figured that we would suffer worse than the enemy."

When the fighting had subsided, a BAR (Browning Automatic Rifle) man named Cox, who had occupied the foxhole next to Hoge's, called Hoge to come over to his hole. "I want to show you something."

A Wehrmacht corporal with the top of his head blown off lay, only a few feet from Cox's position. Cox had heard a slight noise and popped up with the BAR at the ready, firing from the hip. If Cox hadn't gotten the corporal just at that instant, the man would have killed both Cox and Hoge with his Schmeisser machine pistol. He also had a "potato masher" grenade stuffed in the top of each boot.

"If Cox had been eliminated," Hoge said, "I think they could have rolled up that end of our position and maybe seized the whole ridge."

Meanwhile, the Germans down in the beech woods to the left had nearly broken the American platoon that confronted them. They had brought up a platoon of Mk V Panther tanks to assist them, and some of the Yanks started to pull back. Luckily, just at the critical moment, the Panthers got confused and shot up their *own* infantry. That was the end of the threat. Late in the afternoon, Hoge watched through his binoculars as the German infantry slipped away through the beech trees toward its own lines.

The next morning, February 25, Bill Hoge was hit just below the knee by a fragment of an artillery shell from a 77mm field piece. The incoming round sounded like light artillery. He had seen the batteries of horse-drawn field artillery on the first evening he was on the ridge and had tried to take them under fire, but they had disappeared into a pine forest before he was able to do any serious damage. The hit became a costly one for Bill Hoge. Although the wound itself was not serious, the nerve in the back of his leg had been severed, and his foot was partially paralyzed. The war, for him, was over.

Happy Birthday, Major Pickett

By the first week of March 1945, the 11th Armored Division had captured Andernach and the 4th Armored Division had captured Coblenz, both of which were *on* the Rhine, but neither of which had an intact bridge *across* the Rhine. Meanwhile, immediately to the north, the 9th Armored Division had captured the Remagen bridge on March 7 and was crossing there.

Eisenhower then decided that instead of pushing everyone across the Rhine at Remagen, he would have the Third Army cross the Nahe and several other intermingled rivers. The 11th Division Armored would cross the Nahe River and head south toward Worms. At the same time, the Seventh Army would be attacking northeast from the Alsace area. Eisenhower wanted to form a pincer to destroy the German army groups that were in the Saar-Moselle Triangle.

The first units of the 11th Armored Division crossed the Nahe River at daybreak on March 19 and headed to Worms. Once the 11th Armored broke through the German line and went on to Worms, it began to encounter enemy positions with no depth to them, places from which the Germans would fire a few rounds and retire.They were beginning to see that the end was coming, and the fighting became a lot softer and easier.

Task Force Ahee was sent to clear a German roadblock north of Simmern, and by midnight they were on the banks of the Nahe River about five miles from where they would finally force a crossing. They knew that they had to get across the Nahe, which was well defended. March 20, 1945, Major George Pickett's 27th birthday, was the day it would be up to him to get Task Force Ahee across the Nahe. Pickett's 42d Tank Battalion had sent the recon out and had discovered that one of the bridges at Martinstein was still intact.

Joe Ahee consulted with Pickett.

"We'll have the lead tank company ready to rush it," Pickett recommended. "I'll take the recon platoon, and we'll go down and rush the bridge. When I get to the bridge, I'll peel off and the recon platoon will charge across the bridge. Then, when they get across, we could have a couple of engineers go in with the recon unit, immediately jump out, and see if they can find the demolition charges before the Germans can use them to blow the bridge."

Just before daylight, Pickett charged the bridge. The platoon sergeant was in the lead jeep, and Pickett was in the second jeep.

As they came to the river where the bridge was, they noticed a little German railroad station. The arm was up, so Pickett jumped out to check it, while the jeep with the engineers stopped and the sergeant continued across the bridge. They found out later that the Germans sentries were not in the train station but rather behind them in a house on a bluff, about 50 feet up and 50 feet behind Pickett. There was nobody at the bridge itself. When they heard jeeps running and saw a jeep on the bridge, the Germans blew it. The sergeant, who was two-thirds of the way across at the time, was badly wounded.

Pickett knew that he couldn't call for the tank company now because the bridge was blown, so the scheme fell through. He took what was left of his recon unit and headed south, parallel to the river, to look for another crossing. Taking one of the tank companies with him, Ahee went all the way down to Sobernheim, about ten miles downstream, but when they arrived, the bridge there was also blown.

At this point, Ahee told Pickett to go back to Martinstein "and see if you can get across with the other company."

For Pickett, the next half hour was like a tale from the Old West, where a couple of soldiers on horseback try to get through the Sioux Nation. When Pickett got back to the bridge that had been blown that morning, the Germans were shelling the area with *nebelwerfers*. Pickett realized that he had to find a place to ford the Nahe, and he had to do it quickly. By the side of the little railroad station there was a push broom. An engineer said he would recon along the bank, and Pickett told him, "I'll go out in the water and see how deep it is."

Pickett then took the broom and waded out into the Nahe.

"They call these things rivers, but I'd call the Nahe a creek," Pickett said later. "This thing couldn't have been more than 60 feet across and had maybe 15-20 feet of water. At the deepest point, it was only up to my waist."

Pickett, broom handle in hand, walked up and down the river, measuring the depth, until he found a place where it was shallow enough not to rise above the exhaust stack of his tanks. "As long as you can keep the exhaust above water, you can ford a stream," he told the major.

By this time, there was a small outpost of Americans behind Pickett who were laughing. "This is a heck of a time to go swimming, Pickett!" they hooted.

The Germans were shooting from what appeared to be a factory building across the river, and once the Americans got down next to the river, they were protected by the opposite bank. The dangerous part was getting down the friendly bank without getting hit by *nebelwerfers*.

Pickett drove back down to Sobernheim and told Ahee that he had found a place to ford, so Ahee turned the rest of the force around. They had almost gotten all the way back when Brigadier General Hunk Holbrook, commanding Combat Command A, arrived on the scene.

Just as Holbrook reached the Nahe, the Germans—who had brought up a panzerjaeger—started shooting vehicles on the American side of the river. Holbrook decided to shoot back, and after 20 minutes, it was over. No one was sure whether he hit the panzerjaeger or whether it just pulled out.

By 11 A.M., the lead tank company started to wade the Nahe through the ford that Pickett had mapped out. Once across, Task Force Ahee went south to link up with the 55th Infantry Battalion. A 96-foot treadway bridge was completed across the Nahe River over "Pickett's ford" at Martinstein at 9:40 P.M., and wheeled traffic moved across it throughout the night.

For Task Force Ahee, it was clear sailing for the next 20 miles, until they linked up with the 55th Infantry in the afternoon. They never heard a shot fired, which was unusual because they were in a mountainous area with high, rolling hills that had sharp drop-offs, where they would not have been able to get off the road if they had been attacked. For a tank outfit, this is the worst place in the world to get caught. Fortunately, the Germans hadn't defended the area.

When they arrived in the town of Breitenheim, where the 55th Infantry had set up an outpost, Pickett soon learned that some of the fellows had found a building in the town that was filled with a platoon of what seemed at first to be Wehrmacht enlisted women. Pickett went down to investigate and discovered what must have been 20–30 women between the ages of 18 and 22, in German uniforms. They all had the word *feldhur*—"field whore"—tattooed under their arms, along with a serial number. This specialized platoon included Poles, Greeks, and young women from other occupied territories who had been "recruited to service the troops."

"This is the first time that I even knew of the fact that an army did things like this," said Pickett, shaking his head. "We Americans are sort of naive. What we do is we turn our fellows loose, and the first guy who goes into the aid station wants a prophylactic. Americans always managed to get to women, one way or another, but the Germans were more organized. They had organized the women into battalions and had given them to their troops to use when they wanted to go see them."

The biggest problem Pickett had after the task force had captured the *feldhur* battalion was guarding the men who were guarding the prisoners. It became a running joke around town that more and more guards had to be detailed to guard the women because guards were needed to watch the guards who were watching the guards who were guarding the women. The women were later evacuated.

Beyond the Rhine

Using boats and landing craft, Patton's troops crossed the Rhine at Oppenheim near Mainz on March 22. In the north, Montgomery's 21st Army Group crossed the Rhine at Wesel on March 23. The First Army broke out of the Remagen bridgehead on March 25, and within three days all three armies were thundering across Germany, just as they had across France in September.

There was great optimism in the ranks of the Allied armies sweeping across Germany at the end of March. Berlin and victory were within sight.

But on March 28, much to Winston Churchill's chagrin, General Eisenhower secretly confirmed the decision, forged at Yalta, to let Stalin's Red Army take Berlin.

By this stage of the war, the whole German defensive operation in the Rhineland was in shambles. Once the US Third Army crossed the Nahe River and headed for the Rhine, with the Seventh Army attacking northeast, and once the initial German positions were penetrated, the whole German defense network fell apart.

"However, it was still dangerous because you never knew if you might come upon four or five guys who still might want to fight," George Pickett observed. "You didn't know what was going to happen next, as far as an organized resistance. It became very thin until we got to the Rhine."

Task Force Ahee had received orders to cross the Rhine at Worms and head directly for Hanau near Frankfurt. The entire area was about 50 miles long and 10 miles wide, but it contained only one first-class road running in a favorable direction. The terrain, although initially flat Rhine valley land, soon rose abruptly across wooded hills. The only good corridor was up the Kinzig River Valley on the south flank. It ran as far as Schluchtern and then down the Fleide River Valley toward Fulda. The only choice for a plan of attack was to send a single column up this corridor, with another column prepared to branch off and operate to the north if determined resistance developed.

At the first light of morning on March 29, the first elements of Combat Command A slipped across the Main River at Grossauheim, and, as planned, passed through the 26th Infantry Division at 7:05 A.M. Task Force Ahee made first contact with the Germans at 7:15, when they ran into a roadblock in a heavily wooded area just northeast of Hanau. The roadblock was mined and defended by infantry and *panzerfaust* (antitank) teams, but the GIs overwhelmed the defending troops a half hour later, and the tanks continued to press ahead while engineers removed the roadblock.

Before the war, the Germans had built a training area for antitank guns just outside of Hanau on the road to Rüchlingen. When the American tanks started to cross the field, it was like shooting fish in a barrel for the Germans, because the troops were literally firing on the range where they had been trained to shoot. Lieutenant Alfred Fermanian, who was only 20 years old, was the point man, in the lead tank when Task Force Ahee cleared the woods. Suddenly the first two tanks went up in flames. Very frequently, those at point on the way out never come back.

"I'll always remember that boy," George Pickett said later, "because he was the first one of the young sergeants that we gave a battlefield commission. We had pinned the second lieutenant's bars on him only a week before. By this time, we had been at the front for quite a while, and attrition had taken almost all of our platoons leaders, so we needed some more officers and there weren't any trained tank officers for them to send us, and Division gave us the option of commissioning some of sergeants."

After losing a couple of tanks, Pickett selected some infantrymen and moved out across the German artillery range. Task Force Ahee was fairly low in terms of infantry strength, but the Germans were pounding the training field, so there was little else that could be done. The Germans seemed to have the range with the antitank guns, but when Pickett started across the field the mortars fell in front of his men or behind them.

After the Germans killed a couple of Pickett's men with machine guns, everybody hit the dirt. The last hundred yards or so, Pickett and his soldiers were crawling. By the time they got into Rüchlingen, the Germans had begun to put up a vicious fight, but when they saw that they were grossly outnumbered they pulled back.

"I don't really know if we chased them out of there with artillery, but the last half of the process of taking the village was like a piece of cake, compared to the first half," Pickett recalled.

After they got into Rüchlingen, Pickett walked back to where the main elements of Task Force Ahee were waiting and told them to "bring in the tanks."

When Pickett reached the rear echelon, he noticed that there was a conference going on among some of the senior officers in the 11th Armored Division and the officers in another division. He said to one of his tank commanders, "There's a tremendous amount of rank standing around at the edge of these woods, with a firefight going on in town, but they seem more interested in what they're talking about than what *we're* doing."

Pickett just looked back at all the senior officers with their maps and said, "These guys are crazy. If the Germans who brought those antitank guns up could see this mob, they'd blow them away. Compared to most of the us who spend our time at the front, the senior officers live a charmed life. Here's a whole bunch of them with maps and aides and vehicles, and the Germans have just knocked out two tanks about a hundred yards up the road. They could have just as easily thrown a few rounds in here, but they didn't."

"I don't know whether the Germans saw them or not," the tank commander laughed. "Probably they were too busy running from 'Pickett's charge' across the gunnery range."

"I felt more like I was playing shooting-gallery ducks," Pickett told him. "I was running back and forth trying to keep everybody off their ass and moving."

Pickett's "trying to keep everybody off their ass and moving" did not, as he had complained, go unnoticed by the senior officers in attendance at the edge of the woods. Indeed, Pickett was to take home a silver star from Rüchlingen. The citation noted that he was "directly responsible for coordinating the attack on Rüchlingen, by infantry elements of both the 11th Armored Division and the 26th Infantry Division."

From Rüchlingen, the 11th Armored Division moved into the Fulda Gap and up to Meiningen and Zella-Mehlis. The division was also ordered to clear the enemy from its zone west of the Oberhof-Suhl line and protect the corps right flank east of Fulda. It then undertook, on its own initiative, the task of securing the crest of Thüringia Wald on the south flank. Pickett in Combat Command A was ordered to seize Stutzerbach.

By April 3, the 11th Armored Division reached the city of Suhl. The Germans by this time were using the Volkssturm, an army of ordinary people—old men, kids, veterans of World War I, in short, almost any man from the age of 16 to 60 who could carry a gun. To the Americans, the Volkssturm looked to be 16 to *80*, and by now the division tried to avoid killing people.

One of the things George Pickett resorted to was putting a white flag on the antenna of his jeep and just driving down the road toward a village. He did not run into any trouble unless he met with SS troops. Only the SS would fire on a white flag because they didn't want *anybody* coming in who would offer the poor Volkssturm a chance to surrender. The SS knew that most of the people would quit fighting if they got the opportunity.

On the morning of April 3, Pickett drove to the southern edge of Suhl with his white flag flying. Nobody ever came out to meet him, so he went around the corner and found that the Germans had defended the high ground just to the east of Suhl. About a half mile farther north was the huge Walther Arms Works, one of the world's biggest and most renowned makers of small arms and infantry weapons.

Pickett had just gotten out of his jeep and had walked up to a low hill to reconnoiter the German positions relative to his own when a mortar went right through the turret of one of his assault-gun tanks and blew it up, killing everyone in it. Among the tanks in the 42d Tank Battalion was the assault gun platoon with 105mm howitzers, which often used white phosphorus shells. The explosion set the phosphorus on fire, and in the searing white-hot hell of this intense inferno, George Pickett got phosphorus smoke in his eyes. He wound up with them bandaged for nearly two days. There wasn't any permanent damage, but, as he complained, "The burn itself was so bad that any time I took the bandage off and opened my eyes, it would burn like the devil."

With the capture of Suhl and the famed Walther arms works, the 11th Armored Division found itself with a dizzying array of high-quality weapons and machine tools. Among the arms, armament parts, and matériel seized in the area were 1,600 P38 pistols; 4,600 7.65mm guns; 325 .22 caliber guns; 2,210 sniper rifles with complete scopes; 4,420 with incomplete scopes; 1,140 partly assembled rifles; 113 lathes; 97 milling machines; 41 drill presses; nine punch presses; two hydraulic presses; and 40 grinders.

At least 500 new model carbines and 2,500 automatic assault guns—with sufficient parts for an additional 5,000 were also uncovered at Suhl. Over a million rounds of small arms ammunition were included. This matériel was found in several large plants and in over 50 small, decentralized plants. In addition to weapons, several of these factories had produced parts for robot aircraft.

Victory in Italy

When spring arrived, Horace Brown's 88th Infantry Division—and indeed most of Mark Clark's Fifth Army—was rested and retrained. The replacements were now well qualified to fill the holes caused by the horrendous losses in the fight through the Gothic Line and the northern Apennines. The attack was set, and again the 88th Division was part of the main effort.

There was initial fierce fighting to roust the Germans out of their defensive positions, then the 88th Infantry Division began to roll into the Po Valley. It was reminiscent of the march on Rome after the Allies had punched through the Gustav Line a year before. The Fifth Army was now moving into beautiful country, Italy's breadbasket. On April 24, it reached the Po River in the vicinity of Ostiglia. The bridges had been destroyed by the retreating Germans, so there was a pause to make the crossing. Brigadier General Lewis wanted to go to the river and observe, and Horace Brown went with him.

They were on the bank several feet above the waterline when they heard mortars coming in. Brown hit the dirt. He looked up and saw that Lewis was still on the bank.

"Get off there!" Brown hollered.

"You've got to make a show," Lewis said.

"Show, hell! They're going to *kill* you!"

The general came down, but he did not fully accept Brown's warning. Later, as he again went up on the bank, a sniper who had scoped him fired off a round. General Lewis collapsed to the ground.

The round had entered the front of his helmet near his general's star, scraped along his almost-bald head like a strong fingernail, and gone out the back side. He was stunned and had a huge headache for a while but otherwise was unharmed. After that, Lewis became much more cautious.

The 88th Infantry Division crossed the Po River on April 26 and headed toward Verona, as did the 20th Mountain Division on a parallel road to the left. On the radio, 88th Division Commander "Bull" Kendall sent word: "*We* had better get there first."

It was not to be. Fifth Army headquarters diverted the 88th northeast to Bassano del Grappa. This was probably just as well, because when the

20th Division finally reached Verona, it faced a short but strenuous fight with SS troops.

"From the breakout of the Anzio beachhead to the capture of Milan and the subsequent surrender of Axis forces in Italy, the war in Italy seemed like a steady retreat of half-hearted Italian troops and badly thinned-out German forces," Paul Skowronek said. "The Italians were obviously not dedicated to the war, and the people were extremely cooperative as soon as the combat zone moved past them and their region became occupied by American troops. In the last year of the war, the Germans must have decided that Italy was a lost cause, and they left only enough troops there to keep the retreat from turning into a rout. The steady Allied advance from one temporarily defended line to the next appeared to proceed like a massive field exercise, only with live ammunition and real casualties."

Horace Brown, however, disagreed. "The Germans in Italy remained a formidable force until the last successful Allied offensive. Their numbers were about equal to ours. We had air and artillery advantage, whereas they had position advantage."

Ernie Durr was transferred to the 34th "Red Bull" Division, joining it in September 1943 shortly after it landed in Italy. For the next two years, the 34th Division fought its way through Sicily and through the grinding Italian campaign. In December 1943, because of his experiences on the rugged terrain, Ernie Durr had authored a forceful study recommending the attachment of an additional engineer combat battalion to infantry divisions operating in difficult terrain, a practice that had become common by 1945. In July 1944, shortly after the liberation of Rome, Ernie was awarded the French croix de guerre with silver star for his service as liaison officer to the 3d Algerian Infantry Division. That same month he was promoted to major as assistant G-3 of the 34th Infantry Division.

On April 26, 1945, in the Po Valley, Ernie was reconnoitering for a new 34th Division command post when he came across some Italian civilians who told him that they knew of a group of Germans "ready to surrender." Durr took a jeep and went to check it out. It proved to be an ambush. Ernie managed to push his jeep driver out of the way as he returned fire. The next day, the 34th Infantry Division overran the ambush site and found his body. One of the men who had been taken prisoner in the previous day's battle explained: "The last I saw the Major . . . he was in a heavy firefight and completely surrounded."

Major General C. W. Ryder, commandant of cadets during the Class of 1941's cadet days, was the 34th division commander during the Sicilian campaign and during part of the Italian campaign, and it was he who wrote the grim letter to Ernie's widow, Edie. "I know of no young officer who showed equal promise of a brilliant military career to come. Everything he did he did well, and when he was with me, he was one of my select few upon whom

I could depend at all times and under all circumstances. As you know, I had noticed him as a cadet at West Point, and when he came to me at Oran and asked for a job, I was more than glad to arrange things so that he could come with me. My one idea thereafter was to give him a rounded-out experience in the part of the military profession that really counts—command and staff of combined arms, in war, so as to assure his going to the top during the rest of his military life. It is almost impossible to realize that what I had planned for him will not happen."

As the war in Italy ground to a halt, Bizz Moore found himself facing down a German armored division that desperately wanted to get out of Italy through the Brenner Pass. The Germans had been rough on the Italian populace and wanted to escape Italy before they were thrown into prison camps. The Germans had ammunition but little fuel for their tanks and armored artillery. If they had had the fuel, they might have made a fighting run for Germany. The United States 1st Armored Division convinced them that the casualties were not worth running the gantlet.

The German armored units were positioned in the hills close to the French border. The 1st Armored Division was west of Torino, spread out from north to south, blocking the roads running east to the Brenner Pass. Both the Germans and the Allies had heard that the armistice would be declared "any day."

On April 29, three days after Ernie Durr was killed, German general Pemsel and his staff met at Fifth Army Headquarters to arrange for the surrender of all German forces in Italy. On May 2, the 88th Infantry reached the mountains around Feltre-Fonzaso, and late in the afternoon they received word that the Germans had surrendered in Italy and were about to surrender Berlin. Overjoyed, Horace Brown and the others Americans at Feltre-Fonzaso thought they would be there for a while and would get some rest.

This was not to be. The 88th immediately received orders to move north and west to close the Brenner Pass to the retreating Germans and at the same time make contact with Allied forces in northern Europe.

They moved out immediately in what was to be a one-division drive into an area where the remaining German troops had retreated. Wehrmacht, Luftwaffe, and SS headquarters were in the Bolzano area, so the Germans were heavily armed. Brown's old 349th Infantry Regiment and the 337th Field Artillery Battalion were the units that actually closed the Brenner Pass and made contact with the 103d Infantry Division coming south from Austria. They established positions in the area of Brenner Pass, while two other regiments with their artillery moved into the Merano and Bressanone areas.

The 88th Division headquarters, division artillery headquarters, the 339th Field Battalion (a medium 155mm howitzer battalion), and the 88th Reconnaissance Troop—around 1,000 personnel, including Horace Brown—moved into Bolzano. The city, which was then known to the Germans by its old Austrian name—Bozen—was "defended" by 20,000 still-armed Germans. It stayed that way for almost three weeks because the 1,000 Americans, 450 of whom were from Brown's battalion, had nowhere to put the 20,000 Germans in the city they had "captured" and no way yet to feed them. The only Germans they kept under heavy guard were 2,500 SS troops.

The Americans almost immediately started to enjoy the same restaurants and drinking places their former enemies did. The Germans sat at their tables and sang; the Americans sat more quietly at their own tables, enjoying the songs. One evening, Horace Brown came upon a very handsome German paratroop major—fully dressed with all his decorations, including the Iron Cross—surrounded by rear-echelon American soldiers who were taunting him about who had won the war. Brown sent the men packing and directed the German major to leave the vicinity and get back to *his* own area.

The second night in Bolzano, General Lewis had told Brown that he wished to have some guests for dinner and wanted the mess to be so informed. Brown did not at first catch on to what the general was talking about, but he soon realized that Lewis was inviting some high-ranking Wehrmacht officers—no SS—to dinner. This gesture of reconciliation and respect seemed strange from a man who had been fighting the Germans for over three years with II Corps Headquarters in Africa, the Fifth Army Headquarters, and the 88th Division in Italy.

After about two to three weeks, the rear area was prepared to accept prisoners, so the Americans in Bolzano gathered up the Germans and sent them back. The move was unannounced and started with the 2,500 SS who were under guard.

"The one thing that surprised me was how well some of the SS officers had been living," Brown said. "When the SS general's German mistress, an actress, came down, she took my breath away. She was Hitler's pure Nordic type and a stunner. After 45 years, I still remember how she was dressed."

The Wehrmacht and Luftwaffe officers were permitted to take only what they could carry. This meant that they left a great deal of gear behind. Brown later picked up the winter overcoat and flight jacket that had belonged to the commanding general of the Luftwaffe in Italy. He gave the overcoat to the USMA museum but wore the sheepskin flight jacket until it was in tatters.

After Bolzano was turned over to the Italian authorities, the 88th Division was assigned to take over the POW command. The division artillery was to move back to Livorno (Leghorn). Since Horace Brown was traveling back independently, the Catholic chaplain, a good friend, asked to ride along with him. It was a very hot, steamy June day, and as they were driving past Lake Garda, they decided to have lunch. There was no wine available. After a moment, the good Father said, "I do have a bottle. It has been blessed, but I can bless another."

They had a fine lunch.

Götterdämmerung

During the Rhine crossings, the US Ninth Army—much to its consternation—had been assigned to Field Marshal Montgomery's 21st Army Group. On April 4, however, the Ninth was reassigned to Omar Bradley's 12th Army Group where it would remain for the rest of the war. Hodges' First Army and Simpson's Ninth swept into the Ruhr River Valley, and by April 14, they had captured or surrounded half a million German troops in Germany's industrial heartland. These two armies continued to work more closely within the 12th Army Group during the war's final days than they did with the third element of the Group, Patton's Third Army.

Having conquered the Ruhr, the First and Ninth Armies, now numbering nearly a million men between them, were preparing for the final push across the north German plain. Eisenhower's decision to stop short of Berlin had yet to be publicly announced, so the prospect of marching triumphantly into the battered German capital was now a prime motivating force for the north wing of Bradley's 12th Army Group.

Ironically, on the night of April 12, when news came through that President Roosevelt had died, Eisenhower, Bradley, and Patton were together at Patton's command post near Ohrdruf, discussing the final meeting place between Allied and Soviet forces. Roosevelt's death gave the German leadership—especially Adolf Hitler—a momentary boost in morale, but there was really nothing left that the German armies could do to reverse the hopeless tactical situation that they were in. Occasionally they offered stiff resistance, but by this time they lacked the armored and mechanized forces necessary for anything but point defense.

After crossing the Rhine River a day earlier than Field Marshal Montgomery's 21st Army Group, George Patton's Third Army was ready for a classic blitzkrieg across Germany's waist, from Oppenheim to Frankfurt to Leipzig, and then on to Czechoslovakia and Austria. Patton intended for his Third Army to be the fastest moving of the Allied Armies—on the Western or Eastern Front—and he was not to be disappointed. Three days after crossing the Rhine at Oppenheim, the Third Army opened a second bridgehead at Saint Goar, roughly halfway between Mainz and Remagen. Having seized control of the bridgeheads, Patton directed engineers to start constructing actual bridges. Patton was determined to have two Rhine crossings to supply his blitzkrieg.

Patton's 4th Armored Division captured the vast German underground command center at Ohrdruf on April 4. Built in 1938, it had never been used, but it was being prepared for use by Hitler as a possible final retreat even as the 4th Armored Division's Shermans arrived at its threshold.

From Suhl, the 11th Armored Division and Task Force Ahee attacked toward Meiningen. The battle near the sprawling hospital center in Meiningen was not a major one, so Task Force Ahee was ordered to cut south to Coburg.

Saxe-Coburg-Gotha was originally the home of Prince Albert of Saxe-Coburg-Gotha, who married his cousin Queen Victoria of England in 1839. However, when Task Force Ahee arrived on April 10, George Pickett was the only man in the unit who knew what the castle was. Coburg Castle was a huge stone structure that was actually more of a fortress, so the Germans generally referred to it as *Festung Coburg* (Fortress Coburg). Because the German troops inside had heavy machine guns and had been ordered to fortify themselves in order to maintain control of the town, Joe Ahee directed his eight-inch howitzers to begin bombarding the walls of Festung Coburg. An early morning dive-bomber strafing attack contributed to the softening of the city. Around 10 A.M., a flight of P-47 fighter-bombers arrived at Coburg. It circled continuously over the city, ready to strike in the event the Germans refused to surrender, and a bomb was dropped through the roof of the castle, starting a fire. The enemy was given until 10:30 A.M. to capitulate.

Finally, emissaries from Coburg approached the 42d Tank Battalion command post, seeking surrender terms for the castle and the city. Joe Ahee told George Pickett, "George, I want you to get a white flag and go on up there and parlay with those Krauts. See if you can't get them to surrender so we don't have to blow the whole place up and hurt somebody."

Through his binoculars, George Pickett could see a great deal of confusion at Festung Coburg as he walked toward it. Just outside the castle gate, he was met by a German officer wearing the gray-green uniform of the Wehrmacht rather than the black uniform of the SS. Shortly afterward, the German officer in charge surrendered.

As they walked through the castle gate, the first person Pickett saw was the Duke of Saxe-Coburg-Gotha, who was also the president of the International Red Cross. Seated about 20 feet away on a little terrace, having an aperitif with a German captain, was one of the most beautiful women Pickett had ever seen in his life. She was the duke's granddaughter.

The duke, speaking perfect English, asked Pickett what he wanted.

"I wish we could stop the fighting," Pickett told him.

"Yes," the duke nodded grimly. "I think that's a good idea." He then looked at the German officer, said something in German, and the officer clicked his heels, looked at Pickett, and told him in German that they now had a truce. At 10:30 A.M., both the castle and the city surrendered. Under terms imposed in the capitulation, the civilians commenced to clear the streets of roadblocks and other obstacles.

Later, the duke, in a commanding voice, asked Pickett, "Could you now get some of your people up here to help us? Your airplane dropped a bomb into the area where we have put all of the art treasures and museum pieces, and the museum itself is on fire."

Pickett turned to his driver, who had followed him up the hill. "Busch, go back and tell Colonel Ahee that I want to get 100 men up here as fast as possible."

Within an hour, they had a bucket brigade, with about 50 Germans and 50 Americans furiously passing water buckets, and they were finally able to put out the fire.

While Pickett was securing the Coburg Castle, Combat Command A had continued to block all roads north and east of the town, and a verbal order was issued to all major units at 1:00 P.M., outlining the plan for renewal of attack. George Pickett's Combat Command A was ordered to bypass Neustadt and advance in its zone to capture Kronach and establish a bridgehead across the Hasslach River. Similarly, Combat Command B was directed to advance in its zone and secure a bridgehead across the Hasslach near Marktzeuln.

Going into Kronach, there was very little exchange of fire and virtually no German resistance. From Coburg to Kulmbach, it was almost like a road march. For George Pickett, the last firefight of the war—and nearly of his life—came on April 16 in the vicinity of a little village on the road between Kronach and Kulmbach. He was leading a small force consisting of two companies down the road to see if it was clear, and, just as he turned the bend, the Germans opened fire on his jeep with a machine gun. A bullet went through the windshield between Pickett and the driver. "A fellow could still get hurt in this war!" Pickett said to his driver.

"Let's get the hell out of here!" the driver replied.

They drove into a cul-de-sac formed by haystacks and took cover. When an American tank arrived, the Germans evaporated into the maze of buildings in the village.

Suddenly, considerable artillery fire started falling around Pickett and his men. An infantry outfit was coming across country from west to east while the 42d Tank Battalion was going from north to south, and the artillery barrage was American. Pickett, his driver, and the tank were being attacked by friendly fire.

Nobody in the line *ever* fails to perceive the irony of that term.

Pickett had no communication with the other unit, so the only thing he could do was get into the jeep and race back toward battalion headquarters and find somebody who had a radio that would connect up.

"I don't know what they're shooting at, but they're putting down pretty heavy fire," he said. "They would rather shoot first and ask questions second."

Machine gun bullets through the windshield and friendly fire were only the beginning of George Pickett's troubles on April 16, 1945. When he got back to the battalion command post, he was met by Brigadier General John L. Pierce, commander of the 16th Armored Division. Pierce looked him over for a minute and finally said, "Are you Major George B. Pickett, 023932?"

"Yes sir."

"You've been absent without leave from my division for six weeks."

Pickett was aghast. For the past six weeks, he had been dodging bullets and capturing dukes. Suddenly, a general from *another division* was telling him he had been AWOL!

"General," Pickett stammered, "I've been here fighting for six weeks."

"I can't help that. I have orders on you. You were supposed to report to *me*."

It turned out that for six weeks Pickett had been assigned as a battalion commander in the 16th Armored Division, but he didn't have any orders. The 11th Armored Division had moved so fast that the orders never caught up with him. He had fought for six weeks as executive officer of the 42d Tank Battalion after he was supposed to be commanding a battalion in the 16th and didn't know it.

The story was that the 16th Armored was brand-new from the States and had never seen a day of combat. It was all the way back on the Rhine at Mainz. General Pierce called someone at 11th Division Headquarters and they told Pickett, "Pack your stuff and go with him. You belong to him. You have for six weeks. We just didn't know about it until today."

As the 16th Armored Division was coming in, George Patton wanted to have as many experienced battalion commanders in a new outfit as he could, so several men were promoted out of the 11th and into the 16th. George Pickett's name had come up in General Patton's "little black book" on a list of potential battalion commanders, where the general had recorded it during that cold day in January when they had crossed paths in the Ardennes. Pickett went around to say good-bye to everybody. "It feels sort of funny to be leaving an outfit that you've fought with all this time, especially when we know that the war is almost over," he told his pals.

Joe Ahee shook Pickett's hand and said solemnly, "You have a US Army .45 caliber pistol. You'll have to leave it with me because my next exec will need it."

"You mean you're going to turn me loose without even a gun?" Pickett gasped.

"You don't need a gun," Ahee laughed, knowing that George had two or three Walther P38s already stashed away.

General Pierce had an aide and another major with him, so Pickett became the fourth man in the jeep. With all the general's baggage and that of the two majors, things got rather cozy. When they reached division headquarters at Bayreuth, Pickett finally met the 11th Division Commander, Major General Holmes Dager, the general that he had been fighting for over the past two months but had never seen. But Dager was leaving the division. At Bayreuth that night, Pickett also saw General Hunk Holbrook and Colonel Virgil Bell, who invited him to a farewell dinner in Dager's honor at the

Senior Officers' Mess. The next morning, they loaned Pickett a jeep. He and Pierce then made their way back to Third Army headquarters, which at that time were at Fulda.

When they reached Fulda, General Pierce stopped over at the Third Army Headquarters command post to get some paperwork and told Pickett, "Go around and find out as much as you can. I want you to be my assistant operations officer until we get back to the 16th Armored Division headquarters."

Pickett realized that being an operations officer is a staff job usually held by a major. He had just been assigned to command a battalion and he wanted the lieutenant colonel's silver leaf that went with it. On two occasions he had already missed opportunities to be promoted. This was a bird in the hand, and Pickett didn't want to lose it, but he quickly said, "Yes sir," and went to the operations office. Most of the people working there were men that he had known before. They were glad to cooperate and told Pickett everything that Pierce needed to know.

At breakfast the next morning, Pickett was directed to the Officers' Mess, which was in an old German barracks. Pickett wandered in alone and looked around until he saw a room that had tablecloths and silverware. "These people at Army Headquarters know how to live," he thought.

Pickett sat down, dirty uniform and all. He hadn't had a bath for God knows how long, and he knew he must look and smell horrible. "At least I've shaved, washed my hands, and combed my hair. That's all I can do for now," he thought.

He seated himself and a man in a little jacket took his order. The waiter had just placed a steaming plate of corned beef hash and eggs in front of him when a well-dressed woman walked into the room. Pickett, recognizing her at once, nearly dropped his fork. It was Marlene Dietrich!

"I'm in deep trouble," he thought to himself. "Marlene Dietrich isn't here to be eating with any ordinary majors."

Nevertheless, Marlene Dietrich strolled over to where he was eating and began chatting with him.

"I knew I had blown it," Pickett recalled later. "I realized that I must be in the commanding general's mess, which you *don't do*, not as a dirty major who still smells like he's lived in a medium tank for the past six weeks!"

Suddenly, they heard a voice coming down the hall. It was George Patton—all four stars! Pickett got up to leave.

When he saw Pickett, Patton said, "The war's almost over. How would you like to be my junior aide?"

Pickett nervously replied that it was a great compliment, then excused himself explaining he had orders to report to the 16th Armored. Patton and his aides enjoyed an enormous chuckle at Pickett's expense.

When he got to Mainz, Pickett took his post as commander of the 64th Armored Infantry Battalion. From then on, his life became part of John Pierce's brand-new 16th Armored Division.

On April 15, General Simpson, whose Ninth Army was now on the banks of the Elbe River only 50 miles from Berlin, flew to Bradley's headquarters to present his plan to "enlarge the Elbe River bridgehead to include Potsdam," a suburb of Berlin. Bradley listened to the plan—a perfectly workable tactical plan—and decided to telephone Eisenhower one last time to see if there was any chance it could be implemented. Overhearing Bradley's end of the conversation, Simpson had his answer.

"All right, Ike," Bradley said. "That's what I thought. I'll tell him. Goodbye."

Simpson was told to wait at the Elbe River for the Red Army.

On April 19, Soviet forces, under Field Marshal Georgi Zhukov, arrived on the eastern outskirts of Berlin and began their encirclement of the city. On April 25, even as Soviet tanks were shooting their way through Berlin's suburbs, the Soviet soldiers were beginning to make contact with men of the First Army's 69th Division along the Elbe.

The official handshakes between American and Red Army division commanders took place on April 26, and for all practical purposes the war was over.

Eighty percent of the men of the Class of 1941 served overseas in World War II, and nearly half of these took part in the great drive across *Festung Europa* that began on the Normandy beachhead on June 6, 1944, and ended in the heart of Germany 11 months later.

General Patch's Seventh Army—part of Devers' Sixth Army Group, along with the French First Army—had the responsibility of crossing the Rhine River in the south and sealing off possible routes by which German forces could escape into the Bavarian and Austrian Alps for a last stand in this easily defensible terrain. Using assault boats, the Seventh had crossed the Rhine at Worms on March 26, and by the end of the next day a pontoon bridge was under construction, one that would soon be named for General Patch.

The 65th and 71st Divisions of the Third Army reached the Danube River at Regensberg on April 24, and the 86th, 99th, and 14th Armored Divisions arrived two days later, as the 65th and 71st were undertaking assault crossings of the river. Vanguards of the Third Army now raced toward Czechoslovakia and Austria. So rapid was the army's advance, that artillery and even tanks were allowed to lag behind. Rubber-tired recon cars and jeeps, capable of faster speeds than tanks, often led "armored" columns. Weather, and the masses of surrendering Germans clogging the roads, were the only obstacles that remained.

The idea of an alpine redoubt—a heavily defended mountain fortress the size of the state of New Jersey in which die-hard German forces could hold out for years—was the subject of speculation since the early days of Allied planning for the war in Europe. There was little hard evidence that it existed, but Allied planners didn't want to take any chances.

The Seventh Army surged into Bavaria, the largest and southernmost of the German states, and took Nuremberg, the spiritual center of the Nazi party, on April 20, Adolf Hitler's birthday. From there Patch turned south to seal off all possible routes into the mysterious, and as yet unconfirmed, alpine redoubt.

Between April 27 and April 30, the 42d and 45th divisions of the Seventh Army liberated the concentration camp at Dachau and captured Munich. Resistance in Munich—especially from the SS—was intense, but a pro-Allied civilian uprising helped the Americans secure the city.

On May 4, the Seventh Army's 3d Division took possession of Hitler's "Eagle's Nest," the octagonal fortress at Berchtesgaden that had served the führer almost as a second capital. If there *had* been an alpine redoubt, Berchtesgaden would have been its nerve center.

There hadn't been. In southern Germany, the war ended when the American flag was run up Hitler's personal flagpole.

On May 4, Pete Tanous, formerly of Cadet Company F and now with the 7th Army, was driving out of Degendorf with his master sergeant, heading south into the Bavarian Alps toward Berchtesgaden. As they rounded a curve, they saw a German officer in the middle of the road, waving a white handkerchief.

The sergeant covered Tanous with his gun as he got out of the jeep to speak to the officer.

"We want to surrender," the German said.

"Who are 'we'?" Tanous asked hesitantly.

The German turned and waved his hands toward the forest on both sides of the road. At this signal, out came about 150 people in Wehrmacht uniforms—doctors, nurses, enlisted men, and other officers.

Tanous gulped. "I sure as hell don't want to become a statistic on the day before VE-Day," he said to his sergeant, "so let's keep our weapons ready until we're sure that this small battalion really means only to surrender."

Into Czechoslovakia

The initial orders to George Patton's Third Army were to not go into Czechoslovakia, but Eisenhower made the decision to take Pilsen, and he conveyed this to Patton by way of Omar Bradley, the Twelfth Army group commander, on the evening of May 4, 1945.

That same day, George Pickett's fast-moving 16th Armored had passed through the 97th Infantry Division salient on the road from Nuremberg to

Mies on the border between Germany and Czechoslovakia. Simultaneously, other elements of the XII Corps of Patton's Third Army had entered Linz, Austria—which was virtually undefended—and had succeeded in securing a bridgehead across the Danube with far less difficulty than American forces had experienced two months before at the Rhine. The next day, they passed into the Sudetenland, the ethnic German region that Hitler had annexed in 1938 and that had been the first domino in the line that had ultimately led to World War II.

For Pickett and the rest of the 16th Armored Division on XII Corps' north flank, the drive into Czechoslovakia was such that even the army's official review called the action an anticlimax. The fighting was unreal, a "comic-opera" resistance by German forces who *wanted* to surrender but felt somehow obliged to fire a shot or two in the process.

The Sudetenland was neither German nor Czech. The little towns near the border, with houses linked by fences and decorated arches over the gates, had the look of Slavic villages, but the population was unquestionably hostile. On May 6, the 16th Armored, along with the 2d and 97th Infantry, reached Czechoslovakia's "Little Maginot Line," the breakwater of fortifications built before the war to defend the country against the invasion from Germany that never came. By the time the panzers had reached this point seven years before, Czechoslovakia had already been defeated by the English, French, and German politicians sitting around a table at Munich.

Now it was the Americans' turn. George Pickett's M26 tank chugged past these silent, undefended forts of the Little Maginot Line, again untested in battle. The Yanks then burst suddenly from the Sudetenland into a riotous land of colorful flags and cheering citizenry. As the tanks of the 16th Armored rumbled past the abandoned antiaircraft guns that had protected the big Skoda industrial complex on the outskirts of Pilsen, the people shouted their buoyant welcome: "*Nazdar! Nazdar!*"

As if they had stepped across some unseen barrier, the Americans now found themselves in a land of frenzied delight. War and nonfraternization lay behind them. Germany was an "enemy nation" where the Americans had been forbidden to fraternize with civilians, but Czechoslovakia was *not* an enemy, so the same rules did not apply there. It was like Paris all over again— though on a lesser scale and with different flags—but with the same jubilant faces, the same delirium of liberation.

There had been a small skirmish—a few rounds were exchanged—at the Little Maginot Line, but by the time the Yanks reached Pilsen, the German resistance had crumbled and was all but forgotten. The Third Army had faced an old foe, the 11th Panzer Division, but this time the panzers were bent not on attack, but on surrender. With an odd conglomeration of tanks and other vehicles, the remnants of the division marched out of the Bohemian woods with their commander, General von Wietersheim, and directly into prisoner-of-war cages.

The official American entry into Pilsen was made by Task Force Blue, which consisted of the 64th Armored Infantry, of which Pickett was the commander, minus one rifle company. Pickett had a recon troop with an attached tank company, which he always insisted on having before he would move. Since none of the other men had ever been on the front line, and Pickett was the only man there who had ever been in real combat, General Pierce had given him the lead battalion.

No one had known what to expect because intelligence reports showed a ring of 88mm guns surrounding the town and a heavy aircraft defense system near the Skoda plant. As things turned out, the Germans had already decided to give up. The 64th Armored Infantry Battalion was later credited with capturing 87 airplanes, roughly 40 tanks, and 5,000 prisoners in Pilsen.

"They had already stacked their arms, you understand." Pickett laughed as he described the scene later. "That's the way you build up your box score. It's like a prize fighter who fights a bunch of bums to run up his box score. We captured a fantastic number of enemy, but it was mostly an inventory of people who were ready to surrender. The only real firefight was in the cathedral square in downtown Pilsen."

Pickett set up outposts in the cathedral square, at the airport, and at the former German barracks. He had no sooner set up his command post than a man in a Czech Army uniform came running up to him, shouting excitedly, "You've got to do something. The SS troops are slaughtering civilians in Prague."

The man, a member of the underground that had risen up against the Germans, had identified Pickett as the commander of the liberation force, so he wanted him to also liberate Prague.

"The Germans are doing the same thing in Prague that they did in Warsaw," Pickett fumed silently.

The memory of the Warsaw massacre was still fresh in everyone's memory. In October 1944, the advancing Red Army deliberately waited outside the Polish capital as the Germans slaughtered the Polish patriots. Now the same thing was happening—albeit to a lesser extent—in Prague.

Pickett alerted the 64th Armored Infantry that it would soon be receiving orders to move out. He ordered his recon platoon to go as far as it could on the highway to Prague and still be within radio range, and the platoon took off.

When General John L. Pierce arrived with the main force of the 16th Armored Division a few hours later, Pickett told him that he had sent the recon out. Pierce asked him where.

"They've reached a village called Horovice, which is 20 miles this side of Prague."

Pierce flew into a rage. "Get them *back* here!"

Pierce then gave Pickett a lecture on obeying orders, and Pickett responded by telling Pierce about how General Patton paid off on initiative.

"Sir, there's a fight going on in Prague, and I can't understand why we're standing here with a whole division, and a second division is moving in right behind us. Why don't we get off our butts and get down there?"

General Pierce just told him that the orders came from a higher authority than Patton, and that Pickett had better get the recon platoon back to Pilsen on the double.

As Pickett was about to learn, the liberation of Prague itself was secondary to politics. The story, now well known but then hard to comprehend, was that during the conference of Allied leaders at Yalta on February 4–9, 1945, Soviet Premier Josef Stalin had drawn a line on the map of central Europe, and that line went two miles outside the city limits of Pilsen. As the story went, Stalin told Roosevelt not to come any farther east than that. Roosevelt had conveyed that message to Eisenhower, who had in turn told Patton, who had told Pierce that the 16th Armored Division should not pass the line at Pilsen.

When the lieutenant from the recon platoon on the road to Prague called, Pickett told him to "haul his ass back." By late afternoon, the lieutenant had not yet returned, so Pickett got a jeep and driver and, when nobody from Pierce's staff was watching, he called John Curtin, who was now his executive officer, and drove up Highway 14, the road to Prague. He had gotten almost to Horovice when he ran into his recon platoon coming back through a tiny Czech village.

As he stopped his jeep, he saw something such as he had never seen in his life. Nearly 200 Mongolians—terribly dirty but all wearing Red Army hats and carrying Russian submachine guns—were in the village. They were the first Soviet troops Pickett had encountered. They were hauling their supplies in a couple of ox carts, and there was not a motor vehicle in sight. With them was an officer Pickett described later as "a thing that looked like a cross between a woman and a rhinoceros." She carried a bullwhip in one hand and a Luger pistol in the other, using the whip to keep her troops moving forward and the pistol to shoot Germans. Every now and then she would give an order, and if it wasn't obeyed fast enough she would snap her whip. The Soviet troops appeared to be scared to death of her.

The Mongolians seemed to be in the process of tearing the entire town apart. They were knocking down the walls of buildings and ripping mattresses apart. The Yanks could tell that they were still in the process of raping the women of the village. Pickett and the other Americans stared in stunned silence. The whole scene reminded southern boy George Pickett of Sherman's March to the Sea during the Civil War, when the Union Army went through the South looking for buried Confederate silver.

At last, the Soviet troops noticed the Americans. The Russian officer addressed Pickett in Russian, which he could not understand. She began shouting at him, and for a moment it looked as if she was going to take the

Americans prisoner. However, she must have known the English word "tank," because when Pickett started trying to make her understand that he had a whole division of American tanks behind him, she quickly lost interest in taking him or his men prisoner.

Pickett and the recon platoon drove back to Pilsen. The incident was never reported to General Pierce. The recon had been to within 25 miles of Prague, but no units of General Patton's Third Army ever advanced into Prague officially. George Pickett, West Point Class of 1941, and his recon platoon may have come the closest of any American troops.

A Guest Goes West

On April 19, from the Luckenwalde POW Camp complex near Berlin, Tuck Brown and Jim Forsyth of Black '41 observed Red Air Force dive bombers, so they knew the Soviets were getting close to the German capital. Finally, on April 21, the German guards, who had been wearing civilian clothes under their uniforms for several weeks, vanished, and the Red Army entered the camp. Tuck Brown broke down and cried, he was so happy. One of his friends, a man named Pinky, had a real breakdown and in the summer of 1947 was still in an institution.

When the Russians first came to Luckenwalde, the food situation got a little better with the addition of horse meat from the battle near Jüterbog. However, the Soviet units doubled the guards on the camp. It dawned on the American and British officers that they were *still* prisoners. As if to underscore this, the Russians pulled them out of the camp and marched them east for a day. Then, for some unexplained reason, they marched them back to Luckenwalde and doubled the guard yet again.

On April 25, the Red Army had a mass funeral for the dozens of Soviet POWs killed in the last week of the fighting. Many of these deaths had, however, been caused by the Soviets themselves. At the meal call for Soviet POWs shortly after the Red Army had arrived on April 21, each man was checked by a Russian doctor. Those with no teeth or with tuberculosis were taken aside. Although no one knew it at the time, each of these men was taken out and shot later that night.

Once Luckenwalde was taken over by the Red Army, the citizens in the surrounding towns had to put out small flags on their houses and apartments. A Communist put out a red flag, a member of the Nazi Party had no flag, and everyone else had a white flag. The rules for Russian soldiers were then posted:

- Be respectful of people in a home with a red flag. Take what you want but pay for what you take.
- People in homes with white flags may not be killed or raped. However, you may take what you want without paying.

- People in homes with no flags may be killed, raped, or beaten. You may take what you want.

Soon trucks with Russian soldiers were driving up and down the streets of Luckenwalde, firing into the windows and doors of homes with no flag. One woman came to the camp desperately wanting to join the POWs because she had been raped and beaten.

While the Americans and British remained in custody, the French prisoners were soon released by the Russians. They were allowed to "liberate" what they wanted from German homes and were encouraged to get to France by May 1 to vote Communist in the election.

Soon after the French prisoners were released, Tuck Brown happened to hear that one of the Frenchmen had liberated a German truck but could not get enough fuel to go to Paris. Brown got in touch with him and agreed to help the man get fuel from American forces if he would take a load of people to the American lines at Torgau.

Brown rounded up some men who were willing to risk the chance of being shot at the gate or along the road. The group included four Americans, three Britons, one Australian, and two Norwegians. They pooled their cigarettes and marched to the front gate, where Brown saluted the four Red Army guards and the men presented them with over six cartons of cigarettes. Brown then marched the men out the gate and loaded them into the back of the French soldier's truck. Brown got in the front seat with the Frenchman and started waving the French flag. It all happened so fast, and the guards were so preoccupied with their cigarettes, that the plan worked.

The truck sped away, and Luckenwalde disappeared in the distance. All the way to Torgau, Brown continued to wave the French flag. The men were halted three times in their 40-mile dash, but never by anyone who spoke French, so Brown would just wave the flag, smile broadly, and say, "*Bonjour, nous som Français.*" Near Torgau, they crossed a pontoon bridge and finally came under the control of the US Army. Tuck Brown told the American major in charge that he was a captain in the US Army, although he was wearing a British flying officer's uniform and carrying a French flag.

"I promised the driver fuel to get his truck to Paris," Brown told the major.

"I don't know," the major said. "That sort of request would have to go through proper channels."

"Look here," Brown said emphatically. "This man probably saved our lives. Give him the gas!"

Brown supervised the fuel transfer, thanked the driver, and finally realized that he was sick. He was taken to a first aid tent, where the medic said that he had a temperature of 103.5 degrees and called for an ambulance. Soon Brown was in an American hospital bed with a penicillin shot in his arm. Two days later, he was flown to a larger hospital and put on six meals a day.

When he entered this tent hospital, Brown weighed only 132 pounds, which meant that he had lost 50 pounds in prison. A hospital train finally took Tuck Brown to a hospital complex near Reims, only about three blocks from where Eisenhower accepted the final German surrender.

Most, if not all, of the American POWs held at Luckenwalde were eventually released by the Red Army. When Tuck Brown saw Colonel Gold in October 1945, Gold related that he and the rest of the Americans were not repatriated to the United States forces until early June, about five weeks after Brown's escape.

VE-Day

World War II in Europe did not end the way World War I had: "On the eleventh hour, the eleventh day, the guns fell silent."

On April 30, as house-to-house combat raged in Berlin the führer killed his dog, gave poison to his new wife, and ended his own life with a pistol. On May 2, all German opposition in Berlin collapsed, and Eisenhower ordered Patton to halt the advance of the Third Army in Czechoslovakia.

The government of Admiral Karl Dönitz—who had succeeded Hitler as führer—had only one objective: surrender. In four days of foot-dragging, his sole goal had been to allow time for as many German troops as possible to surrender to British or American units and avoid the brutality and torture the Soviet troops were sure to inflict upon their captives.

On May 5, a German delegation arrived at Eisenhower's headquarters at Reims in France, hoping to surrender the German state to the supreme allied commander. Ike told them that surrender would have to be unconditional and must be a surrender of *all* German forces to *all* Allies—including the Soviets. There were still Germans trying to fight Anglo-American forces in order to be captured by them. The Germans were desperate not to surrender to the Soviets. On May 6, Dönitz sent General Alfred Jodl to plead the case of a separate surrender, and Eisenhower simply told him that he would not accept the surrenders of any more of the Germans who wanted to give themselves up to British or American units while other units were still fighting the Soviets.

An agreement was finally reached, and at 2:41 A.M. on May 7, in a red brick schoolhouse in Reims, the surrender documents were signed. The next day, the procedure was repeated in Berlin for the benefit of the Russians. World War II in Europe officially ended at 11:01 P.M. the same night.

Fox Rhynard, who was in London, wrote, "Great joy. Great relief! We survived."

Jack Murray was put in charge of all German forces that surrendered to the 87th Division. There were about 40,000 troops, including 30 generals, 400 women, and an entertainment troupe. After a month, the Army decided to discharge them all and Murray was told to pay each officer 20 German marks and each enlisted person 10 German marks. He set up a plan with the help of a German financial officer and drew the 2.5 million marks the plan required from a bank in nearby Plauen—later part of East Germany.

A great many German units fully expected the Americans to join with them to fight the Red Army. When the 11th Panzer Division surrendered to the Third Army, General Wietersheim was truly surprised that, instead of putting the division in an assembly area, the Americans started disarming his men and putting them in prison camps. Someone had started a rumor among the Germans that the Americans would join with them and everyone would fight the Russians together.

Like everyone else, George Pickett's 16th Armored Division received official word that the war would terminate on May 8. The major impact of the "line" east of Pilsen was that Pickett could not accept the surrender of any Germans from the other side of the line. The Germans still had a large force—at least 85,000 troops, including four corps and their entire Seventh Army—massed in Czechoslovakia between the Soviets and the Americans. They didn't want the Red Army to get them, yet the 16th Armored Division had orders not to accept the surrender of anybody on the west side of the Pilsen line.

"The truth is," Pickett confided later, "every chance we had to capture a German when nobody knew it, we took him. These weren't SS units. If they were SS, we would have let the Russians have them because we couldn't shoot them and most of us thought that the SS were little better than butchers, and should somehow be punished. The fellows that were coming were all members of the army, the Wehrmacht. They were ragged, hungry, and half-starved."

On the night of May 8, the mayor of Pilsen threw a huge party in the cathedral square with bands, beer, and food. Afterwards, the GIs took two days to sober up. By May 10, General Pierce ordered the 16th Armored Division to move out of Pilsen and back to Tschemin, an hour west. Several of the men asked George Pickett why they were moving back, and he told them that, as he understood it, "The old man wants to get us the hell out of there because of the wine, women, and song, so he's moved us out into the country."

Because he had caught "holy hell" for pushing his recon platoon up the road toward Prague, George Pickett thought for a while that General Pierce would not want to promote him, but to Pickett's amazement, Pierce recommended him for promotion on May 8, and Patton's headquarters cut the orders on May 19. There was a rumor that Eisenhower had already put out orders against further promotions because the army was cutting everything back to prepare for the Pacific offensive, but Patton was still promoting some of his front-line people on May 19.

Wendell Knowles of Black '41 was with XIII Corps artillery in Simpson's Ninth Army at the end of the war. He watched the surrender of the German Army on the Elbe River, north of Magdeburg. "Looking across the river, one could see the German units come up the river, stack their arms and equipment, and swim to our side," he recalled. "For a short time, we had over 80,000 German POWs bivouacked on a grass airfield at Klotze. We had made some efforts to cross the river, but these projects had been halted on orders from higher authority." At one point it became necessary to throw a bridge across the Elbe and to send POW details back over the river to retrieve mess gear and other items that they were going to need at the POW compound.

Mike Greene was in a hospital in France when the war ended. He had been wounded on April 15, when his jeep struck a mine and he was sprayed with shrapnel. Greene returned to his division six weeks after he was injured, and several weeks after the war had ended.

When he got the news, Andy Evans was returning to England, having just flown a strafing mission over Berchtesgaden, his last of the war. Evans had become group commander of the 357th Fighter Group, and it was only a matter of a few weeks before the 357th was reassigned to Germany at an air base that, coincidentally, had been the group's next-to-last target just before the end of the war.

Immediately after VE-Day, Evans and his deputy commander flew to every German airport they could get into with P-51s to examine the destruction that had taken place. "Frankly, most people wouldn't believe what Germany looked like at the end of World War II. In fact, when you look at it now, it's unbelievable, because it was *flattened*, to the point where I couldn't see how they could *ever* put it back together."

Less than two weeks after the war ended, Evans made his first flight to the 357th's newly assigned base near Munich. He went by himself, just to check it out. When he landed on the strip, an officer in an unfamiliar uniform came up to him as he was climbing out of the cockpit, and, in broken English, asked Evans to "please come with me."

When they got to the barracks area, Evans found that the officer had his men all lined up and dressed up in clean uniform. He ceremoniously took out his pistol and surrendered it to Evans. The officer was Hungarian, and his unit had been in charge of security for the base. Now that the man had accomplished his purpose, he saluted and marched his men off the field. That, for a fighter pilot, was quite an interesting experience.

A few weeks later, after the ground echelon of the 357th Fighter Group arrived, a Soviet pilot in a Yakovlev Yak-9 fighter aircraft requested permission to land at the base. The control tower alerted Evans who along with an interpreter, was there to greet the pilot. They found that he wanted to defect to the United States. There was nothing Evans could do but to pass the word up to higher headquarters, and after a few days he got the message, "Tell him to get the hell out of there because we don't want any part of it!"

The 357th gave him some fuel and he left. Evans never knew what happened to him after that.

Later in May Pete Tanous's XXI Corps headquarters was sent north to Leipzig to prepare for turning the area over to the Red Army. After a few days there, Tanous joined the 30th Division as division quartermaster, preparing for movement to Le Havre followed by shipment to the United States. After 30 days of leave, the division was to go to the Pacific for the onslaught on Japan.

Bizz Moore went into the hills of southern Germany east of Stuttgart looking for "werewolves" or "wolf packs," one of Hitler's last propaganda dreams. Even after combat ceased, able-bodied Germans were supposed to gather in the hills, forests, and mountains and harass and murder the occupying troops. "I never met anyone who ever encountered a wolf pack. However, we searched for them and, unfortunately, confiscated the farmers' hunting shotguns and any other firearms we found. It made a hungry winter for many."

When he got the news, Merritt Hewitt was in Paris on three days of leave from the Sixth Army Group headquarters in Heidelberg, Germany. "I was aware that the end of the war had arrived and would soon be announced," he wrote. "The mood in Paris on VE-Day was much as might be expected, although many knew it was just a matter of when it would be announced. At 11:00 A.M., when it was announced, I was in the Au Printemps department store, and business shut down there and all over Paris. Great exuberant crowds gathered in every open space. By evening, the crowds were so wild and so thick that I gave up even trying to use the Metro and just walked back to my hotel."

As Hewitt reflected, "The real war was over. Japan was on the other side of the world and obviously finished. Inasmuch as I was regular army and I realized that most of the army would soon be demobilized, I started looking for a new assignment. Military government would be the one expanding field of opportunity. To achieve a transfer, I used contact with one West Pointer in Heidelberg to borrow an airplane and pilot to fly me to Paris and back."

He knew another West Pointer in the personnel sector in Paris and managed to obtain an assignment with the Headquarters of Military Government in Berlin.

Horace Brown had been with the 88th Infantry Division for two years as it battled its way north through Italy. Having liberated Bolzano on the eve of VE-Day, the 88th was sent south to Livorno to take charge of processing German POWs. Suddenly, Brown received orders—effective June 22, 1945—

sending him back to Washington for a special class with later assignment to the War Department General Staff.

The 88th Division was a 15,000-man division, and Horace Brown had been in it over three years, almost as long as he had been at the USMA. It had taken over 15,000 casualties—either killed or wounded—and had over 45,000 on the rolls prior to complete deactivation. Subsequent to the POW command, the 88th went on to serve in Trieste, with one regiment in Austria.

"The impact on me was similar to that of West Point," Brown reflected in later years. "I have always felt that my debt to each was almost the same. My affection for my classmates and almost all my civilian-life comrades is as great as that for those with whom I fought the war in the 88th. Again, now an elderly man, I have to take off my glasses and wipe my eyes when I think about them."

Brown left Casablanca and caught a C-54 transport aircraft across the Atlantic. When he arrived in Washington, he tried to call his wife but was unable to reach her. When he arrived at her door in the middle of the night wearing his field uniform and carrying his duffel bag, she was completely surprised. To this day, she still remembers the shock she felt.

Tuck Brown had been in a hospital at Reims suffering from pneumonia even as the surrender was being signed. As soon as he was well enough to get around by himself, he was sent to Camp Lucky Strike. From there, he was taken by train to a ship named *West Point*, which was about to sail for the United States. As the *West Point* entered New York harbor and Brown saw the Statue of Liberty, his eyes filled with tears of joy. Home at last!

That night, Brown stood in a long line to phone his wife, Dody. "I could hardly talk for the love I felt for her," he recalled later.

The next day, he was sent to Camp Kilmer in East Orange, New Jersey, and shortly afterward, he was reunited with Dody and two-year-old daughter Sandy, whom he had never seen.

Between D-Day in June 1944, and VE-Day in May 1945, 5.4 million Allied troops had entered Europe, most of them Americans. Fully one-third of all Americans killed during World War II—a total of 135,576, including two dozen men from Black '41—were killed during those 11 months. The goal, for which many had fought and some had died, had been achieved.

12

The Final Victory

Back to the Philippines

In late July 1944, General Douglas MacArthur had met in a historic summit conference at Pearl Harbor with George Marshall; Admiral Chester Nimitz, commander of US naval forces in the Pacific; and President Roosevelt. The purpose of the meeting was to decide where next to turn the course of the war in the Pacific.

Nimitz outlined the Navy plan. The Navy was to bypass the Philippines and enter the western Pacific to attack Formosa. For this purpose, all of MacArthur's American forces except a token group of two divisions and a few air squadrons were to be transferred to the command of Admiral Nimitz, who was to continue to proceed at full speed across the central Pacific. By the summer of 1945, Nimitz would be ready to invade Formosa.

MacArthur asked, "Just how will you neutralize and contain the three hundred thousand Japanese troops left in your rear in the Philippines?"

According to MacArthur, Nimitz never clearly explained this. MacArthur left him little opportunity, telling the assembled group that he was "in total disagreement with the proposed plan, not only on strategic but psychological grounds. Militarily, I feel that if I secure the Philippines, this will enable us to clamp an air and naval blockade on the flow of all supplies from the south to Japan, and thus, by paralyzing her industries, force her to early capitulation."

MacArthur was opposed to the naval concept of frontal assault against the strongly entrenched island positions of Iwo Jima or Okinawa. He stressed that Allied losses would be far too heavy to justify any benefits gained by wresting these outposts from Japanese control and that the islands themselves were not essential to the enemy's defeat. He argued further that by cutting the islands off from their supplies, Japanese resistance could be reduced easily with negligible American losses.

MacArthur was also critical of what he regarded as a major blunder in originally abandoning all efforts to relieve the Philippines back in the spring of 1942. He felt that if the United States had "had the *will* to do so, we could have opened the way to reinforce the Bataan and Corregidor garrisons, and probably not only saved the Philippines, but thereby stopped the enemy's advance eastward toward New Guinea and Australia." He argued that there was a moral obligation to release this friendly possession from the enemy now that it had become possible and that failing to do so would result in death to thousands of prisoners, including American women and children, male civilians, and POWs who were being held in Philippine concentration camps. Practically all of the 17 million Filipinos remained loyal to the United States and were a potential source of support. On other islands, such as Formosa and Okinawa, the indigenous population was loyal to the Japanese.

The meeting adjourned with the president making no final decision, but the following morning MacArthur once again pointed out how the recapture of Luzon was necessary to the winning the war and how it would be simple to deny Japan valuable oil, rubber, and rice supplies once Manila Bay and the northern part of Luzon were back in American hands.

The president interrupted: "But Douglas, to take Luzon would demand heavier losses than we can stand."

"Mr. President," MacArthur replied, "my losses would not be heavy, any more than they have been in the past. The days of the frontal attack should be over. Modern infantry weapons are too deadly, and frontal assault is only for mediocre commanders. Good commanders do not turn in heavy losses."

Roosevelt was convinced, and MacArthur's plan was approved and integrated into Allied strategic planning at the Quebec Conference in September.

The plan called for troops to land on Mindanao in the southern Philippines during November and to then invade Leyte on December 20, 1944. However, when Admiral William F. "Bull" Halsey's Third Fleet, operating in the central Pacific, reported that the Japanese naval presence was too weak to resist a landing at Leyte, Nimitz and MacArthur recalculated and altered the plan by moving D-Day on Leyte up from December 20 to October 20.

The invasion force consisted of 100,000 troops comprising the XXIV Corps veterans of the southwest Pacific campaigns and the X Corps, which had been diverted from a now-canceled plan to retake the island of Yap. Both corps, under the umbrella of Krueger's US Sixth Army, went ashore at 10 A.M. on October 20, and General MacArthur joined them at midafternoon to broadcast a message to the Philippine people that he was making good on his promise of two-and-a-half years before. He had, indeed, returned.

However, it was not to be easy going for the naval units supporting the Sixth Army. The Japanese Imperial Navy staged a counterattack, and the Battle of Leyte Gulf between October 23 and 26 became the largest naval battle of World War II. As had been the case at Midway in 1941, the Imperial Navy lost.

The Japanese came close to winning the battle. Compared to Midway, they had better operations and a much better area in which to fight. The layout of the inlets in the Philippine Islands gave them the opportunity for deception, covering, and developing forces, whereas at Midway they had been attacking strictly in an offensive role, and it had been fairly easy for U.S. forces to predict what they were going to do.

After the war, George Pickett and other Americans in Japan talked to many people who had served under Admiral Isoroko Yamamoto, the commander of the Imperial Navy at Midway, who died in 1943. All of them seemed to think that if Yamamoto had survived, the Japanese would not have been defeated as badly as they were. Discussions often turned to the Battle of the Philippine Sea, where Bull Halsey chased the Japanese fleet, leaving the landing beach on Leyte open for MacArthur's invasion. The prevailing theory among the Japanese was that Yamamoto would not have permitted that to happen, and that if he had been present the Americans would have been seriously defeated in the Philippine Sea.

Ashore on Leyte, General Tomoyuki Yamashita, the butcher of Malaya and Singapore, who now commanded the Imperial Sixteenth Army, had given orders to hold at all costs. Yamashita had 260,000 troops in the Philippines to defend Leyte against MacArthur and his troops.

Bob Cummings of Black '41 was a company commander in the 306th Infantry Regiment of the 77th Infantry Division on Leyte. On November 30, near Dagami, the 306th Infantry was pinned down by fierce Japanese fire, and reconnaissance was necessary to determine the exact location of all the enemy positions. Cummings exposed himself time after time to enemy machine gun, mortar, and rifle fire as he worked his way to within ten yards of a Japanese Nambu machine gun and assisted in blowing it up. Slowly and methodically, he collected valuable information that enabled the battalion commander to plan a successful coordinated attack on the entire enemy position at Dagami. Cummings was mortally wounded during his reconnaissance mission.

By December, six weeks after their defeat of the Japanese fleet and two weeks after Bob Cummings was killed, the Americans had achieved complete air superiority and were able to halt all resupply ships traveling between Manila Bay and Leyte.

Having secured Leyte, MacArthur and Krueger were ready to return to Luzon. Like General Homma, MacArthur landed his main force at Lingayen

Gulf. Yamashita had anticipated such a move but was surprised by how soon after his Leyte victory MacArthur was able to accomplish it and by how rapidly the Americans advanced once on land.

The Sixth Army came ashore on January 9, 1945, a little more than 37 months after Homma's Imperial Fourteenth Army had first arrived. The tactics of the two armies were similar, but their roles were now reversed. Recalling how his own forces had held out for over three months on Bataan, MacArthur decided that this strategic peninsula should be taken with the greatest haste. On January 29, he sent two divisions of the Sixth Army's XI Corps to seize Bataan. This time, it fell in less than three weeks.

On January 30, MacArthur dispatched a team of US Army Rangers and Philippine Scouts on a daring commando raid to rescue the American and Filipino POWs who were being held at Cabanatuan Prison. The two men of the Class of 1941 who had been held captive by the Japanese for the past three years were not there. The Japanese had begun evacuating Americans to POW camps on Formosa even before MacArthur's forces set foot on Leyte.

Bob Pierpont was at Batangas Prison for years, but on October 11, 1944, he and 1,789 other prisoners were taken aboard the troop transport *Arisan Maru*, which departed Manila later that day, presumably for Japan. On October 24, in the Bashi Strait between Luzon and Formosa, the *Arisan Maru* was sunk by an American submarine whose captain had no idea that there were Americans aboard. Bob was not among the handful of survivors. He was awarded the Purple Heart posthumously. His parents could not conceive of an accident of war in which death resulted from "friendly" action. They felt Bob was murdered by the United States. His roommate later told Lynn Lee that even many years after the war, Pierpont's parents would not see him or any of Bob's friends from the army.

Hector Polla had been at Cabanatuan Prison, but on December 28 he was among the prisoners put aboard the *Enoura Maru*, bound for Formosa. Also aboard were some of the 900 survivors—many with severe wounds that had gone untreated—from the *Oryoku Maru*, which had been bombed by American planes on December 15 at Subic Bay.

Polla spent his fourth New Year's Eve in the Far East aboard the *Enoura Maru*, knowing that within days Americans would liberate Cabanatuan. At last, the end of his captivity was in sight; MacArthur was on Leyte. He watched the dawn of 1945 in the Bashi Strait. Somehow, his floating hell managed to reach Formosa. Then, on January 9, just as the *Enoura Maru* made harbor at Takao, American bombers—who now ruled the skies over the western Pacific—appeared overhead. The ship was sunk in the harbor, and as a result, neither of the men of Black '41 who had been captured in 1942 would live to taste freedom.

Bob Kramer was the only member of the original Black '41 group who was able to elude captivity and death. After Bob Pierpont was taken prisoner, Kramer made his way to Mindoro Island and survived by living off the land for nearly a year. He joined a band of Filipino guerrillas at the end of 1943 and crossed with this group to Panay Island, where they continued their harassment of the Japanese. By this time, US Navy submarines were making regular supply runs to the guerrillas in the Philippines. In June 1944, Bob Kramer was taken aboard one such ship and made his this way to Australia. After a short leave in the United States, Kramer returned to the Philippines and went ashore on Leyte shortly after the Sixth Army landing.

Edgar Clayton Boggs of Black '41 landed in southwestern Luzon with General Robert L. Eichelberger's Eighth Army on January 31, three weeks after Hector Polla died. Clayt had been awarded the Bronze Star in New Guinea on June 20, 1944. When his company was pinned down by heavy machine gun fire, with "complete disregard to his personal safety and with calmness and dispatch, [he] reorganized his company and moved them under continuous fire to a flank position on higher ground, from where his company succeeded in eliminating a large number of the enemy."

Clayt Boggs went ashore on Luzon commanding Company B of the 61st Division's 20th Infantry Regiment. On February 5, 1945, Company B was advancing on Manila when Boggs learned that his first platoon was being held up by two Japanese tanks that were about 50 yards ahead of them. Boggs dashed forward to the first platoon line and established his observation post in a former Japanese dugout. In the hail of fire, he could see that there was little protection for his troops, so he moved them back 60 yards to the rear and called for mortar fire to knock out the tanks.

A direct hit destroyed one of them, but while Boggs was directing his mortar observer to put fire on the second tank, a large number of Japanese infantrymen suddenly swarmed out of concealed positions around the tanks and started a counterattack. As they did so, the second tank lurched out of its dug-in position and headed directly toward Boggs's observation post. He pumped a few rounds at the enemy infantrymen with his M1 and then grabbed the radio as mortar shells crashed down on the position where the tank had been.

"It's not *there!*" he screamed, giving the mortar crew a new set of co-ordinates.

More mortar shells came down, but still the tank inched forward. All the while, Boggs and his riflemen blasted away at the banzai charge. The charge finally was halted. Many of the Japanese were killed, and those remaining retreated in disorder. Then all at once the tank started firing again, scoring a direct hit on the Company B observation post and killing Boggs.

On February 10, Krueger's Sixth Army converged on Manila itself, which was not—as in 1942—an open city. It took the Americans 13 days of heated fighting to recapture the Philippine capital. American airborne troops filled the skies over Corregidor on February 16, and by the first of March, Manila Bay was at last reopened to Allied shipping.

Within weeks, all Japanese resistance in the Philippines had collapsed. The US Army was at last back where it had been when Japan started the war. The next step was to take the war home to Japan itself.

The End in Sight

After the defeat of Germany and Italy in May 1945, the Allies turned their full attention to fighting Japan. Having won the battle of the Pacific, the Allies focused on the Japanese mainland.

The Battle of Japan had, in fact, already begun, and it was taking place in the sky. The USAAF's Twentieth Air Force was activated on April 4, 1944, for the sole purpose of conducting the strategic air offensive against Japan. The Twentieth was the only USAAF component to receive the giant Boeing B-29 Superfortress. The B-29 was the largest strategic bomber yet built, with a far greater range than any other bomber. It had been a pet project of the USAAF chief, General Hap Arnold, since the late 1930s, when earlier, smaller heavy bombers were barely off the drawing boards. Now it was a reality. Arnold's objective was to use the B-29 to demonstrate that a major world power like Japan could be defeated by airpower.

Arnold believed that a strategic offensive using B-29s would negate the need for the planned invasion of Japan. General Douglas MacArthur, as supreme commander, had control of General George Kenney's Far East Air Force—the Fifth, Seventh, and Thirteenth. However, when the Twentieth Air Force was introduced, it was not placed under MacArthur's tactical control, and unlike the other 15 air forces, it reported directly (via Arnold) to the Joint Chiefs of Staff. This was by design. Arnold intended the organization as a prototype for the autonomous postwar Air Force he envisioned.

The Twentieth Air Force was originally assigned to bases near Calcutta, India, with advanced operating bases around Chengtu in northeastern China, which were the closest airfields to Japan then available to the USAAF. The first B-29 mission to be flown against Japan was a strike against Yawata that took place on June 15, 1944. Missions from Chengtu continued through the summer, but it soon became clear that supplying the B-29 force—which was accomplished via air across the Himalayas and 2,000 miles of contested Chinese air space—was going to be prohibitively difficult.

With the capture of the Marianas Islands—Guam, Saipan, and Tinian— Hap Arnold made the decision to relocate the B-29 strike force to bases there. The Twentieth Air Force flew its first mission against Japan from the Marianas on November 24, 1944, and raids continued on the average of every third day thereafter. Between March 9 and March 18, General Curtis LeMay, commanding the B-29 strike force in the Marianas, conducted a series of five nighttime incendiary attacks on the principal urban areas of Japan. These raids, which averaged 300 Superfortresses per mission, did more damage to Japan's war-making capability than all the previous attacks combined, but they were only the beginning.

Richard Kline wound up as a B-29 squadron commander in the 16th Bomb Group of the 315th Bomb Wing on Guam and led the 15th Bomb Squadron's first five missions against Japanese. The flight time of B-29 missions ran from 17 to 22 hours, so with the preflight briefings, crews were awake and busy for between 20 and 25 hours at a time. The missions themselves exacted a toll of nervous exhaustion, but aside from a few moments over the target itself, the chief concern was fuel conservation and cruise control.

At one point, returning from a successful mission, Kline had a big Superfortress on autopilot at 5,000 feet. Several hours south of Tokyo, he dozed off. This would not have been so bad were it not for the fact that his copilot was also asleep, along with the entire crew. Suddenly, Kline awoke to find the plane about 50 feet off the water! The autopilots of 1944 were not as well developed and refined as those in use today and would not compensate for weight and balance as the aircraft consumed fuel: thus they allowed the B-29 to drift slowly downward.

Kline quickly altered the flying mode, jerked back on the yoke, and the event passed unnoticed. "To this day," he recalls with a shudder, "I do not know if any of the crew really knew what happened."

Meanwhile, George Pittman had been at Tinker Field near Oklahoma City, wondering whether the war was going to end without his getting overseas. He went to the depot commander, who called personnel on his squawk box and told them to assign Pittman to the next opening. Five days later, Pittman joined the 359th Service Group as commander of the 570th Air Engineering Squadron, and a few days later the troop train with them on it moved out for Seattle and the Western Pacific.

As senior officer, Pittman wound up as the commander of troops of the 358th and 359th Service Groups aboard the SS *Extavia*. Other ranking officers managed to fly out to the Marianas as advance parties, but Pittman spent 38 days on the high seas.

Once on Tinian, the 570th Air Engineering Squadron set up camp on the north side of North Field, slated to become the world's largest airfield during World War II. Tinian ended up with four parallel 8,500-foot runways, with individual B-29 handstands between each runway along the long taxiways. It was here the 6th, 9th, 504th, 505th, and 509th Groups were supported by four service groups and six engineering squadrons. Pittman inherited the largest of the two service centers. By May, the Twentieth Air Force had enough B-29s in the Marianas to launch 400-plane raids several times a week. In August, 836 planes took off on a single day.

Most of Pittman's job involved traveling around the squadron maintenance operations and attending various meetings and conferences. His flying was confined to B-24s that carried post-strike photos from Tinian to the 20th Air Force Headquarters on Guam by night. The service groups also had some Piper L-4 Cubs that had been shipped to the 20th Air Force for B-29 panel operator/engineers who were pilots but did no B-29 piloting.

Although Pittman had been a B-29 test pilot at the depot at Kelly Field and was one of the first USAAF pilots to fly the B-29 as aircraft commander/ first pilot, the bomb squadrons had their line pilots do all their flying. "After all," Pittman admitted, "they were doing the combat flying in those aircraft."

Meanwhile, the Seventh Air Force had moved its VII Fighter Command into bases on Iwo Jima to serve as fighter support for the B-29s and to work with them to achieve total air superiority over Japan in anticipation of the final invasion.

By July, with the Twentieth Air Force and the VII Fighter Command in control of the skies over Japan and Okinawa secure, Allied strategists began to formulate plans for the invasion of Japan itself. That would take place in two parts. Operation Olympic—the invasion of Kyushu, the southernmost of the four main islands of Japan—would gain a foothold from which to launch the ultimate assault on the imperial heartland, the island of Honshu. This offensive, scheduled for November 1945, would be followed in March 1946 by Operation Coronet, the invasion of the Kanto plain on central Honshu. The Kanto plain, containing Yokohama and Tokyo, would be the most heavily defended enemy territory yet encountered in World War II, and Coronet would have to focus the most formidable force ever assembled. It would be an operation that would dwarf Overlord.

The Nuclear Genie

Based on the losses that American Forces had sustained on the islands of the central Pacific, Iwo Jima, and Okinawa, the invasion of Japan itself was viewed as potentially the most costly military operation in the history of the world. It was likely that the United States could lose as many men in this single operation as in all the other battles of World War II combined. The total estimate was one million American casualties and up to four million Japanese—both civilian and military—in a showdown that could last into the first part of 1947.

The campaign to capture Tokyo, Yokohama, Osaka, Kobe, and other population centers would be costly enough, but when the Japanese troops withdrew by the millions into the rugged mountains near Hokkaido in northern Honshu, the resulting battles could well become protracted bloodbaths. With these statistics in mind, the difficult decision was made to let the nuclear genie out of its bottle. Only a weapon more terrible than any that had ever been seen before could shock the Japanese into realizing the futility of continuing the war.

The task of building the world's first nuclear weapons had begun in 1942 under the direction of the US Army's Manhattan Engineering District, commanded by General Leslie Groves. The Manhattan Project staff grew rapidly over the next two years to include some of the world's best scientists as well as US Army personnel from both the Corps of Engineers and the Quartermaster Corps. Among the latter was Peer deSilva of the Class of 1941, who had made the move to the Manhattan Project from Fourth Army Headquarters in 1944.

On July 16, as President Harry Truman and Secretary of War Henry L. Stimson were sitting down for the war's final Allied leadership conference at Potsdam near Berlin, the first of the bombs was successfully tested at a place known as the Trinity site in the New Mexico desert.

"The bomb as a mere probable weapon had seemed a weak reed on which to rely," Stimson reflected, "but the bomb as a colossal reality was very different."

Together, Truman and Stimson gave a green light to the chain of command to execute plans to use the weapon.

In June, the 509th Composite Group was assigned to General LeMay's Twentieth Air Force and began training for a mission known to only a few. On the morning of August 6, three planes from the 509th Composite Group flew over Hiroshima. One of them, the *Enola Gay*—so named for the mother of Paul Tibbits, its pilot and the commander of the 509th—carried the first of two bombs that were to provide the war's *coup de grace*.

George Pittman's service group was indirectly in charge of maintenance of the 509th Composite Group and witnessed the departure of the *Enola Gay*. "We did not know exactly what was to happen, but the 509th got whatever they wanted," he recalled. "If they needed something, our service center controllers would break their backs to see that they got it. If they wanted gold plating, they had the priority to get it. We saw the shapes of their training bombs and we saw the radiation-protective gear in the hands of the medical officers, so we knew something special was afoot. They had special military police and the strictest security any of us had ever seen. You could not even stop on the road alongside their parking areas without being told to move along."

The first nuclear weapon used in wartime—a Uranium-235 bomb nicknamed *Little Boy*—exploded over Hiroshima at 9:15 A.M. on August 6. The 9,000-pound bomb, which had the force of 200,000 conventional bombs, incinerated the center of the city and created a mushroom cloud that the *Enola Gay*'s tailgunner could still see when they were 400 miles away.

The decision to target Hiroshima had not been taken lightly. Up to this point, the president and his cabinet left the tactical decisions to the field commanders who had been trained to make them by West Point, Annapolis, and experience. Although Henry Lewis Stimson helped to formulate the grand strategy, he didn't pick individual targets for specific bombers. However, when it came to history's first use of nuclear weapons, the responsibility for target selection had to be made in Washington.

For technical reasons, and in order to better assess the effects of the bomb, the target would have to be one that had received relatively little bombing from earlier B-29 raids. This ruled out Nagoya, Osaka, Kobe, and Yokohama. Tokyo was already ruled out, as the object of the exercise was to force the Japanese to surrender, not to kill the very people who could surrender. General Groves favored bombing Kyoto, Japan's fourth largest city and one that was completely untouched. Stimson was incensed at the suggestion. He had visited the city before the war and considered it to be a cultural treasure on par with Florence and Venice.

A second criterion was that there would have to be some military significance to the target, so Groves put together a list of relatively undamaged cities with major military facilities. The top three were Hiroshima, Kokura, and Nagasaki, in that order.

Following the Trinity test, the United States had three nuclear bombs. The tactical plan was to use two of the bombs—leaving the third in reserve—and to detonate them in rapid succession to create the illusion that the United States had many such weapons available for routine use.

After Hiroshima, the Japanese Imperial government blustered and tried to figure out what had happened and what to do next. The Japanese had an atomic weapons program of their own, but they were still several years away from being able to deliver a nuclear counterpunch.

On August 9, just three days after the bombing of Hiroshima, a plutonium bomb called *Fat Man* was sent to the second target on the list. But fate intervened—there was a heavy cloud cover over Kokura—so *Fat Man*'s fury was unleashed upon Nagasaki instead. This was the straw that broke the camel's back. After more than a decade of war, the Japanese agreed to surrender unconditionally on August 15. There was nothing left but the paperwork.

The Armistice was signed aboard the USS *Missouri* on Tokyo Bay on September 30. The war was over.

VJ-Day

Richard Kline was still on Guam when the news came in. "The men went nuts with happiness. I had a hard time stopping them from firing their .45 automatic side arms into the air in celebration. They all wanted to gas up the B-29s and fly home *immediately.* The demobilization was to be a nightmare."

George Pittman supervised the maintenance support of bomb groups converted to dropping 55-gallon drums of supplies to POWs in Japan and Manchuria.

Wendell Knowles had been redeployed from Germany to the United States in July 1945 with the 381st Field Artillery Battalion. On his arrival in Boston, he was given 45 days' leave and ordered to report to Camp Cook, California, in August to become part of the XIII Armored Corps Artillery. The XIII Corps' mission, as Knowles later found out at Fort Leavenworth, was to land on the south coast of the island of Honshu in November, on D-Day plus five, and "sweep the Tokyo plain." VJ-Day came midway during his leave. When he got to Camp Cook, he began demobilizing and discharging troops.

With the war now over, the Class of 1941 had a chance to reflect on how their experiences at West Point had prepared them for their roles in mankind's bloodiest war.

"Technically, I was ill prepared," Harry Ellis said. "Close-order drill was of great importance at West Point, but that was the first training I eliminated as a troop commander. However, the spirit of leadership, devotion to duty, and loyalty were probably the greatest assets I carried away from West Point."

"I think that we were all very lucky," said Bill Hoge. "Our success depended very largely on our tremendous natural resources and tremendous industrial production. When Roosevelt announced that we would build 100,000 planes, I didn't think it was possible. There was some wonderful staff work at high levels of the Army that permitted us to concentrate our power at decisive points."

Linton Boatwright said, "It didn't matter whether you were Army, Navy, Air Corps, Coast Guard, a factory worker, Rosie the Riveter, or a railroad technician. The population of the United States and Great Britain supported the war effort in every way that was possible. Everyone seemed to understand that the contribution he or she was making, no matter how small, was of importance in achieving final victory over the forces of Hitler and the Japanese Empire."

Boatwright cited an instance of being invited to dinner at the home of an English factory worker and his family in Birmingham, England, in May 1944. The "old man of the family," who was 76 years of age, was in the midst of a bout with influenza. His two daughters were concerned that he was returning to work too early and were saying that, in view of his advanced age and his weakened condition, he should stay at home for a few more days. However, the old man would have none of it. He said, "I am perfectly ready to go back to work—and besides, my factory has a quota to make, and I'm not going to let a trifling head cold stop me from pulling my share."

Says Ben Spiller, "It's not an original observation, but I see World War

II as having introduced Americans from all walks of life to one another, rich to poor, coal miner to college grad, easterner to westerner. It was a great melting pot. The war was a team effort and everyone seemed to appreciate the actions of the *other* members of the team. Never in our history has there been so obvious a striving together to achieve a common goal . . ."

13

Between the Wars

Coping with Peace

Lieutenant Colonel George Pickett volunteered for the occupation of Germany in June 1945. The Army directive regarding discharges said that anybody—except regular Army people—who had 85 points could go home. Regular officers like Pickett, who were not going back to the States, were then transferred to another unit in the army of occupation or shipped out for the war against Japan. Most of the people in Pickett's 64th Armored Infantry didn't have 85 points. They were young and had between 40 and 60 points, so the majority was vulnerable for reassignment. Within a week, the 64th Armored Infantry Battalion was dismantled and the 16th Armored Division was inactivated. "High point men" were transferred into it, and the division was shipped back to the States.

Each man had been asked to fill out a form indicating whether he wanted to go home for discharge, reenlist to fight the Japanese, or volunteer for the occupation army. For West Point men with regular army commissions, there was no choice about discharge. For Pickett, it was a matter of stating whether he preferred occupation duty or going to fight the Japanese. He went to Colonel Noble, the commander of Combat Command A and his immediate superior, to ask for help. "What's the best thing a man can do from a career standpoint—go fight the Japanese or be in the army of occupation? We've all been trained to fight the Germans in an armored division, and MacArthur has no armored divisions. The Navy has no armored divisions. There's also the rumor that MacArthur doesn't like Patton and doesn't like Patton's people."

"I don't know if that's true or not," Noble replied, "but that's the rumor. They say that if any of Patton's people get involved, MacArthur doesn't want them. Whether or not that applies to young lieutenant colonels like you, Pickett, I don't know."

When the war in Japan suddenly ended, there wasn't any other option. The "demobilization plan" was instituted in June 1945, and the Army was soon shipping people out of the old units as quickly as possible and replacing them—theoretically—with people who had just come over in new units. The units that were neither going to Japan nor staying on in the occupation were inactivated and their personnel absorbed.

When the war ended, members of the Class of 1941 were seasoned veterans who found themselves in positions of increased authority and responsibility. Jesse Thompson was assigned to the be the United Nations mediator during the Arab-Israeli conflict in 1948, and Richard Kline took part in the Strategic Air Command show of force that kept the Russians out of Iran. Meanwhile, Bizz Moore took home a Silver Medal from the 1948 Olympic Games.

"Whenever the classmates got together in those years, we told war stories, Air Corps and ground troops alike," Moore recalled. "We partied a lot, sang the wartime songs. We were exuberant to have survived, and sad for those, and their widows, who did not make it."

Mike Greene and Merritt Hewitt were both in Berlin in the fall of 1945, but in different commands. Greene recalled that throughout this time he was in contact with, and served with, many classmates as the Army expanded, because they each held positions of comparative importance within the system. During the war, there was a time when most battalion executive officers in the 11th Armored Division were classmates. They met periodically during combat action or in rest areas between actions, but these contacts usually were brief due to the nature of operations.

Mike had little contact with his brother Larry, however. Although they were both in armored divisions—Mike in the 11th and Larry in the 1st—they had no contact between May 1942, when Larry went overseas to North Africa just one week after his wedding, and early 1946, when Mike returned to the States for his own wedding. It wasn't until Larry returned to the States two-and-a-half years after he first went to England that he met his son.

When Mike Greene was in Berlin, he was a battalion commander of a provisional motor transport battalion in General Clay's US Group Control Council. Also in Berlin at that time was John Lee, who was operations officer for the 82d Airborne Division, and Francis Myers, who was also in the 82d Airborne. Ralph Kuzell served in Berlin with the Office of Military Government, and it was there that he met and married his wife, Margaret, a State Department employee.

The Military Standing Committee of the United Nations—of which Pete Crow was a member—met monthly. The ritual "come to order" was given in four languages. "There being no objection, minutes of the last meeting are approved," was also repeated in four languages. Rotating chairmen vied to establish the record time for completing the process. The record was 90 seconds!

Pete Crow was later assigned to the Pentagon in personnel planning. Next, he went to command and staff school with classmates Clint Ball, who had spent the war with the 351st Bomb Group in England, and Elmer Yates, who had been in the South Pacific as an engineer with the 24th Division. Afterward, Crow went back to the Pentagon and to Harvard Business School. Crow later returned to the Pentagon to spend nearly four years as executive to the Air Force member of the Munitions Board before going into intelligence. The Crows' first daughter, Punky, arrived in 1949 while her dad was serving in this assignment, which he described as his "short tour with the spooks."

Three members of the Class of 1941 lived in the same apartment building in Boston. Pete Crow and Hank Boswell were attending Harvard Business School, and Bill Seawell was at the Harvard Law School, where he shared classes with Jack Murray and Lee Ledford from 1948 to 1951. Also in Cambridge during those years were Gordon Gould, Jim Carroll, Jack Christensen, Malcolm Troup, Ralph Upton, John Redmon, Bill Starr, and Tom Cleary.

The Occupation

George Pickett had been in Pilsen until May 16, and then in Tschemin, which was seven miles away. On August 15, the 64th Armored Infantry was moved up to Graslitz and Falkanov. Although Pickett didn't realize it at the time, the latter was important because it was a known source of radioactive material. Marie Curie had gotten her pitchblende from the mine there in 1898, and after the Americans moved out the Russians moved in and mined the ore to build their first nuclear bomb. Pickett was assigned to guard and protect "coal mines" at Falkanov but didn't know until five years later that he had actually been guarding a uranium ore mine.

When the 64th Armored Infantry moved out on September 15, it was replaced by Czech troops, of which there were two distinct "armies." The "Swaboda Czechs" were a division that had fought with the Russians on the Russian front. Another brigade of Czechs had fought with the British and wore British-style uniforms with a little flash on them that read "Czech." They were the first group that the Yanks did business with, and they had started taking over some of the installations as the American troops pulled out.

When the 64th Armored Infantry Battalion was shipped back to the States as a carrier unit, George Pickett was assigned to the 2nd Batallion, 18th Infantry, in the 1st Infantry Division and he turned his area over to an officer from the "British Czechs." For months, the Americans had no trouble getting into Czechoslovakia because they controlled the border. Then one day, when Pickett had decided to take some people to see the monument that the 1st Division had built, somebody opened fire with a machine gun when they got within 50 feet of the border. Bullets bounced off the front of the jeep.

Pickett went back to get an interpreter and a white flag, and this time the Czechs came out and talked to him. It seemed that the British Czechs had been replaced with Swaboda Czechs, and although they were Czech, they didn't want any part of the Yanks. They had been told by the Red Army's commissars that Americans were the enemy.

For most of the troops still in Europe, the primary concern was getting home. It had taken the United States three years to get its army of a million to Europe. They could not be returned overnight, and there was a great deal of dissatisfaction. The general cry was, "The war's over; I want to go home." There was a great deal of unrest among the troops before the situation would be resolved.

General Lucius Clay replaced Eisenhower as supreme allied commander, but Clay was functioning as a military governor in Berlin and the actual troop command was operating under the deputy commander of USFET (United States Forces, European Theater), Lieutenant General Clarence Huebner, in Frankfurt. George Pickett was alerted to take the 2d Battalion to Frankfurt at one point. There was a demonstration going on and the MPs couldn't contain it, so Huebner decided to move some troops in. Pickett took a squad of ten men with him into the city, leaving the main elements of the battalion outside of town on the highway.

Pickett walked onto the steps of the I.G. Farben Building, which was now the American headquarters for Germany, and found a large crowd of GIs milling around, shouting. One of the GIs got out of hand with one of Pickett's sergeants, and the sergeant just let him have it. The men looked at the 1st Infantry Division patch and the medals on the sergeant's chest and let out a scream. "My God! It's a Big Red One. He'll kill us!"

With this, all of the "rioters" evaporated. That was the end of the riot.

The 1st Infantry Division, whose insignia was a large red numeral, had earned a justifiable reputation for toughness during the war, and most soldiers who were newly arrived in Germany held it in awe.

Two weeks later, the 2d Battalion was alerted to go to Bad Kissingen, where a USAAF major general couldn't control his people. They had cut holes in the barbed wire fence so they could go out and "shack up" with young German women instead of obeying the curfew.

General Milburn called Pickett into his office and told him, "Take your battalion, under cover of darkness, to Bad Kissingen, report to the general there, and clean up this goddamn mess."

"Begging the general's pardon, why does the 2d Battalion always get these kind of details?" Pickett asked.

Milburn said, "Damn it, Pickett. Clarence Huebner *commanded* the 2d

Battalion in 1917–1918, and if there's any dirty job he thinks needs to be done, the 2d Battalion gets it." "You're to surround the town and go through it building by building. Anybody that's not in an official billet is to be arrested."

"Am I supposed to use force?" Pickett asked.

"All I said was you arrest them. If they resist arrest, I'm not going to ask any questions."

The 2d Battalion surrounded the town of Bad Kissingen and went through it with bayonets, digging soldiers and airmen out of brothels, beds, bars. A full colonel who was the principal assistant to the general in charge there was caught in bed with his mistress.

The occupation troops found the Germans to be a most docile, cooperative people. For example, when George Pickett was in Würzburg in February 1946, the city needed to get a hot-water steam plant running, so Pickett went to the prison camp, got a German colonel, and told him what to do. The next thing Pickett knew, he had a former German engineer battalion repairing the city. As Pickett would soon observe, "Once they knew they were licked and got out from under the thumb of Hitler and the SS, they became very cooperative."

There were rumors of a planned resistance to the Allies, the Edelweiss Underground, but it just didn't happen. The German people had had enough, and all they wanted was to survive that winter of 1945–1946, one of the most terrible winters on record in Europe.

By the spring of 1946, the Americans had released all of the Wehrmacht soldiers from the POW camps, so that the only people being held were members of the SS at the rank of sergeant and above. The SS was being held as part of a search for war criminals. The Allies tried to get the ordinary soldiers out of the POW camps as fast as possible because it was the humane thing to do and because they were needed to rebuild the German economy.

In Hof, the US Army wanted to set up an Officers' Mess jointly between the 28th Constabulary Squadron and the 2d Battalion but there were no cooks. It so happened, however, that Field Marshal Von Manstein's pastry chef was in the prison camp. The 28th Constabulary tapped him, and the 2d Battalion had the best pastries in Europe.

"If we hadn't been able to utilize the skill level of the German prisoners, we would have had a hell of a time getting through the winter ourselves. It made things a lot easier, and of course, when you have Von Manstein's pastry cook, it also makes it a lot *nicer!*" said George Pickett.

By the middle of 1946, however, most of the troops who were going home had gone home, and the role of the occupation forces became routine. The occupation troops were involved in such things as guarding POW camps, operating and administering the Displaced Persons program and DP camps, guarding and protecting the border, and operating the Nuremberg War Trials.

"Being a battalion commander didn't take that much time," Pickett laughed. "Outside of seeing that they were fed and housed and clothed and didn't get in trouble and kept busy enough so that they could do their jobs, and then keeping the DP camps operational—there wasn't a heavy load on the officers. It reminded me a lot of the peacetime Army before 1941. A captain went to work at eight o'clock and left at noon. If there was a problem, the first sergeant called him."

If war can be described as "a few hours of unmitigated hell, surrounded by months of boredom," then the same was even more true of the occupation. Because there was really so little for the occupation troops to do, the army gave individual commanders the option of setting up battalion "junior colleges." The occupation army's "junior colleges" could help a man finish his high school diploma or give him courses that most colleges would accept for credit. In October 1945, a GI could take a course in college algebra at the 2d Battalion headquarters, which was in an old school building in Ufenheim near Würzburg. George Pickett set up a complete school that taught everything from high school to college courses that he could find people to teach. In addition to being the battalion commander, Pickett was teaching college algebra and a course in world history.

Pickett gave up command of the 2d Battalion about a week before Thanksgiving in 1946, and after that, things were so quiet the officers managed to move their families up from Amberg. George, his wife Beryl, and three other couples spent the week before Christmas 1946 in Nice on the Riviera. It was a far cry from his Christmas two years before during the Battle of the Bulge. George Pickett left Europe in June 1947, and two months later he joined the faculty of the infantry school at Fort Benning, Georgia.

From March 1948 until July 1950, Horace Brown served in Germany with the Big Red One as division artillery intelligence officer and later as the executive officer. Three months after he arrived, Soviet forces suddenly cut off all access to Berlin except for aircraft. It was to be a period of great stress: the 11-month Berlin Blockade had begun.

When Brown had arrived, the 1st Division was in very poor condition because it had been scattered over Germany on an occupation mission with no emphasis on combat training. Within days, however, the division was assembled in the Grafenwohr Maneuver Area, where it would remain for six months, conducting firing and maneuvering exercises.

The only other American force of consequence at this time was the Constabulary, a lightly armored police-type occupation force. No tanks were available when the American forces in Germany went on the winter maneuvers of 1948–1949. The tank build-up began soon after, and by the spring of 1950, when Horace Brown looked at all the tanks rolling around the maneuver area, he "felt a lot better and more confident."

Before the war, Horace Brown had married Lucia "Chick" Sloan—nobody ever called her anything but Chick—a biology instructor at Dreher High

School in Columbia, South Carolina. When they moved to Muskogee, Oklahoma, where Brown was assigned to the 88th Infantry Division stationed at Camp Gruber before he went overseas, Chick took a substitute teaching position that evolved into a full-time job. While Horace was in Europe, she worked in civil service at Fort Sam Houston, Texas, but after his return the roles of army wife and mother became her primary occupations. She was also president of the Army Wives group in Erlangen, Germany.

While Horace was assigned to 1st Division Artillery in Erlangen in 1949, a large turkey shoot was scheduled. Marksmen, including Horace himself, came from around the area with specially made rifles with scopes. Almost all of them had qualified as experts. Chick, who was about six months pregnant with John Sloan Brown (who would graduate number eight out of 729 cadets in the West Point Class of 1971) decided to participate as well. She had been a member of a women's rifle team in Hawaii, but hadn't had the opportunity for practice that the infantry and artillery men had. She drew a carbine, got an old master sergeant (who had served for her father), went to the range, and zeroed her carbine.

There was a host of red faces when she won the turkey shoot. Her photo, with a write-up of the event, was published in the army newspapers in Europe. "One good side effect ensued," Horace laughed. "Our home was one of about two in Erlangen that was never robbed."

The Olympics

The cross-country meet in Van Corlandt Park in New York City in the fall of 1939 had pitted Army against the favored Minnesota team. George "Bizz" Moore, who would run a 4:21 mile in the spring of 1940, spearheaded Army's victory. Nearly a decade later, Moore backed into the idea of the 1948 Olympic Games quite unexpectedly while he was teaching at West Point. After classes each day he went down the hill to the quarter-mile track where he was the officer-in-charge of the track team and assisted the full-time civilian coach. In 1947, more than a year before, the United States had begun to get ready for the first Olympic Games since Berlin in 1936. Coaches all over the country were asked to nominate likely athletes in order to form a training squad for each sport.

The civilian track coach at West Point said he would nominate Moore for the 1,500-meter run. It was a nice compliment, but Moore pointed out that, while he had been a good college miler in 1940 and 1941 and had won most of his races, he had never been an Olympic-caliber miler. He and the coach then listed track-and-field men who had a chance to make the United States team.

Later, the swimming coach and track coach were asked to nominate men for the Modern Pentathlon team, which had always trained at West Point.

The Modern Pentathlon was a five-day event with one sport per day: a 5,000-meter steeplechase ride; an épée (dueling sword) competition in which every one of 45 to 60 competitors fences every other competitor, with one touch resulting in a victory; a pistol shoot at bobbing targets with scoring rings; a 300-meter swim; and the finale, a 4,000-meter cross-country run over difficult terrain of hills, fences, streams, soft loam in the woods, and usually an uphill finish.

The Twelfth Olympiad opened amid great splendor in London in July 1948. For Bizz Moore and his teammates on the Modern Pentathlon team, the 12-hour competition on the fencing day was probably the highlight. They knew from all the history of the Olympic Games that they were at a great disadvantage. The long, grueling fencing competition begins with the men from each country fencing with their own countrymen, to prevent two men from "throwing" their bouts to a third countryman with a chance to be first among the fifty fencers.

Moore knew that if he gave the experienced Europeans too long to look him over, most of them would find some way to beat him. He decided that he had to attack early and keep his opponents busy defending themselves. Not too long after the competition began, European experience came to light. When Moore showed up on the scoreboard as winning several bouts, competitors not active on the strip came drifting over to see what was going on. After about six hours, the Europeans had drilled themselves to avoid Moore's violent attacks, and touches became harder. On the other hand, he was in better physical condition than some of the older competitors, and as they wore down, their wisdom was less effective against his speed.

Once Moore had established his reputation for violence, he was able occasionally to sneak in a quiet point under the bell of a tense opponent, bringing joy, he was sure, to his fencing coach's heart—a real "finesse" touch!

When the fencing round ended, the top Swede and the top Brazilian had tied for first, winning 28 of 43 bouts, or 65 percent. Moore and the top Finn won 26 of their 43 bouts, or 60 percent. By this time the Swede had two first places in two events, and Moore had a second and a third place and was ranked second in overall competition.

Prior to the shooting event, all teams agreed that normal international shooting rules would apply. If a target was found with one hole too many, the best hole would be canceled and the other five scored.

In the middle of the event, while Moore was shooting, a Swedish officer loaded six rounds instead of the legal five, and the British soldiers operating the range faced the targets six times instead of five. Everyone was startled, but each brought up his gun and squeezed the trigger before he finished the count in his head, "One thousand one, one thousand two, one thousand three."

There were clicks of empty guns all down the firing line. The Swede's gun, however, went off, and his target had six holes.

"No problem. Throw out the high hole. The low five holes count!"

"Hah! Not so! There will be a meeting of the International Jury of Appeals."

This group was presided over—and dominated by—its Swedish president, so in the end the range was made quiet. The "offending competitor" was allowed to step up to the firing line alone, no other guns firing ahead of him or after, and shoot his five rounds over again. Having had a chance to rest and concentrate, he made a good score. Only then, after a tension-ridden hour, were the other shooters and the Swede allowed to complete their 20 rounds.

George Moore ended the day with a score of 183 out of 200 possible, far lower than the 190 he had fired when the US team was selected. In his case, the Olympics were much more nerve-wracking! Still, when the standings after three days of competition were posted, he was in second place overall.

The swimming on the fourth day looked like Moore's Waterloo. It had been his worst sport. He tried hard for a good start and splashed his way down the 50-meter pool for all he was worth. When that day ended, the scoreboard revealed that he was *still* in second place overall.

Finally, the running day arrived. Here Moore knew what he was doing. He had been captain of the West Point cross-country team and the Academy's best miler during his Second and First Class years in 1940 and 1941. There was a complication, however: Three days earlier, Moore had sprained his ankle on a training run. He had had a worse sprain in 1940 and had learned how to use two-inch adhesive tape to strap his ankle. The tape allowed the ankle to flex front to rear, but stopped any tendency to roll over left to right. It was a slight handicap, but respraining the ankle before the end of the Olympics would have been disastrous!

The top Swede was also a pretty good runner. His lead in overall points was so great that unless he broke a leg on the run, he was sure of the Gold Medal no matter how much the nearest few competitors bested him on this final day.

Moore and the number-two Swedish athlete had a much different situation. Each had beaten the other in two events. Their scores after four events were only a single point apart. Thus, whichever runner had the better time on this last day would win the Silver Medal. They were far ahead of the next few men in total points, and both were good runners, so for themselves, they were the only two men that mattered.

With 30 seconds to go, Moore approached the starting line, ankles freshly taped. He had studied a chart of the course, a rough loop. As the official starter counted down the last seconds, Moore got down in a sprinter's starting position. A few spectators laughed at the sight of a cross-country runner in such a position, but it may have made a vital difference. He went out fast to release his pent-up nervous energy and then settled into his familiar pace. In the mile run as well, Moore had never thought that his initial burst from the starting line took away any energy he would need at the end.

The race led up a short, steep, sandy hill that was best assaulted by a sprinting rush followed by a scramble on all fours up the final, almost vertical, bank near the top. Then the course wound down a long, scrubby slope, over a barbed-wire fence, and down through a stream. The soft loam of the woods was especially tiring and discouraging, but once out of the woods, Moore could see the finish line only a quarter mile away, uphill across a rough field. Through his wind-teared eyes he could see much arm-waving, and he heard indistinguishable shouts. Finishing a race was one thing Moore had always been good at. He began digging. He was exhausted but pleased the end was near.

Moore finished, gasping too hard to speak. Hearing sketchily, he felt an old general thrust a stopwatch into his hand. Not understanding, and needing only more oxygen, he handed it back. Later he was told that General Guy V. Henry had wanted him to accept, as a gift, his favorite stopwatch with his running time still set on it, showing that he had won the Silver Medal by one-and-one-half seconds.

The crowd at the finish line knew that Moore and the Swede had come out of the woods at exactly the same time. Thus, whoever did better up that last quarter mile would win. This accounted for the excitement and arm waving and the offered watch. Upon reflection, however, the general decided to keep his favorite stopwatch.

At last came the ceremony on the victory stand. The band played the national anthems of Sweden and the United States. Later, a group picture was taken of the 200 people involved in the event. Finally, there was a formal dinner, a tattoo by the Guards and their pipe band, and the 1948 Olympic Games were over for the Modern Pentathlon teams. Moore managed to watch a few other events in progress, and then the team headed home.

The Army Loses an Air Force

In 1944, the Joint Chiefs of Staff had appointed the Special Committee for Reorganization of National Defense to look into the question of overall postwar command structure. In April 1945, after a year of extensive investigation, the committee recommended—as had many others over the years—that a US Air Force be established on a coequal basis with the US Army and US Navy, under a single department of "National Defense."

In November, Eisenhower returned from his triumph in Europe to take the place of retiring Army Chief of Staff George Marshall. Eisenhower was a firm believer in air power and wasted no time in presenting his view that a separate and independent US Air Force should be created.

On July 26, President Truman signed into law the National Security Act of 1947. The Act, effective on September 26, 1947, provided for three "executive departments" under the new Department of Defense. These subdepartments—Army, Navy and Air Force (the Marine Corps remained in the Navy Department)—would each be presided over by a civilian secretary and comprise a branch of the military commanded by a chief of staff. The chiefs of staff would be members of the now-institutionalized Joint Chiefs of Staff, and its chairman would be an advisor to the president and the National Security Council.

As Black '41 classmate Jim Laney saw it, the Army–Air Force separation "evolved naturally. Both were too large and too different in mission orientation to any longer function as one entity."

A by-product of the Act was that after September 26, about one-third of the Class of 1941 was no longer in the US Army. They now wore the new, blue uniform of the US Air Force.

In 1947, after his assignment with the Japanese occupation forces, Richard Kline, who had flown B-29s with the USAAF's Twentieth Air Force during the war, returned to the United States, joined the new US Air Force's infant Strategic Air Command (SAC), and was assigned to Davis-Monthan AFB in Arizona as a squadron commander in one of the few squadrons then equipped with nuclear weapons. Meanwhile, President Truman was having trouble convincing Stalin he should remove Soviet troops from Iran. During World War II, British and Soviet forces had occupied Iran to forestall German attempts to take over its government. When the war was over, the British withdrew. The Soviets did not.

In the late summer of 1948, Truman ordered three nuclear-armed B-29s to make a round-the-world flight to demonstrate to Stalin the wide-ranging capability of the Strategic Air Command's nuclear force. Richard Kline was the commander of that flight. Stalin quickly got the point and removed his men from Iran.

14

Korea

Drawing a Line

World War II was the challenge that absorbed the Class of 1941 upon graduation, but five years after it had ended America once again was at war, this time on the rugged Korean peninsula as the Communists swarmed south in an attempt to swallow the fledgling Republic of South Korea.

The Korean War presented a situation that was the antithesis of that in which the United States had found itself in World War II. It was the first time in the twentieth century, and arguably the first time in American history, that the US Army found itself fighting a major war with limited political goals. In World War II, the goals and objectives of the two sides were symmetrical: The Axis wanted to dominate the world, while the Allies demanded unconditional surrender.

In Korea, the goals and objectives were politically asymmetrical: The North Koreans wanted both political and military domination over the entire Korean peninsula. The United Nations, the self-proclaimed protector of South Korea, wanted only the southern half of the peninsula to remain under South Korean control. One side had unlimited goals, while the other side had a limited goal.

For the United States military, which constituted the overwhelming majority of United Nations forces actually in combat, the Korean War was a very disillusioning experience. After having defeated the powerful Axis powers in World War II, the United States found itself expending mountains of matériel and tens of thousands of American lives to preserve a stalemate against an enemy keen on victory.

At West Point—as at Annapolis—America's officers had been drilled repeatedly on the dictum that there was no substitute for victory. Stalemate was no better than defeat. In World War I, the stalemate of trench warfare had killed millions and demoralized Europe. The American entry into that war had been precisely the event that had ended the stalemate and had made victory possible. In World War II the object was a total, unconditional end to the war. The United States seemed invincible because it *knew how* to win. The Korean War taught that the United States also had to decide it *wanted* to win.

In January 1950, with China "lost" to communism, the United States had drawn a line in the Far East that did not include Korea. Dictator Kim Il Sung, who had been installed in North Korea by the Soviets when that region was still in the Russian zone of occupation at the end of World War II, wanted to control all of the Korean peninsula in the name of communism, and he was willing to go to war to accomplish his goal.

Jack Murray served as a deputy battalion commander in Korea between 1946 and 1948 during the occupation. When he was pulled out in 1948, he said, "It wouldn't surprise me if the North Koreans moved south. Our training of the South Koreans during the occupation had not prepared them for what they had to face."

Kim's North Korean Army was trained and equipped—thanks to the Soviet Union—to sweep the South Korean way of life out of Korea in six weeks. There is no doubt the North Koreans would have done it were it not for the fact that the United States had decided that this cold, rocky, wind-swept peninsula was the place to make its first stand in the postwar confrontation between Western democratic ideals and the perceived threat of Communist totalitarianism.

In January 1950, in the wake of the resounding Communist victory in the Chinese Civil War on October 1, 1949, Secretary of State Dean Acheson made a speech in which he outlined American priorities in the Far East and pointed out the places where the United States would stand firm to resist the advance of Communism.

Formosa, the island to which Chiang Kai-shek's defeated Nationalist forces had retreated, was on Acheson's list. The Philippines, site of America's colossal defeat in 1942, was included, as was Japan, the great world power that had rendered that defeat, but which was now occupied by victorious American armies. Korea was *not* on Acheson's list. Acheson, one of Henry Lewis Stimson's protégés, had clearly not benefited from the counsel of his mentor, who would die later that year. Stimson, with his depth of experience and his first-hand knowledge of the Far East, would probably never have made such a profound miscalculation.

Surprise

The Korean War started without warning on June 25, 1950, when a massive North Korean invasion force overwhelmed the South Korean forces guarding their side of the demarcation line at the 38th parallel.

On June 27, two days after the invasion, President Harry S. Truman announced that the United States would send equipment and airpower to support the South Korean government of Syngman Rhee, whose army was then in a complete rout. Meanwhile, the United States sought to have this counteroffensive against North Korea's invasion given the imprimatur of the United Nations. Because the Soviet Union, as one of the five victorious Allies in World War II, had a permanent seat on the United Nations Security Council, it could have vetoed the resolution that designated the defense of South Korea as a United Nations peace-keeping action. As it turned out, the Soviet Union was boycotting the Security Council over its refusal to unseat Chiang Kai-shek's Nationalist Chinese government and let Mao Tse-tung's Communists be seated in the United Nations.

The counteroffensive did receive the official blessing of the United Nations, and although the United States—and, to a lesser extent, South Korea itself—provided the lion's share of the forces in the field, the armed forces of a great many United Nations members, including Canada, France, Britain, and Turkey, took part in the effort.

As Truman considered the American response and lobbied the United Nations, events in Korea itself were moving with breathtaking speed. North Korea's army was bigger, better trained, and better equipped than South Korea's. The Soviet Union had furnished it with 150 T–34 medium tanks (the backbone of the armored force that had defeated Hitler on the Eastern Front) and 100 combat aircraft. By contrast, the South Koreans had no such weapons, nor any antitank or antiaircraft artillery. Only 72 hours after they crossed into South Korea, the North Korean forces were able to occupy the city of Seoul, the capital and largest city in South Korea.

General Douglas MacArthur, commanding American occupation troops in Japan, was immediately named to command all United States forces and was designated United Nations commander once the United Nations resolution passed. MacArthur sent American tactical airpower into South Korea just before the fall of Seoul, but the aircraft soon had to be withdrawn to Japan as bases were overrun. On July 5, he also sent in the 24th Infantry Division, under General William "Bill" Frishe Dean. Based in Japan, it was an occupation garrison force and not up to strength as a combat unit.

The men of the 24th Infantry were overconfident, under-equipped, badly trained, and very surprised. Unable to stop the North Korean advance at any point, by July 14 they had stumbled back in virtual disarray to Taejon in the southern part of Korea, where Dean hoped to make a stand. In only two weeks, the 24th Infantry lost 5,729 men. The remaining troops were exhausted, but it was imperative that the North Koreans be stopped—or at least slowed—at Taejon in order to buy time for MacArthur to plan a counterattack.

The enemy slammed the Americans holding Taejon with repeated, ferocious, head-on assaults, pinning them down, infiltrating their rear, and cutting off their retreat. By the morning of July 20, the situation was hopeless. Even General Dean himself was hunting T–34s with a 3.5-inch bazooka. The 24th Division had sustained 30 percent casualties, with an unusually large percentage of officers who were experienced veterans of World War II. The 34th Regiment, which was perhaps the most intact of the 24th Division's components, had lost its commanding officer and all contact with two entire battalions.

Dean finally gave the order to withdraw. By nightfall, Dean himself left the area, leaving elements of the 34th Regiment to cover the tail of the retreating Yanks. However, the North Koreans had completely surrounded Taejon, and there was no way out. Dean's jeep was destroyed, but he managed to get away on foot and elude capture for a month.

William Thomas McDaniel was the first member of the Class of 1941 to see combat in Korea. He went to Japan with his family in 1949, assigned to the 34th Infantry Regiment of the 7th Infantry Division. When the war came and the 34th was sent to Korea, McDaniel was its operations officer.

McDaniel and his men were overrun and captured on July 20 at Taejon, and along with almost 376 other American prisoners of war, they were taken north to Seoul. From there, they were consolidated with other prisoners and marched over 300 miles to a POW cage in Sunchon, North Korea.

It was North Korean policy to shoot men who could not maintain the pace for the grueling march of 22–25 miles a day. To prevent this, Tom McDaniel, who was a major and the senior POW officer, argued and cajoled the guards into giving him ox carts for the men who couldn't keep up. At one point there were 15 carts in the column, carrying 60 men. However, with the scant rations of only a few balls of rice a day and virtually no medical support, the men grew weaker, and more and more of them fell out and were gunned down. It has been estimated that only 296 men were still alive when the two-week Death March finally reached Sunchon.

On October 20, near Sunchon, with the American forces now closing in on North Korea, the POWs were promised their first meal in several days. Tom McDaniel was among those told to make up a detail to a nearby village to obtain food. He never returned.

The remaining POWs were taken to a railway tunnel, where their captors turned machine guns on them. United Nations forces, which overran the area later that day, discovered the slain Americans. Of the POWs under McDaniel's command, less than 30 had survived the entire ordeal that had begun in Seoul.

For his actions in helping to inspire and preserve the lives of his men, McDaniel was posthumously awarded the Distinguished Service Cross and given a promotion to lieutenant colonel. He was carried on Army records as missing in action until July 1953, but his death was not officially verified until 1955.

"Two By Two"

The first United States troops sent into Korea were in what was called the "two by two" units. As a result of Defense Secretary Louis Johnson's cost cutting, the US Army had two battalions in a regiment instead of three, and two rifle companies in a battalion instead of three. In cutting "fat," Johnson had not only cut out one-third of the combat strength, he had cut out one-third at every level.

Beginning in the spring of 1949 after President Truman was inaugurated, Secretary Johnson introduced a policy of selling off the enormous stocks of supplies left over from World War II. This included tents, canteens, belts, field equipment, officers' bedrolls, and every imaginable type of equipment. The idea was that instead of having this matériel rotting in warehouses, the government was getting some kind of return on its investment. The operative theory was that the supplies would never be needed again. This was Louis Johnson's way of "cutting the fat and sparing the muscle." Except for the tanks that were left on Johnson Island in the Pacific, most of the US Army's tanks had been turned into scrap. Johnson Island was a centralized location that the Army shipped tanks to from the Pacific theater to save money by not having to ship them back to the States. Many tanks that wound up in Korea came from Johnson Island, where they had been rusting for five years.

The muscle of an infantry outfit is in the machine gunners or mortar crews. The muscle in a tank outfit is in tank crews and the number of tanks available. The muscle in an artillery outfit is in the number of guns in the battery and the number of batteries in a battalion. When the 1st Battalion Task Force went into Korea in July 1950, it had two rifle companies and a composite heavy weapons company that had only half the heavy weapons it should have had.

"As long as we can still carry all the regimental flags on parade, everybody thinks we have the right number of people," George Pickett said in retrospect. "You can say the same thing that George Scratchley Brown of our class said about the British that got him in trouble: 'They have nothing but generals, admirals, bands, and flags.' That was about the situation we were in during the occupation of Japan—generals, admirals, bands, and flags."

George Pickett's tour at the infantry school had been extended for an extra year, so June 1950 found him still at Fort Benning, Georgia. On June 19 he was given a 10-day leave, so he and Beryl packed up the kids and went down to Daytona Beach. On the way back on June 25, they heard on the car radio about the attack by the North Koreans.

"What do you make of this, George?" Beryl asked.

"Some Americans are going to be fighting over there within 90 days," he said simply.

"I hope it's not *you*," she said.

"I wouldn't be a bit surprised *who* it was, because this thing is going to blow up."

The radio went on to say that President Truman had said the United States was going to use air and sea power to stop the North Koreans. Hearing this, Pickett commented, "Horse shit! Air and sea power are not going to stop a land invasion. No matter what these Air Force fanatics say, they can't stop an enemy in the advance!"

Of course they didn't, and Truman quickly decided to send in ground troops. The ground troops that were available within the occupation forces in Japan—the 24th Division—weren't sufficient. More troops would have to be sent, planning for this led to the necessity for a corps headquarters, and corps headquarters needed an experienced armor officer. So in the last week of July, George Pickett found himself with orders to be at Fort Bragg, North Carolina, within 48 hours. He got the phone call at 9 A.M. on a Saturday morning.

After he got his orders, Pickett called Fort Bragg and asked them if they had any special instructions. They told him, "If you have any equipment left over from World War II, you should bring it with you, because we're short on field gear."

"What sort of equipment are you 'short on'?" he asked increduously.

"Bring anything like a helmet, a canteen, webbed belts, or bedding rolls," they told him. "You know, field equipment."

George went through his things and pulled out stuff that had been rotting in his footlocker for five years. It had smelled like the end of the war when he had stored it five years before, and it didn't smell any better in July 1950. He threw the gear into a barracks bag, and Beryl drove him to Fort Bragg on Monday morning because there was no train or plane that would get him there at the right time.

Once he got there, the first thing the army did was to inventory the things he brought with him and pick them up on the government property books again. "I really don't consider this stuff combat-serviceable," Pickett told them.

"It will have to do. You're still short about seven or eight items, so you'll have to pick those up when the purchasing/contracting officer gets back."

"Where's he gone?" George asked. "If I've been told to be here at the crack of dawn, where's the purchasing officer? By the way, why do we have to have a *purchasing* officer? Isn't there a *supply* officer?"

"There is a supply officer, but he doesn't have any supplies, so the purchasing officer had to go down to Friendly Joe's War Surplus Store to purchase the supplies that Friendly Joe bought at about four cents on the dollar a year ago, when Louis Johnson decided to sell them. Now he's selling them back to us for a big profit."

Friendly Joe wasn't in the program to "cut the fat and save the muscle." He had bought it and he got his profit out of it. The men who came in at the last minute on Monday, June 26, had to get their field equipment from Friendly Joe's War Surplus Store in Fayetteville, North Carolina, rather than from the US Army supply system.

The Pusan Perimeter

By August 5, 1950, only five weeks after they had crossed the 38th parallel, the North Korean invaders were able to occupy nearly all of South Korea. South Korean and United Nations forces had been pushed back to the area immediately adjacent to the port city of Pusan. It was a situation reminiscent of the British at Dunkirk in 1940, and for MacArthur it brought back haunting memories of his own predicament in the face of the Japanese onslaught in Bataan in 1942.

Despite the similarities with Dunkirk and Bataan, the dissimilarities favored the United Nations forces. At Dunkirk, the British had neither air nor sea superiority. At Bataan, MacArthur had no airpower whatsoever and the Japanese had naval superiority. In the defense of the Pusan perimeter, however, the US Air Force had achieved complete air superiority, and the North Koreans could not touch United Nations shipping going into the port of Pusan. However, the North Koreans held 95 percent of the land area of the Korean peninsula, and the United Nations troops had their backs to the sea. If the North Korean advance continued as it had since June 25, Pusan would soon be overrun, and Syngman Rhee's government would become a government in exile, like that of Chiang Kai-shek.

But the Pusan perimeter held fast. United Nations forces—mainly the US Eighth Army under Lieutenant General Walton H. Walker (Class of 1912)—consolidated their position. Newly mobilized forces from the United States began to arrive, which helped to resupply and strengthen existing troops.

Among the men sent in to help reinforce Walker's Eighth Army perimeter at Pusan were two classmates from the Class of 1941. Joseph Knowlton had served with the First Army in Europe, and Frank Benton "Ben" Howze had served two-and-one-half years in the Pacific theater during World War II. Both Joe and Ben had been at Fort Sill in 1949, but the next time they got together was at Camp Drake, a training and processing center on the outskirts of Tokyo. They spent four days there during the first week of August 1950 before getting orders to "go South"— to Korea.

By September 15, the 25th Infantry was just west of Taegu and was anticipating a move west across the Kumho-gang, towards Waegwan. That morning, Howze and his driver took a jeep and went across the river to reconnoiter for new positions in an area that was supposedly clear of North Korean forces. They accomplished their reconnaissance, and on the return trip they stopped to test radio communication with the battalion command post.

Suddenly, North Koreans were everywhere. Howze grabbed his rifle and took aim. Although he was able to empty his ammunition clip, he was fatally wounded before he could get out of his jeep.

Howze was not the only casualty on the Pusan perimeter. During a break in the action of a combined United States–Korean infantry attack, which the 73d Tank Battalion was supporting, Merritt Hewitt, then the battalion operations officer, was hit by machine gun fire while he was trying to disengage one of his companies. He was taken to Pusan by helicopter in an outboard plastic pod and from there to Tokyo on the British hospital ship *Maine*.

"The demobilization of our armed forces after World War II had been so widespread, and the loss of discipline was such a traumatic experience for many professional officers," Hewitt said with disbelief. "There were just so few people in the Pusan perimeter, and supplies were not available. For instance, the Eighth Army armored officer was at the dock when we arrived, seeking spare fan belts for M-26 tanks that had already chewed up all those in the theater. It was obvious that we were a very limited force. We had the feeling that we were the only ones at war."

As Walker's Eighth Army continued to hold the North Korean advance at Pusan, MacArthur, as overall United Nations commander, had already begun to assemble a second force, which he would call X Corps. The purpose of X Corps would be to launch a counteroffensive that would destroy the North Korean army's capacity to wage war and thus would fulfill the United Nations mandate to expel it from South Korea. Where and how this was to be accomplished was up to MacArthur. To attack through Walker's position at Pusan would mean going up against the majority of the North Korean forces on an established line, while striking elsewhere would mean facing a smaller opposing force in the initial engagement and in turn forcing the North Korean army, which was almost entirely committed to the Pusan perimeter, to divide its attention and fight on two fronts.

Inchon

Having experienced repeated triumphs in World War II with amphibious landings, MacArthur was confident of the success of such a maneuver. For a landing site, he picked the port city of Inchon on the northwest coast of south Korea, near Seoul. Because of its confined approaches and treacherous tides, Inchon presented numerous potential difficulties. On the other hand, the probability for surprise was great in so unlikely a place.

Such was the case on September 15, 1950, when MacArthur put X Corps ashore at Inchon. The gamble paid off. The landing was an astounding success and is remembered as a milestone in American military history.

Inchon changed the whole complexion of the war. Almost overnight, the United Nations forces went from being almost pushed from the end of the peninsula to chasing the remnants of the North Korean forces back into North Korea. From that point forward, all the American soldiers—regardless of their predisposition to MacArthur before September 15—were "MacArthur people."

The Pusan perimeter had stabilized and could have been successfully defended indefinitely. Without Inchon, however, the breakout from Pusan would have been bloody.

The following day, only 24 hours after Ben Howze had given his life, Walker's reinforced Eighth Army broke out of the Pusan perimeter and headed north. On September 26, the X Corps, under the direction of Major General Edward "Ned" Almond, recaptured Seoul and linked up with the Eighth Army at nearby Osan. The 24th Infantry returned to Taejon the following day, singing, "The last time we saw Taejon it was neither bright nor gay. / Now we're going *back* to Taejon, to blow the goddamn place away!"

In October, when the Eighth Army went north, X Corps was pulled out and went around to the east coast, where it landed at Wonsan. Unlike IX Corps and I Corps, X Corps remained a separate operational group until the end of December.

Joe Gurfein was among the men involved in the planning and execution of the Inchon landing. After World War II, Joe had gone to Harvard and earned his master's degree in engineering in 1947. In the spring of 1950, now a major, he was ordered to go to Okinawa. By the time he got to San Francisco to ship out, however, the Korean War had started and everyone destined to be shipped *anywhere* was shipped directly to Japan. He would never see Okinawa.

No sooner had he reported for duty in Japan than he ran into another classmate and close friend, Ed Rowny, who was one of MacArthur's staff planners for the Inchon landing. Rowny brought Gurfein into X Corps as assistant operations officer for plans, and they arrived at the invasion of Inchon on the third day after the landing. About a week later, X Corps assigned Gurfein to Pusan to help straighten out the logistical problems being encountered by units that had been on the defensive for nearly three months, but that were now making rapid advances.

Among the frustrations of logistical organization that Joe Gurfein found was the great shortage of parts for generators throughout Korea in 1950, which persisted until as late as 1952. For example, when he requested 100 parts, he was lucky to get four or five. As a result, most generators were not functioning. When the division commander asked Gurfein to make a study of why this great shortage had occurred, he discovered that the division had been *authorized* 30 generators but had 250 on hand. As a result, it got *parts* for only 30 generators.

On September 15, as Joe Gurfein and Ed Rowny were preparing to go to Inchon, George Pickett arrived with the IX Corps to take over a sector on the Pusan perimeter along with the 24th Division and 2d Division. By September 28, when they moved up to Taejon, the rest of IX Corps headquarters had arrived from the States, and instead of a few people pitching their tents and working around the clock, the corps headquarters suddenly had the luxury of a complete staff. Pickett had a staff of five men and equipment.

When IX Corps arrived in Taejon, however, they noticed that the water system had an unusually foul smell. In trouble-shooting the cause, the Yanks discovered the American prisoners taken at Taejon had been killed. "The North Koreans had tied their hands behind them with telephone wire, knocked them in the heads with picks, and thrown them into the water wells to contaminate the city water supply. That was the first atrocity I ever saw— when we were getting the bodies out of those wells in October and trying to identify the Americans they had killed," George Pickett recalled. "They had ruined the water supply by throwing American bodies into it."

For all practical purposes, the entire nation of South Korea was under the control of United Nations forces by the end of September, and MacArthur proposed that those forces now be used to occupy all of the Korean peninsula. The rationale was that the United Nations had called for free elections to form a government for a united Korea, and Kim Il Sung had refused to cooperate. On October 6, the United Nations General Assembly approved MacArthur's proposal, and three days later X Corps and the Eighth Army plunged across the 38th parallel and into two weeks of fierce opposition.

When IX Corps reached Taejon on October 1, the war had swept far into North Korea. George Pickett and Linton Boatwright had already been given the job of planning the postwar withdrawal from Korea to Okinawa, which was to have been the residence of the IX Corps headquarters. The Army intended that I Corps would stay in Korea, IX Corps would go to Okinawa, and probably the Eighth Army would have gone back to Japan. Whatever Americans were left in Korea would have been under I Corps. The war had been won and everyone was planning to be "home by Christmas."

Although he had broadcast the text of the United Nations resolution to the North Koreans on October 9, and the concept of a reunified Korea was often discussed, MacArthur issued the tactical orders piecemeal. The decision to capture Pyongyang, the North Korean capital, was actually made after

MacArthur's famous meeting with President Truman at Wake Island on October 15.

After October 6, the Eighth Army attacked North Korea in the west and started going north, and as they did, the whole mission had changed. The US Army now had to *occupy* North Korea.

Immediately, the planners sat down and drew up maps of North Korea, dividing it into provinces and setting it up for the Civil Affairs people, who were already coming in and were talking about how X Corps was going to *govern* North Korea.

Into the North

On October 20, Pyongyang, the capital of North Korea, was seized by the Eighth Army, and North Korean resistance had all but collapsed. On October 26, X Corps, which had been pulled out of Inchon after the Eighth Army swept through, landed at Wonsan on the east coast and began expanding its perimeter northward. Simultaneously, several units of the South Korean Army reached the Yalu River, which divides the Korean peninsula from China. The unification of Korea by the United Nations forces appeared to be close to a reality.

Meanwhile, however, the Communist government of China had let it be known through diplomatic channels that it would intervene in the conflict if the United Nations troops entered North Korea.

Despite the fact that numerous Chinese "volunteers" were already being apprehended during the operations in North Korea in October, MacArthur considered Chinese threats of a full-scale intervention a bluff, and he proceeded with his plans accordingly. By November, however, the Chinese, under General Lin Piao, had infiltrated 300,000 troops into North Korea, under the noses of United Nations forces that were thinly spread over the rugged mountains. The Chinese would soon quickly defeat many of the units that, only recently, had assumed the war was all but over.

During the two weeks prior to Thanksgiving, Joe Gurfein had been in General Almond's office every day. Previously another major had been in Almond's office adjusting the lines on the general's maps, when Almond suddenly said, "*That's* not where the troops were. I just got back from the front and they're not *there*."

Instead of keeping his mouth shut or saying, "I'll find out," the major replied, "But sir, this *is* where they are."

Almond said, "Get Gurfein. He's an engineer, a map man."

When Gurfein arrived, Almond barked, "Gurfein, correct that map." He made the other major stand at attention while Joe corrected the map.

Orders had come from MacArthur's headquarters indicating the location of the dividing line between X Corps in the east and Eighth Army in the west. As an engineer, Gurfein was used to the regular coordinates of a map, so he plotted the map and showed General Almond where X Corps was in relation to the Eighth Army.

Almond then sent out a patrol to go to that line and coordinate with Eighth Army, but the patrol didn't find the Eighth Army team. The X Corps patrol returned to camp and said the team never showed up. When X Corps wired Eighth Army, the latter replied, "*We* sent a patrol out, but they couldn't find *you*."

The next day, another patrol went out, but the patrols still didn't meet up. On the third day, Almond told the patrol, "Keep going. Don't stop at the line. Just keep going."

They kept going, and they met the Eighth Army patrol about three miles father west. They had stopped three miles short because there was a three-mile error on *their* map, which had not been corrected as Gurfein had corrected Almond's.

Not only had a three-mile gap existed for three days, but the entire North Korean Army that had been surrounded was actually able to escape through it, and the US Army didn't catch it.

George Pickett left Taejon at the end of October, helping to relocate the IX Headquarters first to Pyongyang, then to Sunchon, and finally to Kunuri on the Chongchon River deep inside the western side of North Korea. On the morning before Thanksgiving, Pickett heard a report that a JS-3 "Stalin" tank had been identified just outside of Kujang in a railroad tunnel, where tanks had been parked to keep them out of the sight of American recon aircraft. All the tanks that the North Koreans had were Russian tanks, and the one of which the Americans were deathly afraid was the JS-3, the most powerful battlefield tank in the world at the time. It had a 120mm gun. All of the tanks in the IX Corps were M4A3 Shermans dating from 1944, which had with 76mm guns.

The appearance of JS-3s in any quantity would constitute a tactical crisis, so the Eighth Army armor officer, Colonel Withers, sent Pickett with a recon patrol to identify the tanks.

Pickett and the recon platoon drove up the road to Kujang and spotted the five tanks, but it turned out that they were not JS-3s, but rather T-34s, the same kind the North Koreans had been using since June. They had 85mm guns, which were better than the guns that IX Corps had on its M4A3 Shermans but not as formidable as the 120mm gun on the JS-3.

It was around 11 P.M. the next night before Pickett got back to IX Corps headquarters. Part of the headquarters was packing up.

"What's going on?" he asked.

"The corps commander is going to send a group back to Sunchon," they told him.

Another Surprise

The weather was now bitterly cold, in stark contrast to the sticky, humid heat of July and August. It was Korea's worst winter in a decade. Snow was in the air, but it only served to remind everyone of what MacArthur said at Wake Island when he met there with President Truman on October 15—that the American troops in Korea would be "home for Christmas."

While some South Korean and American units had, in fact, reached the Yalu River and could see the mountains of northern China beyond the opposite bank, the majority of the United Nations forces were on a line roughly halfway between the Yalu and the 38th parallel. The US Eighth Army, under General Walton Walker, incorporating I Corps and IX Corps, was on the west side of North Korea from Pyongyang to Anju, concentrated along the wide and shallow Chongchon River. The 24th Infantry Division was positioned near the mouth of the river, while the 25th and 2d Infantry Divisions were about 50 miles upriver, between Kunu-ri and Tokchon.

Meanwhile, the United States X Corps under General Almond, incorporating the United States 1st Marine Division, controlled the east coast between its beachheads at Wonsan and Hungham. The Republic of Korea (ROK) II Corps was located on the Eighth Army's right flank. Between them and X Corps was a mountainous gap as much as 60 miles in width. This is often cited as the Chinese invasion route but, in fact, the areas inland from the two coasts were so thinly held that Chinese infiltrators using mountain trails were able to move through United Nations lines throughout the breadth of the peninsula.

When the Chinese attacked in force on November 25, their overall troop strength was roughly equal to that of the United Nations force, so their strategy was to exploit their own strengths and their opponent's weaknesses. Because the United Nations force was "road bound"—dependent on vehicular transport to move men and supplies—it was particularly vulnerable in the rugged terrain of North Korea, where there were few roads. The Chinese, on the other hand, were accustomed to moving on foot over rocky areas. The Americans could move faster when they had a road to move on, but the Chinese could move in any terrain.

With this in mind, Lin Piao decided to concentrate all of his forces against only three American divisions: the 25th and 2d Infantry in the west and the 1st Marine Division in the east. Lin's assault echoed the one inflicted on the 24th Infantry at Taejon four months earlier, except that now the onslaught was on a larger scale, a *much* larger scale. In the west, the 2d Division was about to endure the worst pounding of any American division during the war and perhaps the worst suffered by any American unit since the bloody days of the American Civil War.

General Coulter told George Pickett he had decided to withdraw the 2d Division through Kunu-ri and form a new line at Sunchon, and this plan was already approved by Walton Walker at Eighth Army Headquarters. When Coulter gave orders to the 2d Division to withdraw to Sunchon, it went back organized for a motor march, not disposed for battle. The result was a fiasco. The Chinese were waiting and opened fire.

The intelligence had people thought that the enemy in the hills were guerrillas rather than a division-strength Chinese tactical unit with 40 machine guns. Within ten days, the 2d Infantry Division was no longer considered combat-serviceable. Three men from the Class of 1941 died during the 2d Division disaster: Thomas Hume, Donald Driscoll, and William Hickman.

Hume had gone into the field artillery, and in 1943 he had received his wings as a liaison pilot. The following year he was sent to England, and he ultimately entered combat in Europe. After VE-Day, he remained in Germany for the occupation. While there, he became engaged to Hjordis Faber, and they were married on February 18, 1947, in Muskegon, Michigan. After his marriage, Tom returned to Germany to complete his tour of duty, and on November 15, 1948, their son Thomas Faber Hume was born. When the Korean War broke out, Hume and his family were stationed at Fort Lewis, Washington, where he was executive officer of the 37th Field Artillery Battalion of the 2d Infantry Division. He continued to serve in this capacity with the 37th during the tense days of the Pusan perimeter, the breakthrough to the Yellow Sea, and the drive to the Yalu River.

Hume—known at West Point for his gun collection—was frequently seen visiting battery positions with his ever-present sporting rifle slung over his shoulder, dodging intense mortar fire while directing the fire of 105mm howitzers against enemy mortar positions. Even during those trying days, he was remembered as cheerful and businesslike, never giving a second thought to his own safety.

One one occasion, in the vicinity of the Naktong River, he and a small group of men formed a battery composed of enemy artillery pieces that had just been captured by forward infantry elements. From this hastily occupied battery position, which was exposed and subject to small arms, mortar, and artillery fire, they pummeled the retreating enemy.

On November 25, when the Chinese struck, the 37th Field Artillery was well forward within the 2d Division position and immediately came under enemy infantry attack. Tom undertook the defense of the battalion position by forming a group of men consisting of cooks, mechanics, and clerks to fight as infantry. His personal courage no doubt contributed much to the fact that the integrity of the battalion was maintained all that night and during the five trying days that were to follow.

About noon on November 30, as the division started through Kunu-ri Pass, the 37th Field Artillery was one of the last units in the long column of vehicles that undertook the heartbreaking withdrawal.

Hume was last seen entering a small village at the end of the roadblock, where enemy snipers had been reported. It was apparently there that he was captured. It has been officially reported that he was presumed to have died of malnutrition while a prisoner of war some time before July 31, 1951, and to have been buried at Pyoktong in North Korea, but there has never been any official confirmation.

Don Driscoll was with the 9th Infantry Regiment of the 2d Division, and he was declared missing in action on December 1. Don had been with the 99th Division during the Battle of the Bulge and was wounded and then evacuated in December 1944. He went to Fort Benning as a student in 1948, when George Pickett was on the faculty, and late in 1949 he was assigned as an instructor of ROTC at New York University. It was here that he met and fell in love with Laemore Cawley from Pen Argyl, Pennsylvania, who was a secretary in New York. They were married in March 1950, and there followed a few short, happy months before Korea exploded. George Pickett last saw Don in Taejon, heading north.

To this day, nobody really knows what happened to Don Driscoll. One story is that he was killed by the Chinese when his position was overrun. Another story is that he was captured alive, then died on the way to a prison camp. The most devastating story is that he was trying to stop the battalion from running away and one of his own men shot him. On January 1, 1954, the Office of the Adjutant General of the United States officially notified Don's family that since no additional evidence had been uncovered to indicate he was still living, he must officially be declared dead. His body has never been recovered.

William Hickman, like Jonathan Harwood—who later died in Normandy—had entered West Point in 1937 but didn't graduate with his classmates. He did, however, take up a career in the Army, and November 1950 found him in Korea as a battery commander with the 503rd Field Artillery of the 2nd Division Artillery on the line above Kujang. He may have died there, or he could have been killed going back into the pass. Officially, he was listed as missing in action on December 1. Some say that he was killed in the initial attack before the division started back through the mountain pass. Other information points to the possibility that he was killed on the hillside above Kujang. Alternatively, he may have been captured. As was the case with Hume and Driscoll, no one knows. Their bodies were never recovered.

As the 2d Division was in the mountain pass taking heavy losses, General John B. Coulter decided to move the IX Corps command post.

The next morning, the Chinese were firing down at the edge of the airstrip and there was a firefight at the MASH hospital next to the ammunition dump. Coulter, Bill Kunsig, the chief of staff, and one other officer got into the planes and left.

John Greco told Pickett and about 20 others who remained, "General Coulter asked me to tell all of you that it was nice to have had you serving with him, and if he doesn't see us again, he wishes us good luck. So right now, we should form a convoy and leave right away, if we're going to get the hell out of here."

"We were licked," Pickett recalled later. "There was no question that we were defeated on the Chongchon River. To put it all in perspective, it was a completely new experience for me because I had never had the job of pulling anybody out and running, or holding a delaying action or making a retreat, or whatever you want to call it. Going across Europe, the only thing I can ever remember is either holding a front line where the enemy never attacked or being in an attack and counterattack situation with the enemy. The idea of being defeated and having to pull out and trying to reorganize and fight someplace else was a completely new experience for us."

They loaded up six jeeps and headed out. Nobody really knew whether they were driving toward or away from the Chinese forces.

When they finally reached Anju, which was about halfway between Kunu-ri and the coast, they met the 89th Tank Battalion's recon platoon, which had two light tanks, a half-track, and a couple of jeeps. They were coming up from the southeast, and the lieutenant recognized Pickett and stopped. Pickett asked him where they had been.

"We were trying to see if the road was open and we ran into a whole lot of Chinamen about two miles down the road. They had the mountain pass blocked. Now we're trying to figure out if we can get part of the 25th Division out by that route, instead of having everybody go along the coast."

There were two MPs still directing traffic, so Pickett asked them what their orders were.

"We're supposed to stay here until everybody else has pulled out," they told him.

"Get in your jeep and come with the rest of us," Pickett directed them, "because the only people left around here now are the recon platoon from the 89th and us. The *next* group of people that you guys are going to be controlling traffic for will be be speaking Chinese."

One MP wanted to go, but the other insisted that they couldn't go unless the provost marshal relieved them, so both finally decided that they would stay.

"Okay," Pickett drawled, "but if the tank platoon pulls out, you really should go with them because there isn't going to be anybody else between you and the Lord and the Chinese."

They agreed to stay until the tank platoon pulled out.

As Greco and Pickett drove on toward Sinanju on the coast, they passed through what had been the I Corps Headquarters. Pickett could see signs, tent pegs, and equipment where the old bivouac area was.

"They must have left in a helluva hurry in the middle of the night," Pickett commented to Greco. "There are still shoes and shaving kits lying around."

They didn't see any guns or helmets, but they decided to check to see if there were any documents left behind that might compromise critical information. Finding no such items, they continued on toward Sinanju and turned south. It had gotten dark, and they hadn't seen anybody on the road in the last 20 miles.

About midnight, a voice rang out of the dark. "Halt! Who's there?"

It was one of John Throckmorton's men. Throckmorton (Class of 1936) was in command of the front line regiment of I Corps. Greco and Pickett had made it back to the American lines again. They had probably been saved by the fact that the main Chinese effort was launched against the 2d Division in the pass between Sunchon and Kunu-ri. The Chinese had the equivalent of five or six divisions facing I Corps, but they weren't advancing as fast as troops opposing IX Corps near Kunu-ri. The I Corps pullout was a lot more orderly and better organized than the IX Corps pullout because I Corps was not surrounded on three sides by the enemy.

The American Army had never been taught how to make a withdrawal. At West Point, the men had been taught the theory of how you did it, but in practice, a "withdrawal" like that of the British at Dunkirk or the Germans on the Eastern Front—large-scale tactical withdrawals—was not taught. At one point when Pickett and Boatwright were studying the maps with their planning group, a remark was made to the effect that "we've never done this before."

When IX Corps was chopped up, the I Corps flank and rear were exposed, and Walton Walker had to order the entire Eighth Army to withdraw. When I Corps pulled back, they pulled back at eight times the speed with which the Chinese could follow their withdrawal. It was this incremental difference in mobility that enabled the Eighth Army to escape destruction on the Chongchon.

"If the Chinese had had tanks and airpower, we would have been history," Pickett said. "It was bad enough as it was, but if they had had a tank battalion to throw into Kunu-ri at the right time, it would have been utter chaos."

As the IX Corps fell back to Pyongyang and beyond, there were comparisons to Stalingrad, the grueling battle during the winter of 1942–1943, which had become a synonym for "hopeless situation."

Everyone seemed to be saying, "We'll pull back into Pyongyang and it'll be the American Stalingrad." However, while the weather conditions were certainly similar, Pyongyang had a port at Chinampo from which Eighth Army units could have been evacuated, just as the X Corps would soon be pulled out of Wonsan on the east coast of North Korea.

It was a terrible winter, probably the lowest ebb in the morale of the United States armed forces. It got so cold some nights that the tanks' tracks would freeze to the ground and the machine guns wouldn't fire. The M1 rifles jammed because their oil had congealed. The men discovered that hot urine freed the guns' mechanisms, so they peed on their guns to make them fire. It sounds ridiculous, but it worked, and it saved lives.

The winter of 1950 was far more disastrous than the winter of 1944. When the Ardennes campaign ended after 30 days, the Germans were completely defeated. Thirty days after the Chinese attack on the Chongchon, the United Nations forces still faced another Chinese offensive that would ultimately push them out of Seoul.

From a tactical and strategic standpoint, the Chinese success against the Eighth Army was far greater than that of the Germans in the Ardennes. It lasted longer, shoved the Americans back farther, and produced a stalemated war.

The Chosin Debacle

In the X Corps sector on Korea's eastern coast, the Chinese offensive slammed into the 1st Marine Division and elements of the US Army's 7th Infantry Division on the icy banks of the Chosin (Changjin) Reservoir. Located about 50 miles north of the port of Hungnam, the Chosin Reservoir stretches nearly 30 miles across the top of a 4,000-foot plateau, which is surrounded by formidable mountains.

As the leading elements of X Corps had advanced in the days leading up to the Chinese offensive on November 15, the topography and the sub-zero weather had presented more of a challenge than the enemy. The men stomped their feet on the truckbeds in futile attempts to keep their limbs from becoming numb. A typical soldier wore long, woolen underwear, two pairs of socks, a woolen shirt, cotton field trousers over a pair of woolen trousers, a pile jacket, a wind-resistant parka, and wool-insert trigger-finger mittens. To keep their ears from freezing, soldiers tied wool scarves around their heads underneath their helmets. Still, the cold seeped through.

The main Chinese attack had hit the men of the Eighth Army on Saturday, November 25, but through the weekend the men of X Corps saw only an occasional enemy soldier in the hills. They didn't realize yet that over a quarter million Chinese Communist troops were nearly on top of them.

A 7th Division task force, led by Lieutenant Colonel Don Faith, was the first to encounter the enemy in force. On Monday night, November 27, Task Force Faith was assaulted and partially overrun and sustained high casualties.

During the day of November 28, General Almond arrived by helicopter to confer with Faith. He shrugged the enemy attack off as remnants of North Korean units fleeing north. The following day, his theory was thoroughly discredited when the Marines and Army units on the Chosin Reservoir found themselves encircled and subjected to repeated, savage night attacks. By twilight on November 29, it had become clear to both Faith and to X Corps Headquarters that the Chosin Reservoir positions were surrounded. A relief column, consisting of the 2d Battalion and the 7th Division's 31st Regiment, was organized to be sent to Chosin from Hungnam.

Ed Rowny and Joe Gurfein had gone north in late October to help set up the X Corps headquarters at Wonsan. They flew in because ships still couldn't get into the harbor, which had been mined. By the end of November, they had moved north to Hungnam.

Gurfein was sent to the 2d Battalion of the 31st Infantry with orders for the battalion commander to reinforce the Marines. It was after dark by the time they got moving. Near midnight, a booby trap on a bridge in front of the column exploded, wounding one man. A rumor started that it had been an antitank gun and that the Chinese were coming.

Soldiers in front had started to break ranks. The tail and lead companies turned to the rear and started to overrun the battalion commander. Jeep drivers turned their jeeps around and headed to the rear. The driver of a three-quarter ton truck started to unhitch his trailer to turn around. The battalion commander, pushed aside by his troops, stood by silently.

The situation reminded Gurfein of an incident in Africa in 1943, when he had been training a group of paratroopers in mountain warfare. One day as a group of Arabs was bringing some mules down a steep trail, the animals stampeded. About 30 of them ran down a narrow road as Gurfein was walking in the opposite direction with two friends. He had been taught at West Point that to stop a riot, you run toward it. He ran toward the mules and they turned around and ran the other way.

Gurfein attempted the same maneuver again. He physically stopped the soldiers by blocking the narrow road with his body, saying, "We're going *that* way. Move!"

Later, Gurfein said, "The first guy with a loud voice who takes command is the leader. It can even be the private in the last row."

Immediately, Gurfein got the staff together and gave orders. He sent one man forward to get the point moving again, and then the whole 2d Battalion began to roll forward once more. Joe Gurfein had taught Psychology of Leadership at Fort Belvoir in 1942—they even used to quote him in the West Point textbook on the subject—but he hadn't commanded a battalion since he was in Italy in World War II.

"The first, most important thing is, you've got to know your mission," Gurfein observed in retrospect. "You've got to know what you are going to do, because everything else revolves around that. Secondly, you've got to give this information to your troops so they'll know *why* you're doing it. If they don't know why they're attacking, everything is lost. You've got to know your job thoroughly, and you've got to have confidence in yourself."

As he watched the column move north again through the cold, the snow, and the Chinese bullets, Gurfein remembered more ambush situation, 13 years before, when he was a Plebe at West Point. He had been out on maneuvers with blank ammunition when Company B was ambushed in the middle of the night with flares and machine guns. He had stood paralyzed with fear, not knowing what to do. Now, with live ammunition and in a real war, he knew what to do.

"People talk about bravery," Gurfein would say later. "It isn't bravery. You've been trained. You just know what to do."

The wind was howling and it was snowing hard. The enemy could not see the 2d Battalion because the trucks all had their blackout lights on, but they could hear their motors. As the 2d Battalion made its way through the narrow pass and approached the middle of the draw, the enemy opened fire.

Again, the troops began to panic. Gurfein again ordered them forward, and this time the staff was behind him.

The men were glad that somebody was running the show. The tail end of the last company formed itself into a hedgehog perimeter and set machine guns up to protect the rear, while the rest of the group kept on moving. Gurfein kept driving them.

Gurfein continued to yell his lungs out to try to attract enemy fire, because he knew that if the enemy fired toward him, the rest of the men could get around behind the hill. He was reasonably well protected, so by making a lot of racket, he could help the men get around. It worked beautifully, and almost no one was hit.

Finally, they broke through the pass. There was still a platoon left behind with a machine gun, but Gurfein couldn't risk going back for them. "It's tough," he told the staff, "but we'll have to get them in the morning."

About 15 minutes later, the point man returned to the column. "We've contacted the marines!"

"Good," Gurfein replied. "Just keep moving. Everybody keep moving."

The battalion commander turned to Gurfein and said, "I think we've gotten through here now."

Gurfein said, "Yeah," and then he faded into the background so the battalion could report to General Lewis B. "Chesty" Puller, the commander of the 1st Marine Division. When they found the Marine Corps commander, the battalion commander introduced himself and Gurfein, and Joe simply said, "Sir, I'm from the X Corps staff."

By now, the battalion commander had regained his composure. The battalion staff knew what he had failed to do, and how Gurfein had run things, but they kept their mouths shut. The commander was a lieutenant colonel, and Gurfein was only a major. Gurfein told him, "I'll attach myself to your staff and do whatever you want me to do, colonel. I'm here."

The following day, General Almond flew into the little airstrip that had been built at Kunu-ri. He gave the battalion commander a Distinguished Service Cross. Nobody said a word. About three weeks later, the battalion commander was relieved of his command and sent back home, where he died of a heart attack.

Five days later, General Almond returned. He recognized Joe Gurfein immediately. "Joe, what have you been doing here?" he asked. Joe Gurfein told him and was awarded a Silver Star for his role in the action.

The 2d Battalion finally reached Kunu-ri on December 1, where it coordinated with the Marines and with what was left of Task Force Faith. The troops had to fight their way out, an action far more easily said than done. The troops did it in good order, though, carrying their wounded and never breaking ranks. The Chinese forced them out but did not achieve the same kind of victory they had against the 2d Division in the west. On December 11, the battle-weary troops finally reached the port of Hungnam, dug in, and waited for the Chinese to spill down out of the hills, hoping they would not be swept into the sea before the evacuation boats could take them aboard.

At this point, General Almond believed that he could hold Wonsan and Hungnam indefinitely, but such a defense would eventually prove to be tenuous and difficult against the Chinese Army. The situation was like that of Pusan in August, but twice as far from Japan and against a force many times greater.

The complexion of the war had changed dramatically. Three divisions had been badly mauled, and the Eighth Army, in retreat, had been evacuated from Pyongyang.

Home for Christmas

The United Nations had sent forces to chase the North Koreans out of South Korea, and when the going had been easy in October, they had allowed MacArthur to talk them into liberating all of Korea. The Chinese intervention and the resounding American defeat left the United Nations with no stomach for occupying North Korea. Now, they only hoped that MacArthur could hold onto South Korea!

MacArthur proposed a massive offensive to retake North Korea, combined with a strategic air offensive against China—possibly using nuclear weapons—to cut supply lines and in general to make the war very costly for the Chinese. The second choice, advocated by men like George Marshall and Dean Acheson, was to hold only South Korea and avoid bombing China at all costs, thereby keeping the Soviet Union from entering a conflict that many people thought could escalate into a global nuclear war.

Truman sent a wire MacArthur on December 3 stating: "We consider that the preservation of your forces is now the primary consideration."

The troops, who had planned to be home for Christmas, spent the holiday in a far more gloomy habitat. Many had been taken prisoner and were being herded north to confinement in Manchuria, while those who had escaped were still on the run. Walker's Eighth Army had withdrawn from the capital of North Korea, and, doubting it could hold the capital of South Korea, had fallen back. On the eastern coast, the last of Almond's X Corps were boarding evacuation ships at Hungnam on Christmas Eve.

Walton Walker himself did not live to see Christmas. He was killed in a jeep accident on December 23 while IX Corps was at Uijongbu, just east of Seoul. By this time, the United Nations forces had formed a battle line across roughly the 38th parallel.

General Matthew Ridgeway (Class of August 1917), who had commanded the 82d Airborne Division during World War II, came in from Washington, D.C., by way of Tokyo and took over command of the Eighth Army on New Year's Day, 1951. When Ridgeway arrived, General Coulter was appointed Deputy Army Commander, with the rank of lieutenant general, and placed in charge of what would be called "military government."

Like Patton, MacArthur, and other such colorful generals, Ridgeway had a dramatic personality. As Commander of the 82d Airborne Division, he had jumped with his men during the Normandy invasion and had been at the front to the end of World War II, so he was no desk-bound administrator. However, there were two schools of thought about him. The airborne troops admired him, while the men in the armored units—who often derided the airborne troops as worthless drunks—did not.

Symbolism is, of course, often larger than substance in such cases, and so it was with the attitude of the troops toward Ridgeway. In Patton's case, his trademark consisted of a pair of pearl-handled revolvers that he wore in a gunbelt, while with Ridgeway, it was his hand grenade. Of course, the armor men saw the hand grenade as only pompous posturing and said, "By the time he got the tape off to use it, everybody would have been dead 20 minutes. He never could have gotten it off there in time."

But as Ridgeway explained it, "I've been told that the grenades I wear on a harness at my chest are worn for picturesque effect, as a trademark like George Patton's pearl-handled pistols. There is no truth in that either. They are purely for self-preservation. I learned long ago in Europe that a man with a grenade in his hand can often blast his way out of a tight spot. And in Korea, I was determined that, in my prowling along the battlefront, I wasn't going to be ambushed and captured without a fight."

As Linton Boatwright wrote, "I can't imagine any knowledgeable person characterizing General Ridgeway's leadership of Eighth Army as anything but brilliant. He took over a battered and dispirited army, which had a bad case of 'bug-out' fever, and turned it into an effective fighting force in approximately 30 days. We had retreated almost continuously for 30 days when he arrived. I can remember the first time he came to IX Corps headquarters and talked about turning around and fighting. We thought he was a little goofy. When he first arrived, he wanted to counterattack, but soon he saw that he first had to change ways of thinking within his command."

When Ridgeway arrived, he sent for all the officers, addressing them in a group with a stirring speech that there would be no more defeat, that the army would fight and win.

In his own memoirs, Ridgeway wrote he that told the troops: "The power and the prestige of America was at stake out here, and it was going to take guns and guts to save ourselves from defeat. I'd see to it they got the guns. The rest was up to them, to their character, their competence as soldiers, their calmness, their judgment, and their courage."

On January 2, 1951, Ridgeway made the decision to let Seoul fall to the Communists for the second time in five months. He decided to pull the Eighth Army back to the Han River and set up the United Nations defenses there. The Communist Chinese took Seoul on January 4, and suddenly the strategic situation looked exactly like it had when the war first began in June 1950. The great success at Inchon and the bitter sacrifices made north of the 38th parallel seemed for naught.

The advancing Communist troops continued to push the Eighth Army south for the next six weeks, until the Battle of Chipyong-ni on February 14. This was the first tactical defeat that Chinese Communist forces had endured in the almost three months that they had been in the war, but even with improved weather and an enormous influx of fresh troops, it took the United Nations forces until March 18 to finally drive them back far enough north to recapture Seoul.

The Stalemate

On April 11, 1951, the debate over the strategy by which the war would be fought reached a turning point when President Truman fired MacArthur from his post as United Nations Commander and replaced him with Matthew Ridgeway.

It was not until June 1951, when the United Nations forces were at last able to push the Chinese back to the 38th parallel, that the Chinese finally agreed to talk. On June 23, 1951, two days before the first anniversary of the original invasion when the battle lines stood at roughly the 38th parallel, Jacob Malik, the Soviet ambassador to the United Nations, let it be known that now the Chinese were ready to talk truce.

On September 26, a tank-infantry task force opened a hole in their corps front. It would have been possible to drive 25–30 miles into the Chinese lines, but, according to George Pickett, "The generals were worried about going too far. The corps commander wanted to go farther, but the division commander convinced him that we might take too many casualties. They were both worried that the people at Panmunjon might not like it, so after one day, they called off a successful attack that could have led into a major victory."

Pickett went to the IX Corps commander, General William M. Hoge—father of his classmate Bill Hoge—on September 27 and asked to be sent home. He told him, "My usefulness in Korea is over when we plan and conduct an armored operation that has such potential and then find that the top leadership has chickened out for fear that the State Department might get upset."

He got his orders around November 1 and went back to Japan the day before Thanksgiving to become the senior advisor to a new Japanese Military Academy.

For the next 22 months, the Korean War continued on two fronts. On the diplomatic front, the two sides sat across from one another at a table in Panmunjon and argued. On the stalemated battlefront, men fought and died for obscure geographic features whose true names—if they ever had names—are now forgotten but whose nicknames—"Bloody Ridge," "Heartbreak Ridge," "Pork Chop Hill"—have become a permanent part of the lexicon of American military history.

Heartbreak Ridge

In March 1951, Joe Gurfein had taken over as the X Corps operations officer in charge of coordinating US Air Force air support of army forces in the field. Over the next five months, he helped to arrange and coordinate more than 12,000 close air support missions, working with the Marine Air Wing, the Air Force and, of course, the carrier-based Navy Air as well. He worked close air support through the middle of August 1951, when a personnel officer from the 2d Division contacted him, saying, "We have a vacancy for an engineer battalion commander. Would you be interested in coming up to discuss this with General Boatner?"

Gurfein, now a lieutenant colonel, became the regimental executive officer of the 23d Infantry Regiment of the 2d Infantry Division.

Colonel Jim Adams (Class of 1935) was the Commander of the 23d Infantry at that time, and he and Gurfein subsequently became very close friends. They were together at the beginning of September 1951, when the 23d Infantry became involved in the battle at Heartbreak Ridge (Hill 931). Adams got double pneumonia but refused to be evacuated. He lay in his tent and directed the regiment from there, but Gurfein had to do all the footwork, both as commanding officer and executive officer, and he ended up on *top* of Heartbreak Ridge.

Heartbreak Ridge was a long, north-south ridge with spurs coming off on each side. At the very top there was a peak 931 meters high that the North Koreans held. In early September, the Eighth Army ordered X Corps to capture Heartbreak Ridge, and X Corps in turn passed the job along to the 2d Division. By September 13, the division was ready to attack.

The 9th Infantry Regiment came up from one side of the ridge, but because the North Koreans held the high ground, the 23d Infantry Regiment sent a second battalion up to reinforce the first and to protect its north flank. However, the 23rd soon discovered that it was still taking enemy fire from the rear, from the north.

When Gurfein arrived on the scene to see what the problem was, he found the battalion commander of the 2d Battalion in a concrete bunker 400 yards from the battle line. Gurfein relieved him of command and crawled up the muddy hillside in the face of North Korean gunfire to order the senior officer of 1st Battalion to press on with the attack.

For about a month, the North Koreans had been able to fend off the attacks of the division, the main thrust of which was being made by the 23d Infantry Regiment. Major General Robert Young, the 2d Division's new commander, developed a new scheme of maneuver for the division, emphasizing a simultaneous use of all its combat power as well as that of the French forces under General Ralph Monclar in a night raid on the enemy positions.

The final offense by the 23d Infantry, which ended the battle of Heartbreak Ridge, lasted from October 5 to October 15, when troops and tanks ultimately succeeded in getting a company around the ridge from the south. Joe Gurfein took home his second Silver Star. Both Gurfein and Linton Boatwright, who commanded the 37th Field Artillery Battalion at Heartbreak Ridge, were awarded the French croix de guerre by General Monclar.

The duel for possession of the hills along the 38th parallel continued through 1952 and into 1953. The battle of Pork Chop Hill during April 1953 was the last of the memorable confrontations, but by no means the end of the bloodshed.

The Classmates in Korea

As the next three years wore on, many of the other men of Black '41 found themselves assigned to Korea. Lynn Lee arrived in Korea more than a year after George Pickett had left. He came in as commanding officer of the 79th Engineer Battalion maintaining Eighth Army Headquarters. At this time, the Commander of the Eighth Army was General Maxwell Taylor (Class of 1922), who had commanded the 101st Airborne Division during World War II. He would later serve as chairman of the Joint Chiefs of Staff (1962–1964) and, after retiring, would become the United States ambassador to South Vietnam.

One of Lee's more amusing moments as commander of the 79th Engineers was the critical task of picking up General Taylor's refrigerator at the airport and delivering it, with all due respect, to his hut. It was not an ordinary refrigerator, but a refrigerator with a *history*.

When the US Army acquired a block of land in Orlando, Florida, it had come with six civilian houses. The commanding general lived off post, but as chief of staff, Harry "Light Horse" Wilson, the former all-American football star of the 1930s, ranked the best, so he got one of the houses—*sans* refrigerator—and asked Lynn Lee to get him one. The post engineer reported that he had one that the hospital had rejected, but there was one small problem: inside it was all ice trays. Except for the ice trays, there was no room—not even for one ham sandwich. However, Harry, who was a bachelor at the time, was delighted with this assured ice supply for his parties. Presumably, so was Maxwell Taylor!

From 1952 to 1954, George Pittman commanded the 581st Air Resupply & Communications Wing, involved in covert operations of many different types of aircraft—including B-29 bombers, Douglas C-118 transports, Fairchild C-119 transports, Grumman SA-16 rescue seaplanes, and Sikorsky H-19 helicopters—as well as different types of personnel specialties involved in ground operations.

"We received good guidance from personnel involved in the same type of operations in Europe during World War II," he said later. "We had many unsung heroes who could not be recognized because of their covert missions. For example, the H-19 pilots of the 3d Rescue Squadron at the K-16 Air Base rotated home with 75 mission credits. Meanwhile, the 581st had H-19 pilots (including the man who later became the first presidential helicopter pilot) with over 150 mission credits and no scheduled rotations because of the missions they flew. Anything spectacular that was done was credited to 3d Rescue Squadron, not to the 581st, which was the only unit that crossed the North Korean main supply routes in the last several years of the Korean War. We had pilots flying all sorts of missions almost all over the world in situations deserving, at the minimum, the Air Medal, which was not approved on my recommendation because of the covert nature of the missions and the lack of knowledge on the part of the awards people."

While assigned to the Far East Air Forces headquarters in Tokyo, Hamilton Avery made numerous trips to Korea. On one such trip, classmate George Stillson invited him on a clandestine operational intelligence-gathering jaunt in a motorized Chinese junk with Stillson's 1126th Field Group. It was reminiscent of Clark Gable's famous foray against the Chinese Communists in an armed private junk in the movie *Soldier of Fortune*.

Andy Evans was sent to Korea in October 1952, just after completing training as a jet pilot at Nellis AFB in Nevada. He went over as a replacement pilot in a squadron flying North American F-86 Sabre jets. However, just prior to his arrival, a deputy fighter bomber wing commander was shot down, so "Lo and behold, I had to take that assignment in F-84s, a type of aircraft for which I really hadn't trained. It wasn't really all that difficult, except I much preferred to do the air-to-air missions than the air-to-ground, which became my assignment," Evans wrote later.

The North American F-86 was the top air-to-air fighter of the Korean war, with a 14–1 kill ratio over the Communist MiG-15s and was ultimately responsible for over 90 percent of the United Nations aerial victories in the war. The Republic F-84 Thunderjet, on the other had, was a heavily armed "bomb truck" that was ultimately responsible for the most effective ground support missions of the war.

Andy soon discovered that Korea was a different war than the one he had helped to win in Europe seven years before. "We were fighting with one arm tied behind our backs," he grumbled. "We were unable to pursue the enemy to their base as we had done in World War II. We weren't allowed to have that sure 'if necessary' or 'victory at all cost' as we had in World War II."

He had flown 67 missions and was on his third mission of the day when he was shot down by Chinese ground fire and taken prisoner. He weighed about 150 pounds when he was captured and came back home weighing about 88 pounds. For the first six months of captivity he was completely isolated. Kept somewhere along the Yalu River in a hole in the ground, he didn't have room to stand up or lie down and never saw another POW until the day he was released. His Chinese Communist captors knew that he was the senior officer and therefore did not want him to mingle with the other POWs.

The Korean Military Academy

When Tuck Brown arrived in Korea in September 1951, he was ordered to select a site for a Korean Military Academy (KMA) and write an organization chart. With these terse marching orders, Brown went to work. He prepared a table of distribution and a table of allowances, then he made a trip south to look at possible installations. Traveling by train to Pusan, he checked out a jeep at rear echelon headquarters and drove along the coast through Chinhae and Masan before finally selecting a site near Chinhae, where the artillery school was then located. He went back to US Army headquarters and sold this location to both the US Army and Republic of Korea staffs.

To administer the academy, Brigadier General Ahn Choon Saeng was selected as superintendent and Colonel Pak Choong Yoon was chosen as dean. After they accompanied Brown on a visit to the site and decided where to put the barracks and classrooms, Army Engineers began constructing buildings. In early December 1951, Colonel Harry McKinney (Class of 1927) was assigned as senior advisor to the academy, and Brown became advisor to the dean.

Brown procured West Point textbooks for all the subjects and had them translated. KMA would have some courses not covered at West Point, such as "Korean Culture," but otherwise, the curriculum would be the same as at West Point.

The translation process was not, however, without its unforeseen problems. It seems that during the 40-year period from 1905 to 1945, when Japan had occupied Korea, the Korean written language was outlawed, so advanced schooling—high school and college—was available to only a very few. Therefore, there was an acute shortage of teachers. Brown combed the Korean Army for instructors and came up far short, so he made a deal with some of the University of Seoul professors (the university had been moved to Pusan) to teach at the academy in exchange for rice. Brown then arranged to send a truck to Pusan and bring them to the KMA two days each week.

The Armistice

On June 4, 1953, the Communist Chinese at last agreed to a United Nations truce proposal that was essentially a return to the prewar status quo, but it took the South Koreans another month to ultimately acquiesce. On July 27, 1953, the armistice was signed at Panmunjon, and the final adjustment of the border between North and South Korea—roughly along the 38th parallel—was set. Ironically, both Heartbreak Ridge and Pork Chop Hill, where nearly 6,000 Americans were killed or crippled, were turned over to North Korea.

The Korean War cost the US Army alone 37,133 men dead and 77,596 wounded. This number was three-quarters of the total casualties suffered by American forces in Korea. As Wilson Reed wrote in retrospect, "Korea was 'the wrong war, in the wrong place, at the wrong time.' However, the impact of the Korean War profoundly altered the United States, its politics, its understanding of the powers of the presidency, its military position, and its world role. It represented a fundamental alteration of George Kennan's approach to 'containment'—our approach toward the Soviet Union. Kennan defined containment as political and economic; Korea represented the militarization of the Cold War. The president did not seek a Congressional declaration of war, setting a critical Cold War precedent. Congress acceded to this shift in the balance of power between the executive and legislative branches. Anti-Communism, not Isolationism, became the hallmark of conservative foreign policy. All this set the stage for the Vietnam War."

Declaration of war would not be an issue for the United States in Vietnam in 1964, Grenada in 1983, or Panama in 1989. The balance of power did not shift back until the passage of the War Powers Act in 1973, and the issue of a declaration of war was not dealt with again in Congress until the Persian Gulf War of 1991.

As Linton Boatwright wrote, "The principal difference between World War II and Korea was the difference in goals as announced by our government. In World War II, President Roosevelt made it clear that we were fighting to obtain the unconditional surrender of the governments of Germany and Japan. Even the lowliest of soldiers understood the dangers posed by Hitler and the Imperial Forces of Japan. In Korea, the United States government did not appear to know *what* its goals were. In January 1950, Secretary of State Acheson had stated that Korea was not included in the United States defense line in Asia. Yet, shortly after the North Korean offensive of June 25, President Truman dispatched forces from all three services to assist the South Koreans in the defense of their country. To the average soldier in ranks, this doesn't track. He says, 'The Secretary of State has said that this country is not important to my country, yet I'm supposed to come out here to this God-forsaken land and give my life. Why?' As in all wars fought by our country, soldiers did their duty, were wounded and killed, even though in Korea the reasons for their sacrifices were never adequately explained."

Wendell Knowles agreed with Boatwright. "Both wars seemed to have been undertaken for noble causes, but World War II was undertaken with the objective of *winning*, the Korean War was *not*. I will never understand why this was true. After much sacrifice, including 54,246 total United States dead, we brought the war to a standstill without victory."

The last postscript to Black '41's involvement in the Korean War was written on July 1, 1985, when Lieutenant Colonel William Thomas McDaniel was awarded, posthumously, the Distinguished Service Cross—America's second highest decoration—for gallantry as a prisoner of war, almost 35 years after his heroic action and death on October 20, 1950, in the Sunchon Railway Tunnel Massacre. It was the highest award given in the Korean War for action as a POW.

An investigation was instigated by the efforts of many of his friends and his son, William T. McDaniel, Jr., a US Air Force officer. Involving interviews with Americans who were there, the investigation showed that McDaniel's refusal to break under mistreatment by his captors, and his inspirational leadership at a time when the North Koreans were intent upon breaking the morale and spirit of their captives, finally led to his execution at the hands of the North Koreans at the Sunchon Railway Tunnel. The Distinguished Service Cross citation, read at his gravesite in Arlington National Cemetary, said in part, "Lieutenant Colonel McDaniel's courage and unwavering devotion to duty and his men were in keeping with the most cherished traditions and ideals of military service and reflect great credit on him and the United States Army."

15

The Good Years

Settling Down

By 1953, 36 of the 424 men of Black '41 had been killed during World War II and five had been killed in action or declared missing in action in the Korean War. Happily, although none of them knew it, no further members of their class would die in armed combat.

The years that followed the Korean War were good years for the nation and good years for the Class of 1941. For most of the men, it was a time of raising families and pursuing staff jobs. It was a period of great personal and professional fulfillment, years that saw most of the men reach the summit of their careers. George Pittman, for example, headed up the US Air Force safety team in developmental engineering inspection for the Boeing 707/C-135 program at Seattle in 1955. His briefing of Trans World Airlines executives in Kansas City in 1958 on US Air Force safety/accident experience with KC-135s resulted in the overhaul of a training program to convert TWA's fleet from conventional aircraft to jet aircraft. TWA's enviable safety record resulted from Pittman's efforts.

Their careers took many of the men of Black '41 back to faculty jobs at their alma mater. Pete Tanous returned to West Point in June 1956 with his wife Maxine and their son Bruce, who was two-and-one-half years old at the time. "Just viewing the beauty of the landscape, the cadets on their way to classes, remembering that 'first day' in July 1937, impressed me so much, and I felt I was so pleased to be able to spend the next three years at my alma mater. Lumps crowded my throat, and tears commanded my sight," Tanous said.

In February 1952, George Pickett was interviewed and then selected for the job as senior advisor for a new military/naval academy, which was to be known as the "Safety Academy." Admiral C. Turner Joy later appointed John

Cameron, a Navy commander (US Naval Academy Class of 1939), to be Pickett's deputy.

In 1950, the controversial Joseph McCarthy, a second-term senator from Wisconsin, developed an obsession with alleged Communist infiltration in the United States government. He had been reelected in 1952 partly because the idea of rooting out an enemy ideology played well to voters in the midst of the Korean War. However, whatever good might have been done by an investigative inquiry into *real* subversive activity began to wane after the war as McCarthy's crusade turned into a modern-day witch hunt that often crossed over the tenuous border between legitimate investigation and raving paranoia.

In 1954, McCarthy turned his attention to the idea of Communist infiltration of the US Army. Jack Murray of the Class of 1941, and late of Harvard Law School, was assigned to the office of the Secretary of the Army from March 1954 to July 1955 and assisted in the Army's dealings with Senator McCarthy. Murray gathered materials for the Army's case and participated as an advisor to Joe Wilet, the army's lawyer. In the end, McCarthy failed to prove anything, wasted a great deal of time and resources, and was ultimately censured by the Senate.

For the first time, congressional hearings were being broadcast on national television. Most people had never seen anything like these hearings on television, so members of the Class of 1941 and their wives glued themselves to the hearings. Their hero, of course, was Jack Murray. At the time, most people in the army believed that McCarthy was picking on the army for reasons that didn't make any sense.

From February 1951 to February 1953, Mike Greene served in the office of the chief of the Psychological Warfare Department of the army before going on duty with the International Branch, Plans Division, at the Office of the Assistant Chief of Staff of the Army. In March 1955, he was transferred to Germany, where he served as executive officer of the 19th Armor Group in Frankfurt and as commanding officer of the 510th Tank Battalion at Mannheim. In July 1956, he became secretary of the general staff at the headquarters of the Seventh Army in Stuttgart.

Jim Fowler had been the second black man to graduate from West Point in the twentieth century after enduring four years of ostracism, isolation, and torment that would have broken many of his classmates. He had joined the Class of 1941 after graduating magna cum laude from Howard University, and in 1941 he went on to the infantry school with the 366th Infantry Regiment at Fort Devens, Massachusetts. Major General Frederic E. Davison, who served with him at Fort Devens, recalled that Fowler "moved rapidly from platoon leader in Company E to a special company charged with the task of training all new draftees assigned to the regiment. As training for combat intensified, Captain Fowler commanded successfully and successively the Regimental Antitank Company and the Cannon Company."

In the fall of 1943, when the 366th Infantry Regiment was sent to the Italian front, Fowler was moved to regimental headquarters, promoted to major, and assigned as S-3. He was later awarded the Bronze Star. When he left the 366th, Fowler went to airborne school at Fort Benning, and between 1951 and 1954, he commanded the 370th and 373d Armored Infantry Battalions in occupied Germany. In 1958, Fowler was assigned to the Military Advisory Group in Taiwan, where he was responsible for establishing and implementing the four-year academic program at the Nationalist Chinese "West Point."

As Paul Skowronek saw it, Jim Fowler was "a victim of the same racial segregation that we all accepted as the American way of life—in the 2d Cavalry Division and at West Point. The Army and West Point were merely following a system that the nation believed in and was enforcing throughout the country. Jim Crow laws followed Jim Fowler to the USMA. He was avoided by other cadets, and the rumor was rampant that he was being paid huge sums of money by black sponsors to stay at West Point, regardless of the animosity and hardship involved. He stayed, he graduated, he had an average military career but retired early. The Army and West Point moved faster than the rest of America to do away with racial discrimination, and Jim Fowler helped to hurry the process."

Many of the men of the Class of 1941 believed that 1959 was a critical year for their careers. If they hadn't made a War College list by 1959, advancing in rank could become more and more difficult. They had 18 years service by now, and many felt that it was essential to their "stairway to the stars" to get assigned to the War College. George Pickett saw the War College as a prerequisite to flag rank. Without War College experience, he believed he would never rise above the grade of colonel.

When the list of students who were going to the Army War College was published in May, somebody called George Pickett and told him that the list was out. He immediately went over to the adjutant's office at the Armed Forces Staff College, where he was an instructor, to see if he had made it. The major looked at the list, looked at Pickett, and finally shook his head and said, "I'm sorry, colonel, you're not on the list. Here's the list. See for yourself."

George and his wife were eating supper that night when the phone rang. It was an old friend, Hank Middleworth, who had been with George in the 1st Infantry Division in 1945. He said, "George, where are you going to live at the War College?"

"I hate to tell you this, Hank, but the list was posted today and I wasn't on it."

"The hell you're not!" Middleworth said. "You and I are *both* on the list."

"I didn't see either of our names on the list."

"George," Middleworth scolded, "did you look at the *National* War College list?"

"No, I checked the Army War College list."

In those days, under Eisenhower, the National War College was a far better plum than the Army War College because it was a joint-command school—Army, Navy, Air Force, State Department—an integrated service school, and you got a better "ticket punch" out of it. From utter dejection at 11 o'clock in the morning, George Pickett had the wonderful feeling of "being among God's chosen" at suppertime.

When George Scratchley Brown had graduated from the National War College in 1957, he served as executive to the chief of staff of the US Air Force and then as military assistant to the deputy secretary of defense. After four years as a brigadier general and military assistant to Secretary of Defense Thomas Gates and Secretary of Defense Robert McNamara, Brown received a second star and successive commands of the eastern transport segment of the Military Air Transport Service, McGuire AFB, New Jersey, and a weapons testing center at Sandia Base, New Mexico. Promoted to lieutenant general in 1966, Brown returned to the Pentagon for two years as an assistant to the Chairman of the Joint Chiefs of Staff. On August 1, 1968, at the age of 49, he would pick up his fourth star.

Ben Spiller, like George Brown and most of their classmates, found himself serving a tour at the Pentagon during those years. Spiller was on the army rather than air force side. His reaction to this tour was like that of most of his classmates when he decided that he hated Pentagon duty just as much as he had hated Plebe Year at West Point.

Groszaphenstriech

George Pickett reported for duty on August 10, 1960, at Alexander Patch Barracks in a suburb of Stuttgart, Germany. His assignment was to become the chief of plans for the US Seventh Army. Of the six Armies—the First, Third, Fifth, Seventh, Ninth, and Fifteenth—that had constituted American ground strength in Europe during World War II, only the Seventh now remained. It was retained in Europe, first as an occupation force and later as part of the contribution of the United States to the North Atlantic Treaty Organization (NATO).

For most of the half-century that followed World War II, the principal concern facing the US Army was the Soviet threat to Western Europe and the force levels required to stabilize this threat. The North Atlantic Treaty Organization was created for this purpose, but the US Army was NATO's backbone. NATO had no strength without the US Army as its essential striking power.

In 1945, within a few months of VJ-Day, it became clear to members of the occupation army in Germany—if not to the politicians at home—that the Soviet Union, their wartime ally, was going to pose a far greater threat to the postwar stability of Europe than the now-defeated enemy whose territory they were occupying.

For most of the next five decades, containment of potential Soviet aggression in Europe would be the centerpiece of American strategic policy. At the end of the 1940s, there had been a desperate rush to build up the US Army in Germany. Money was no object. Trillions of dollars would ultimately be spent attempting to recreate the power that had existed in April 1945, when there had been 69 US Army divisions in Europe. Four months later, there were only five. The force levels had increased in the 1950s, but by 1960, there still were only five US Army divisions and three armored cavalry regiments in Germany as part of the Seventh Army.

Access to Berlin was the major issue, but it was not a NATO issue, so the United States could not go to NATO for help. Berlin was controlled by the British, the French, the Americans, and the Soviets. The Soviets, though, had control of the railroads and the highway. Routinely, they would shut off access to Berlin, and then there would be a big crisis about what the United States was going to do to make them open it. The first and, in retrospect, the most serious incident had been between June 1948 and May 1949. Only the fact that the United States and other countries had transport planes with which to launch the Berlin Airlift kept Stalin from being able to take over Berlin completely. Another crisis, which began in September 1960, shortly after George Pickett arrived, lasted into the summer of 1961.

The winter of 1960–1961 was marked by heated verbal exchanges and by Soviet troops periodically cutting off surface traffic between West Germany and West Berlin. In June 1961, Soviet Premier Nikita Krushchev announced that the Soviet Union would sign a separate peace treaty with East Germany, the German Democratic Republic (DDR). Soon, word spread that border crossings between East and West were being closed, and desperate East Germans began pouring across the border into the West. In July, more than 30,000 crossed, and during the first two weeks of August, another 48,000 escaped. It would be the last time until 1989 that such mass crossings between the two parts of Germany would be possible.

The crescendo of the 1961 Berlin Crisis came in the early morning hours of Sunday, August 13, when East German Army units and *Betriebskampf-gruppen* (armed industrial groups) began digging up the pavement at the official border crossings around West Berlin. Barbed wire was stretched along the border, and by the end of the day the infamous Berlin Wall was under construction. The border between the two halves of Germany was being closed, said East German party boss and State Council Chairman Walter Ulbricht, "to protect the DDR against threatening West German militarists and all kinds of Western spies, agents, and saboteurs."

President John F. Kennedy responded by ordering 50,000 additional US Army reservists to Germany. George Pickett burned a lot of midnight oil during the summer of 1961, getting up at 3 A.M. and going up to Heidelberg to communicate with Washington via teletype. He laid out the tactical plan as best he could, but from a practical point of view, the military situation was hopeless.

As Pickett summarized it, "The only viable option, in my opinion, was another airlift because we did not have the troop strength to force our way down the *autobahn* [highway] into Berlin. It just wasn't there. The theory behind it was the assumption that if we did make a show of force, the Russians would let us through.

"In my professional opinion, it would have been ridiculous to attack, but we probably would have. You never know what will come out of Washington, D.C., so we were prepared to go if they said 'go,' but we knew deep down inside that it was going to be pretty sticky. We also didn't know if the East Germans would support the Russians or not. The whole thing was a strategist's nightmare. Thank God it was solved politically, because we could have had World War III if the thing would have backfired."

Pickett's biggest job was finding space for 50,000 reservists and their equipment in the existing barracks. The Defense Department hadn't given the Seventh Army any additional money or facilities, so by the end of 1961, the Seventh Army had five divisions, four armored cavalry regiments, and support units in the same facilities that they had used a year earlier for a force that had been a fraction of that size.

In December 1961, General Garrison "Gar" Davidson, the former football coach at West Point and now the Seventh Army commander, told Pickett that as soon as the Berlin Crisis was stabilized, he could command troops in the field. An officer cannot be promoted above the grade of colonel unless he has had a command as a colonel. Pickett was supposed to go to command duty in 1960, but the Seventh Army had needed a war planner, so Pickett's orders had been changed.

It turned out to be one of the biggest breaks of Pickett's career, because in December of 1961 Gar Davidson gave him command of the 2d Armored Cavalry Regiment in Nuremberg, and this post became a stepping-stone for him.

Pickett went to the 2d Armored Cavalry during the first week of December 1961 and served until July 1963. At that time, the armored cavalry regiments on the border between East and West Germany were the "trip wires." If the Soviets attacked, the cavalry regiments were to be the units that would delay them while the army geared up for World War III.

The cavalry regiments had to patrol tremendous distances with a very small force. The 2d Armored Cavalry had one squadron of 900 men patrolling along the Czech border in the south and two squadrons across the East German border in the north, near Coburg. They were constantly monitoring the activities of the Soviet and Warsaw Pact units on the other side of the wire. The biggest problem of the Soviets was keeping people in. During this period, there were still hundreds of people trying to defect to the West. It was difficult to get through because there was an electric fence with guard towers about every 50 to 100 yards. The 20-foot high guard towers were equipped with machine gun crews and searchlights. The entire East German–Czech border was covered by an electric fence. On the western side of the fence—but within East German or Czech territory—there was a minefield. Like a World War II minefield, it had antipersonnel mines and antitank mines. Someone escaping had to get through the electric fence, past the guards with machine guns, and finally through a minefield.

In October 1962, Pickett's 2d Armored Cavalry was involved in an exercise that found it designated as the "enemy" force, maneuvering against the 3d Infantry Division, commanded by Major General Frank Mildren. On the last day of the exercise, Pickett was in the field when he received an urgent call to immediately fly to Nuremberg to meet the corps commander. Even though the exercise wasn't completely over, he turned things over to the executive officer and flew back to Nuremberg. It was here that he was first notified of the Cuban Missile Crisis.

When Pickett had been briefed on the situation in Cuba, the corps commander told him that the Pentagon needed to know exactly what he would do in the event of war.

"We have a war plan," Pickett told him simply. "My people have rehearsed it. We know exactly what to do. If they go to war over the Cuban Missile Crisis, it would be no different than going to war over anything else. We have to execute the mission as written and as coordinated."

A *Zaphenstriech* is a "great tattoo," or outdoor military exercise presented as evening entertainment. The custom comes from the era of Frederick the Great, and it came to be a tradition in the German Army that the ceremony was conducted only on special occasions. In 1963, the German military commander for the Nuremberg area decided to conduct a big or *Groszaphenstriech* in honor of the 125th anniversary of the organization of the 2d Armored Calvary Regiment. It was a huge success. George Pickett and his men marched in and took positions facing the cathedral in the *Marktplatz* in Nuremberg, while the German Fourth Panzer Grenadier Division sent in its band and troops, which marched in by torchlight.

"If you think those Germans had lost any zeal for military ceremony, you're wrong. They had the streets lined!" Pickett said later.

Twenty-four years later, when Pickett went back to Nuremberg, he discovered a huge oil painting and newspaper clippings of the event in the regimental museum. There hasn't been anything like it since. The 2d Armored Cavalry received the biggest honor of any American unit in Germany. Soon after, however, a smaller *Zaphenstriech* was scheduled in Coburg, not for the 2d Armored Cavalry Regiment, but to "honor the departure and give due honor to the friend of the German Army, George Pickett."

The Man in the Gold Stingray

In 1963, Colonel Paul Skowronek presented himself to the Russian commander in chief of the Group of Soviet Forces in Germany (GSFG) as chief of the United States Military Liaison Mission (USMLM), a post he would hold for four years. With a "license to spy," he drove his gold Corvette Stingray more than 80,000 miles throughout the Soviet zone and experienced so many unique and challenging situations that eventually he wrote his Ph.D. dissertation about United States–Soviet military relations in Germany, largely from personal observation.

Skowronek was the tenth chief of the USMLM in the Soviet zone, and the first to begin his term after the construction of the Berlin Wall. His headquarters were in Potsdam, a suburb of West Berlin, in a palace once owned by the Hohenzollern Prince Sigismund but known to the USMLM simply as "Potsdam House."

In the 1960s, the United States did not recognize the German Democratic Republic (DDR)—East Germany—as an independent country, but rather as the Soviet zone of occupied Germany. The Soviets and the Americans had arrived at a reciprocal agreement permitting each to station a 14-man liaison mission in the other's zone of occupation—in Frankfurt and Potsdam respectively—and this arrangement continued even after 1974, when the United States recognized the sovereignty of the DDR. The stated purpose of the respective liaison groups was officially to maintain contact with each other's military commands, but in fact, the gathering of intelligence was a key element of the agenda for both sides.

Under the military agreement signed in 1947, both the Soviet and the United States liaison missions were to be supported in each other's zone with suitable housing, domestic help, and access to food supplies. The Soviet Military Liaison Mission (SMLM) members shopped regularly and extensively at the US Army commissary in Frankfurt, where a wide variety of meats and groceries was available, enjoying better food than Russians could buy anywhere in their own country. In general, food supplies were scarce in East Germany, and the Soviet Army drew on these meager supplies for its subsistence. Since they had nothing comparable to the well-stocked US Army commissaries, they delivered rather mediocre food products to Potsdam House. Skowronek and his staff did not complain, however, because Russian Beluga caviar and Stolychnaya vodka were made available in unlimited quantities, and USMLM members could also shop in the US Army commissary in West Berlin. They paid for their expensive caviar and vodka, as well as some basic food products, in cheap East German marks, which resulted in unbelievable bargains.

Sometimes the difference between the Soviet and American ways of doing things worked to Skowronek's advantage. One night, when several members of his USMLM team were driving near Potsdam, searching for Soviet tank units reportedly conducting a field exercise, their car suddenly hit a stretch of road made slippery by mud deposits from tank treads. The Mission sedan went out of control and skidded into a platoon of Soviet soldiers who had been marching in the dark along the shoulder of the road. Two soldiers were seriously wounded, and one with massive head injuries, who was wedged under the rear wheel of the sedan, lay unconscious.

A Soviet NCO ordered soldiers armed with rifles to surround the vehicle and allow nothing to be moved until an officer could be called to the scene. Fortunately, the USMLM officer, speaking fluent Russian, was able to convince the NCO that the sedan should be moved enough to extricate the unconscious soldier. This was permitted, but nothing else was done until a Soviet captain from a nearby troop barracks arrived in a military truck. The Soviet officer ordered that the wounded soldiers be loaded in the back of the truck, which then drove off, presumably to a hospital. The captain then halted a passing military vehicle and asked the driver to take a message to the Potsdam Soviet commandant requesting that he come to the accident site to discuss the situation with the USMLM team.

Confusion reigned for a considerable time on that dark road before the commandant decided that the USMLM sedan, with its occupants, should be towed to his headquarters for a full investigation. By this time, Skowronek had been notified that his team was being held for investigation, so he went to the Soviet headquarters immediately to do what he could to resolve the matter.

It was nearly midnight by the time he arrived there, and nerves on both sides were becoming frayed. The commandant was demanding that Skowronek's team submit to a blood test to determine if they had been drinking. This had never been done before, and Skowronek had no intention of starting a procedure that in any way limited USMLM's special status in the Soviet zone, so he stalled for time.

Skowronek insisted that he would have to refer such an unprecedented request to the US Army commander in chief in Heidelberg, so his team spent the night in the commandant's office while he drove to Berlin to send a message to Heidelberg. The upshot of the situation was that Skowronek was instructed to return to the commandant's headquarters in the morning and state that the US Army had no objection to a blood test, provided that it was performed by a US Army medical officer.

Skowronek knew that to obtain clearance for a non-USMLM officer to enter the Soviet zone, the request would have to be made at the headquarters of the Soviet commander in chief, and that would take several days. Meanwhile, the now-frustrated commandant realized that his demand for a blood test had gone beyond his level of command and might even be seen as uncalled for by his superiors. Furthermore, because so much time had elapsed since the accident, the delayed blood test would be useless.

As a compromise, the Soviet commandant suggested to Skowronek that if the team officer would sign a report of the circumstances of the accident—which the commandant's staff had prepared—he could release the USMLM personnel and vehicle.

"My officer will have to be advised by an American military lawyer after the statement of circumstances has been examined from a legal standpoint, and that *also* will require reference to my higher headquarters," Skowronek said simply.

By this time, the Soviet commandant was becoming so weary of the entire affair that he finally agreed to release the team to Skowronek's custody and to allow a Mission sedan to tow the disabled "touring car" out of the Soviet zone.

While the Soviets spent a good deal of time shadowing Skowronek and his men, the East Germans also had an obvious interest in them. The East German secret police—the dreaded Stasi—had established a special unit to watch USMLM travel activities and had purchased, with scarce hard currency, several West German BMW and Mercedes sedans with which to follow USMLM vehicles. On the highways, USMLM's normal Ford sedans were not able to outrun the East Germans. However, the Fords had specially raised suspensions, mud and snow tires, protected oil pans, and a cable winch, enabling them to drive cross-country over rough terrain to lose the East Germans. When he first arrived in his post, Paul Skowronek had felt himself greatly stifled, not only by the slow Fords, but by the standard operating procedure that called for the chief of mission to always travel in the Soviet zone with an experienced enlisted driver, so he immediately explored the possibility of cruising around East Germany alone in a faster car.

The car he picked for his personal use was a gold-colored, Chevrolet Corvette Stingray. It disconcerted the East German secret police, who often attempted to follow USMLM sedans, and it intrigued the Soviet officers in East Germany, who were fascinated by the beauty of a flashy sports car. Driving himself in a sports car with which he was thoroughly familiar, Skowronek felt safer than he would have with someone else doing the driving. The idea worked so well that the commander in chief of the US Army in

Europe favored Skowronek's innovation and showed a personal interest in the incidents and encounters resulting from Skowronek's trips in the Corvette. With the Stingray—which had a top speed at least 25 mph faster than Soviet or East German cars—he was able to outrun the secret police sedans quite readily.

In frustration, the East German secret police frequently reported his high-speed getaways to the Soviets, who never took any action against him. The Soviet military commanders were instinctively distrustful of secret police, whether East German or Russian. More to the point, the Soviets probably refrained from interfering with Skowronek's travel activities in order to retain freedom of movement for their liaison mission in the US zone. They wanted Skowronek's counterpart, a Soviet major general in Frankfurt, to be free of West German police control, so they ignored the fact that Skowronek did not observe the 100 kph speed limit in East Germany. The 1947 Hueber-Malinin Agreement, under which the liaison missions operated, stipulated that travel by Mission personnel would be unescorted and free throughout each other's zone of occupation.

The East German secret police in their unmarked sedans did not have radar speed-measuring devices, but Soviet officers told Skowronek that they had complained of his driving at 240 kph (150 MPH) when, in reality, his top speed was only 135 mph. At one New Year's party, the Soviet commander in chief, with a twinkle in his eye, asked Skowronek, "How fast can your Stingray really go?" He seemed a little disappointed when Skowronek admitted that its top speed was less than the high speed that presumably had been reported.

The Russian general thought for a moment. "Let me offer you some friendly advice. Please drive carefully in the Soviet zone," he said. "You are accredited to me personally, and I would not want you to have an accident."

16

Vietnam

Getting Involved

The Vietnam War was America's longest war, an inconclusive and confusing conflagration whose stated purpose seemed small justification for the level of men and matériel invested in it. Its impact on the national conscience made it the most divisive and demoralizing war since the Civil War, and no stratum of American society was satisfied with its final outcome. Ironically, despite the war's having been the most prolonged military involvement of the United States, many Americans ultimately felt frustration that the United States was little more than a weary bystander when the end finally came.

As Pete Crow pointed out later, the conflict in Indochina was even more inexplicable than Korea. "President Roosevelt did not want Indochina returned to French sovereignty," he said. "Our OSS assisted Ho Chi Minh in fighting the Japanese. Had we recognized the free Vietnam in 1945, we would subsequently have had a staunch ally instead of a deadly enemy."

The 1954 Geneva Accords granted Laos and Cambodia their independence and divided Vietnam into two independent countries at the 17th parallel. The northern part of the country became a Communist state with a capital at Hanoi (the old French capital) governed by Ho Chi Minh, while the south ostensibly became a Western-style constitutional monarchy under Emperor Bao Dai. Bao, who could trace his lineage back to Emperor Bia Long, who ascended the throne in 1802, had served as a kind of puppet monarch under both the French and the Japanese between 1932 and 1945. Although the throne was in the ancient imperial city of Hue, the South Vietnamese administrative capital was established at Saigon, a sprawling city on the meandering Saigon River that had been known in colonial times as "the Paris of the Orient."

The United States continued to actively support Bao Dai and his prime minister, Ngo Dinh Diem, while Ho Chi Minh vigorously promoted a Communist guerrilla movement within South Vietnam. These guerrillas, who called themselves the National Liberation Front (NLF), were virtual clones of the Viet Minh guerrillas who had ousted the French and were called Vietcong by the South Vietnamese.

Both the Eisenhower and Kennedy administrations sent military supplies and advisors to South Vietnam, and initially the number of American military advisors was not out of line with the number sent to other non-Communist countries around the world. At the end of 1960, there were only about 900, but a year later there were 3,200. By that time, Vietnam was becoming a critical element in American foreign policy, and the United States military presence there quickly mushroomed. By 1962, there would be 11,000 Americans—mostly US Army personnel—"in country."

The involvement of the United States in Vietnam evolved as a gradual, yet ever-growing, commitment. The only events that could be in the least compared to a Fort Sumter or a Pearl Harbor were the August 2 and August 4, 1964, attacks by North Vietnamese patrol boats on a pair of US Navy destroyers in the Gulf of Tonkin off the coast of North Vietnam. In response, President Lyndon Johnson ordered air attacks on North Vietnamese naval bases on August 5. Two days later, Congress passed the Gulf of Tonkin Resolution, giving Johnson extensive powers to commit American troops to fight in Vietnam. Passage of this resolution paved the way for a progressive influx of American forces, which would eventually number over half a million. The first large-scale deployment came in early 1965, but the numbers escalated rapidly after that.

The period from 1965 to 1967 saw an immense increase in American force levels in Vietnam. In 1964, there had been 47,000 Americans in Southeast Asia. In 1965 there were 184,300, but by the end of 1967 the number had grown to 485,600 and was still growing. About half of the two dozen men from the Class of 1941 to serve in Vietnam were "in country" during this period. These included Charles Cannon, James Graham, Walter Mather, Jack Norton, and Joseph "Sloppy Joe" Grygiel of Army football fame, as well as William Gleason, who served in Edward de Saussure's old job as assistant division commander of the 25th Infantry Division, and Potter Campbell, who was the commander of the 12th Aviation Group and took home a Distinguished Flying Cross, a Legion of Merit, and a host of other decorations, including 17 Air Medals.

Because the enemy was largely a guerrilla force, the helicopter was *the* American weapon of the Vietnam war, just as the terrain of the European theater had favored tank warfare in World War II. As Paul Skowronek has written, "In the rugged, mountainous terrain of Korea and in the jungles of Vietnam, large tank formations would have had limited mobility and vulnerability to enemy infiltration. Further, this terrain drastically reduces the effectiveness of long-range fire from the tank's main gun. Armor was decisive in the open terrain of Europe and North Africa, but it is wasted in highly unfavorable terrain. Helicopter gunships afford unimpeded mobility, freedom from infiltration, and unlimited field of fire above the terrain which is unfavorable for tanks. However, modern helicopters would have been of incalculable value in the Italian Campaign in 1943–1945."

Only one member of the class of 1941 would die in Vietnam, although his death did not occur on the battlefield. Alfred Judson Force Moody, the quiet, friendly, capable man who had graduated at the head of the Class of 1941, died of a heart attack at Camp Radcliff in Vietnam on March 19, 1967.

"Al was brain power without any ostentation at all," Horace Brown recalled. "He readily fit in a group, and you never thought of him as being brilliant because he had more of the everyday mannerisms. He was truly an outstanding, unique, caring friend."

After graduation, Al Moody had married Jean Enwright, whom he had known since childhood, and set off for the "wilds of Kansas" and the horse cavalry. When World War II started, they found themselves back at West Point, where Al joined the faculty. By having been first in the class, Al Moody was cursed with greater perceived value as a planner than as a combat commander, so it was with some effort that he managed to get away from West Point in time to participate in the war as a planner in Europe at General Eisenhower's headquarters in 1944, and on an intelligence mission to the China/Burma/India Theater. Moody ended the war with the 7th Division in Okinawa and Korea, and, upon returning to the States, he acquired a master's degree at Yale in 1948 before being sent to the Pentagon, where he eventually wound up in the army chief of staff's office. He subsequently went on a brief assignment to Paris and then to command a tank battalion in Germany. After a course at the Army War College, he returned to the Pentagon and ultimately became assistant to the Secretary of Defense.

Moody had spent most of the 1960s in Washington. In 1963, he was into his third year at the Office of the Chief of Staff when he and Jean had bought a home in Virginia with "more yard than we bargained for." Suddenly, he was shipped to Korea to command the 1st Brigade of his old 7th Division, and Jean and their four daughters remained in Virginia to "learn more about power mowers." However, Al's tour was cut short by the Chief of Staff's request and he was back in Washington inside of a year. He went immediately to work in what Bradish Smith described as the "ticklish task" of assistant to Secretary of Defense Robert McNamara. The job and the hours were grueling, but in his usual quiet, efficient, and firm manner, Moody became an extremely valued member of the "top-level family," and he was promoted to brigadier general in July 1966.

In March 1967, after a short session at Fort Rucker to learn to fly helicopters, Moody was assigned as the assistant division commander of the 1st

Air Cavalry Division at Camp Radcliff, Vietnam, which was commanded by his old friend and classmate, Jack Norton.

To be again with troops in a command assignment was a job that he dearly loved. He had been "in country" only a few days when the heart attack hit him while he was in his quarters, writing a letter to Jean. It was massive and unexpected, and he died almost immediately in the arms of his aide and the local surgeon. The following day, Easter Sunday, the men of the 1st Air Cavalry placed a plaque in his memory on the altar of the division Memorial Chapel as part of the dedication ceremony at An Khe, Vietnam.

On March 24, 1967, Al Moody was buried in Arlington National Cemetery with full military honors, attended by his family, classmates, and friends by the score, including almost all of the top personnel of the Department of Defense and the Department of the Army, who had come to know and respect him. The last member of Black '41 to die in a war zone was laid to rest.

Into the Johnson Grass

Despite the rapid growth of the American presence, the ensuing years in Vietnam were characterized by a lack of clear-cut strategy and leadership on the part of political leaders in Washington and frustrating, small-unit actions on the ground in South Vietnam. The Johnson administration seemed intent on usurping the decision-making authority usually vested in brigade or even regimental military commanders while abdicating its own responsibility to create a coherent overall strategy that would bring the war to a satisfactory conclusion.

Many American officers advocated a decisive, all-out assault to defeat North Vietnam, an action that could have been completed in a short period of time with fewer casualties then were ultimately incurred. President Johnson, however, insisted on restricting American troops to actions *within* South Vietnam, because he feared Chinese intervention if North Vietnam were invaded—in short, it was a replay of what had happened in Korea. Johnson asked only for a "conditional surrender," but the enemy wanted total victory. As in Korea, the respective goals were asymmetrical.

Pete Crow, who served as director of the budget for the Air Force Department and as comptroller of the Air Force between 1964 and 1973, was among a growing number of military men who had become skeptical of the wisdom of American involvement in Vietnam. Of the hundreds of congressional hearings that he attended during his decade at the Pentagon, one stands out in his mind as being illustrative of the Johnson administration's conduct of the war. Secretary McNamara was the witness, and service budget officers had been requested to attend. Senator Stennis was in the chair. As he banged his gavel, Senator D. Willis Robertson entered. Stennis opened with, "My good friend Senator Robertson has asked permission to ask you a few questions."

Senator Robertson asked, "Mr. Secretary, when you were over here last March, you told us that our boys wouldn't have to go over there. Now what has changed?"

Secretary McNamara gave his usual lucid explanation of the necessity to contain communism.

Senator Robertson then asked, "Mr. Secretary, do you know what 'Johnson grass' is?"

McNamara said he did not.

Then, as Crow recalls, Robertson said, "Mr. Secretary, my friend John Stennis and I hunt quail together. If they flush and go into the 'Johnson grass'—tall, thick, and impenetrable—we leave them alone. Now you are sending our boys over there, where the enemy is *always* in the 'Johnson grass.' It is a national tragedy!"

With that, he slammed his big, hamlike hands on the table and stalked out.

In 1965, as the war in Vietnam was heating up and the Air Force prepared to increase its effort, Pete Crow was among the officers called to a commanders' conference at Hickam AFB in Hawaii. Crow and his staff had worked through the night, and with some 150 projects in hand totaling billions of dollars, they took off from Washington to Hickam, most of them hoping to sleep en route. However, the assistant vice-chief collared Crow to do the briefing.

During the entire 11-hour flight, Crow sorted through a knee-high stack of notebooks, and upon arrival at Hickam in early evening, he was met by the Pacific Air Forces Command Project Officer with *his* stack of notebooks. All through the night they merged projects, eliminated many, beefed up others, and were ready when Air Force Chief of Staff General John Paul McConnell (Class of 1932) called the meeting to order the next morning.

"Crow, how long is this going to take?" McConnell growled.

"Sir, we have about 150 items to cover. At a minute-and-one-half each, it will take nearly three hours, with no questions," Crow replied.

"Make it a minute each," McConnell said bluntly.

It took all day, but at the end everyone had a good understanding of what had to be done.

At a staff meeting at the Pentagon one day soon after the Hickam conference, McConnell opened the meeting with, "I want all of you to read the book *The Battle for Dien Bien Phu.* We just may be getting into more than we bargained for. This is not going to be a picnic. It's going to be tough all the way."

Disillusion

The limitations that President Johnson placed upon US military leaders in Vietnam played a major role in significantly decreasing the morale of the troops in the field, and indeed among troops everywhere, the Black '41 classmates included. Wendell Knowles, who was in Manila in 1965 when dependents in Vietnam were being sent home and the gradual US troop buildup began to get underway, commented that, "Having been disappointed with the outcome of the Korean War, I was very much opposed to the Vietnam War. I felt, early on, that although we had every right to pursue our objectives there, our political leadership would not permit us to *win* the war. If the war was not worth winning, it was not worth becoming involved in." With thoughts about the lack of resolve on the part of administration leadership, Knowles decided to retire at the end of his tour in the Philippines in August 1965.

Some months later, however, the State Department recruited him for duty in Vietnam, and he spent the next nine years there working in the Pacification Program, which he described as "a flawed experiment, but an interesting one."

Pete Crow visited Vietnam twice: once with General McConnell and once with General John Dale Ryan, who succeeded McConnell as Air Force Chief of Staff in 1969. On the first trip, their KC-135 tanker/transport aircraft was so loaded with fuel that they had to take off from the longest runway at Dulles Airport. It had snowed all of the previous night, and at 4 A.M. Crow's staff car, on its way to pick up Major General John Bell, became stuck. Crow commandeered an Alexandria, Virginia, garbage truck to pick up a surprised General Bell. The flight took them through Incilirk to Bangkok, overflying India—where Crow had spent World War II—at night.

On both visits to Vietnam, Crow was struck by the relative comfort, good facilities, and good equipment that the American forces enjoyed. Most of all, however, he was impressed by their pent-up eagerness to do more than attack meaningless tactical targets, while North Vietnam's most important strategic targets in Hanoi and Haiphong went untouched. "Why can't we be turned loose to get this thing over with?" was a universal question.

On one visit, Crow bunked at the trailer house being used by Robin Olds, who had been an usher at his wedding in 1945. Now an Air Force colonel, Olds had been a fighter ace twice over in World War II, shooting down ten of the Luftwaffe's finest. He now commanded the 8th Tactical Fighter Wing—the celebrated "Wolf Pack"—and would ultimately claim four North Vietnamese MiGs. An ace with victories in two wars, and a daring standout among the daring, Olds echoed the universal question, "Why can't we be turned loose?"

At the staff meeting the next morning, General McConnell said that intelligence had affirmed that the Air Force prisoners of war in Hanoi were being tortured. It was a grim meeting. Nearly all present had personal friends in Hanoi. "Could we have had our way, everything would have been turned loose," Crow remembers bitterly.

Tom Lawson concurred in this opinion. "I do not believe the politicians have the right to start a war unless they are prepared to unleash the Armed Forces to go ahead and win it."

On January 30, 1968, the North Vietnamese and Vietcong launched a massive offensive designed to coincide with Tet, the lunar New Year. The northern city of Hue was captured, and parts of Saigon were overrun. A concerted effort was made to also overrun American bases in the Saigon area, Tan Son Nhut, and Bien Hoa.

While the Tet Offensive was eventually beaten back, it was done only at great cost to American and South Vietnamese forces. The price would also be paid in Washington, where the Johnson administration suffered a severe jolt. Tet showed an increasingly disgruntled American electorate that administration policy of the past four years had not worked. The Vietcong had not been defeated as the Johnson administration had claimed, and indeed, they appeared *stronger* despite three years of deep American involvement on a level unheard of since the Korean War. The Tet Offensive showed that the administration's assessment was drastically flawed. During Tet, the "nearly defeated" enemy had been able to strike nearly every American and South Vietnamese unit in South Vietnam, and one commando group even penetrated the US Embassy compound in Saigon.

There was an enormous groundswell for withdrawal from Southeast Asia. In a dramatic, televised address, President Johnson told the nation that he would not be a candidate for reelection in 1968 and that, effective April 1, American bombing raids would not be made in North Vietnam above the 20th parallel. Johnson also announced that he would gradually reduce American strength in South Vietnam, and that he hoped North Vietnam would show good faith and do the same. High-level talks, convened to negotiate an end to the war, began in Paris on May 10, 1968.

Meanwhile, dissatisfaction with the war had grown to the point where Johnson's intended successor, Vice President Hubert Humphrey, was trailing his opponent, Richard Nixon, in opinion polls. On November 1, 1968, four days before the presidential election, President Johnson further deescalated the war by declaring a halt to *all* bombing missions of any kind over North Vietnam. Armed reconnaissance flights would continue, however, and they ultimately revealed massive rebuilding of the supply network that fed enemy forces in the South, as well as a strengthening of antiaircraft defenses. It was not the good-faith de-escalation that Johnson had hoped for, nor did his bombing halt have the intended effect of winning the election for Humphrey.

The Tet Offensive and the transition from the Johnson to the Nixon administration changed the complexion of the war in many subtle ways, but the net result was still the same. To both Johnson and Nixon—both of them outspoken anti-Communists—the Vietnam War after 1967 was more of an embarrassment than the crusade that it had been in the early years. From a strategic viewpoint, it was not a war to be won or lost, but rather a quagmire from which to extricate oneself as gracefully as possible.

The new president continued Johnson's policy of gradually withdrawing troops and attempting to turn full responsibility for the war over to South Vietnamese forces. Although Nixon was not particularly inspired to continue the war, he saw it as embodying a commitment that the United States should live up to, so that future allies would have faith in American resolve.

Among the men from the Class of 1941 who went to Vietnam in 1968 and 1969 were Elmer Yates, Robert Tarbox, and Walter Woolwine, all of whom were involved in engineering or logistics. Mike Greene also arrived to serve a second tour in January 1970, succeeding classmates de Saussure and Gleason as assistant division commander of the 25th Infantry Division, where he remained until December, at which time he became commanding general of Headquarters Area Command, US Army Saigon until May 1971. During the same time, his brother Larry came to Saigon as assistant chief of staff for personnel in the Military Assistance Command in Vietnam (MACV) headquarters, and as circumstances would have it, every Thursday morning at the weekly briefing for the MACV chief of staff, they sat side by side. This was the only time in their nearly 30 years of Army service that they ever served in proximity to each other.

Mike Greene served a total of 44 months in Vietnam. During his first tour, from February 1963 to June 1965, before the United States's entry into a combat role in Vietnam, Greene was a colonel and executive assistant to the commander of MACV. However, during his second tour, from January 1970 to May 1971, Greene once again found himself in combat, serving directly with the officers and enlisted men on the ground.

There was a vast difference between the US Army in Vietnam in 1965 and in 1970–1971. The troops in 1965 were mainly advisors to the Vietnamese, until combat units began arriving in mid-1965. These troops were trained and selected for specific missions. By 1970, the US Army had changed drastically. "There was little unit identity and esprit, morale had sunk, and discipline was lax," Mike Greene said.

The Pentagon policy of limiting tours of duty to one year guaranteed that the best-qualified and most experienced troops were pulled out just as they had achieved the experience needed to make their units, and the new men coming into them, truly effective. Thus, the longest war in American history was fought by men with less than a year in combat.

As Mike Greene pointed out later, during World War II, army units had been organized, trained, and deployed *integral units*, and they fought that way over a period of years. In Vietnam, there really was no sense of unit identification. "The vast majority of officers and enlisted men went into Vietnam on one-year assignments. They went in as individuals and had no real sense of attachment to a unit or idea of duty, nor were there front or rear lines, and no way to distinguish the positions of the two sides."

The Commander

In World War II, the men who commanded armies, army groups, and numbered air forces had been men with 25 and 30 years of service. By 1969, the men of the Class of 1941 had 28 years of experience. Had there been an army group to command in Vietnam in 1969 or 1970, it could well have gone—rightfully—to someone like Potter Campbell, Jack Norton, or Mike Greene—or indeed to Bill Gillis, if he had lived. In Vietnam, however, there was only the US Army, Vietnam (USARV) and MACV, which were under General William Westmoreland and later General Creighton Abrams, both of the Class of 1936.

On the US Air Force side, however, the Seventh Air Force had been reactivated in April 1966 as the umbrella for all US Air Force tactical units in Vietnam. The Commander of the Seventh Air Force was in turn designated as the deputy commander of MACV.

In 1968, when Creighton Abrams succeeded Westmoreland as commander of MACV, the man selected to command the Seventh Air Force was Black '41's George Scratchley Brown. Brown, who had flown with the Eighth Air Force in World War II from Ploesti to the crowded skies over Berlin, had gone on to serve as director of operations of the Fifth Air Force during the last year of the Korean War. When Chief of Staff John Paul McConnell picked him to go to Vietnam, Brown was finishing his second year as assistant to the chairman of the Joint Chiefs of Staff.

On August 1, 1968, George Brown pinned on his fourth star. It had been only six years since he had received the permanent rank of brigadier general, and he was now the fastest-rising member of the Class of 1941. When he arrived at Tan Son Nhut Air Base outside of Saigon, Brown found himself with a curious task. He was taking command of an air force that had downed 67 enemy aircraft in 14 months and had virtually eradicated the entire North Vietnamese Air Force. Meanwhile, the Rolling Thunder bombing campaign, undertaken in 1967 and managed for two years by the Seventh Air Force, had severely crippled North Vietnam's ability to move heavy equipment into the south. Rolling Thunder, despite its tactical success, was now a thorn in Lyndon Johnson's political side, and with the President's bombing halt, Brown found himself presiding over an air war that was now restricted in its geographic scope.

Nevertheless, Brown increased the number of missions flown by the Seventh Air Force—all of them now limited to South Vietnam—by a factor of 18 percent. By 1969, however, American troop strength had reached its peak, and with Richard Nixon's policy of returning combat duties to the South

Vietnamese forces, the number of Americans in Vietnam began to decline for the first time. From 1969 to 1970—when George Brown's tour as Seventh Air Force Commander ended—American troop strength was reduced from a high of 538,700 to a still-considerable 414,900; and the number of missions flown by the Seventh Air Force was down from 966,949 to 711,440.

The Painful Postscript

In the spring of 1972, with American troop strength down to 47,000, its lowest since 1964, the North Vietnamese launched another major offensive against South Vietnam. This time, an unrestricted American air offensive called Linebacker was unleashed against North Vietnam, and it halted the Communist advance. In December 1972, a second air offensive, called Linebacker II and involving B-52 bombers deployed against Hanoi itself, succeeded in compelling the North Vietnamese to sign a peace treaty. This treaty paved the way for the final withdrawal of American forces from Vietnam in March 1973.

The final chapter in the Vietnam War was played out in 1975, three decades after Ho Chi Minh declared Vietnam independent of the French and ten years after the first massive deployment of American forces. This time a North Vietnamese offensive, using conventional rather than guerrilla tactics, rolled through South Vietnam with minimal opposition. On April 30, 1975, Saigon fell and the NLF's Provisional Revolutionary Government was installed—albeit briefly—as a caretaker government until the North Vietnamese could reunify the country under their *own* control. The North Vietnamese had the unconditional surrender of the South.

The United States had entered into the Vietnam War with a noble purpose, but with no explicit idea of how to achieve it. It was a war in which politicians made military decisions and military men were blamed for political compromises. Rules, written by Washington and imposed on American forces, were ignored by everyone but American forces. Even the South Vietnamese people did not perceive the situation in the same way the American politicians claimed they did.

As Linton Boatwright put it, "In World War II, even the lowliest private understood why we were fighting. In the Vietnam War, neither the people of the United States, or South Vietnam understood why we were fighting. Practically all South Vietnamese senior officers were padding their pockets with various sorts of kickbacks and profits. Many owned bars, whorehouses, and drug outlets in the zones they commanded, and their principal efforts were given to the maintenance of income from these activities, as opposed to the prosecution of the war effort. The South Vietnamese people resented this exploitation and, in the main, did not support the war effort."

On the ground, Americans fought and died to capture objectives that were later casually abandoned. As a result, the proportion of South Vietnam controlled by Communist forces never changed appreciably in all the years that Americans were "in country." Enemy body counts became the only means of measuring progress, but they were almost never reliable because of the widespread practice of falsifying reports.

As Black '41's Andy Evans said many years later, "Hopefully, we learned not to get involved where we have no objective worth fighting for. I would like to think our senior military leaders would resign in protest in the future rather than accept such a 'no win' assignment again. The Vietnam War was a farce. There was no political, and therefore no military, objective. The military entered the war without the enthusiasm of World War II. We substituted 'body count' of enemy dead in place of territory taken and held, and/ or bombing the enemy's heartland into submission. Ultimately, it became a place for career advancement and where military accomplishments were distorted. It was our most wasteful war—the wrong war in the wrong place at the wrong time."

For the United States, the Vietnam War was a national trauma. It was said that the United States lost the war, but in fact, American troops had been withdrawn two years before North Vietnam's final victory. It was said that the United States could have decisively won the war at any time, and that is probably true. It was said that the Vietnam War was not worth the casualties expended, and that also is probably true.

In Washington, presidents Johnson and Nixon both became embarrassed by the war and moved the United States from a policy of containing Communism to one of disengaging from Vietnam without "losing face." Ultimately, the United States did extricate itself from Vietnam, but the long-term damage inflicted on American morale, both in and out of the armed services, was incalculable.

We'd lost more than "face."

17

Retirement and Reflection

Black '41's Last Decade in Uniform

The Vietnam War had been a deeply unsatisfying experience for military men, and when it was over, the blame for the disaster fell on the heads of military men. It was a painful burden that would color the rest of Black '41's career in uniform.

George Brown and George Pittman had remained good friends throughout their careers in the US Air Force. They saw each other when both were in Korea with the Far East Air Forces. Later, in the 1950s, when Pittman had needed C-124 transport planes to help airlift large construction equipment to the Thule Air Base construction site, Brown provided him with the assistance from his 62d Troop Carrier Group at McChord Air Force Base in Washington, D.C. The big aircraft operated from Westover Field, where Pittman was the deputy chief of staff for operations, and thus largely responsible for the Thule support airlift operation.

When he returned from his tour as commander of the Seventh Air Force in Vietnam in 1970, George Scratchley Brown began three years as the commanding general of the Air Force Systems Command at Andrews Air Force Base near Washington, D.C. He later moved to the Pentagon as secretary of the air staff, and it was in 1973, while he was working in this capacity, that president Nixon picked him to succeed John Dale Ryan as chief of staff of the US Air Force.

When the surprise attack against Israel by Egypt and Syria occurred on October 6, 1973, General Brown, only two months into his new job, immediately ordered supplies and equipment to be stockpiled at final embarkation ports, ready for rapid shipment to Israel. Once the presidential decision for resupply was reached, the timely arrival of US military equipment and supplies helped Israel mounted a devastating counteroffensive that won the Yom Kippur War.

As George Pittman remembered him, Brown never "threw his rank or position around," even when he was secretary to the air staff before he became chief of staff. "He was always the friendly classmate, cordial and gentle. As I recall, he loved polo, and I had an attraction to horses from my farm days with farm animals, so even though we were in different battalions at West Point, I saw him and we hailed each other frequently."

One of Brown's first priorities as chief of staff was upgrading the air force's strategic bomber fleet by replacing aging Boeing B-52s with Rockwell B-1s—swing-wing aircraft that could carry the latest electronic equipment and twice the payload of the B-52s and yet penetrate deeper into Soviet territory.

The lineage of heavy strategic bombers that began during World War II with the Boeing B-17 and the Convair B-24—the plane that George Brown flew to Ploesti and beyond—culminated in the Boeing B-52 Stratofortress, which became operational with the US Air Force's Strategic Air Command (SAC) in 1955. It was originally intended that the B-52 would be succeeded as a first line strategic bomber by a future generation of bombers in the early 1960s. Both the supersonic Convair B-58 and the remarkable, hypersonic North American B-70 appeared on the scene during the 1960s, but neither aircraft survived the budget-cutter's ax, and the Strategic Air Command therefore entered the 1970s with the B-52 still its leading aircraft, and no successor in sight. A new aircraft to augment and replace the B-52 eventually evolved into the North American Rockwell B-1 bomber. A preliminary development contract was issued on June 6, 1970, while George Brown was in Vietnam with the Seventh Air Force. When George Brown became the US Air Force's chief in 1973, he readily accepted the development of a new strategic bomber as one of his highest priorities.

In 1974, when Admiral Thomas Moorer stepped down as chairman of the Joint Chiefs of Staff, the president decided that George Brown was the best air force candidate to replace Moorer. He officially promoted him into the job of chairman on July 1. Brown was to be the first air force officer to hold the position since General Nathan Twining, who was chairman from 1957 to 1960.

The only man from Black '41 to wear four stars had gone all the way to the top, and his term would ultimately straddle three presidential administrations, Nixon's, Ford's, and Carter's. It was as chairman of the Joint Chiefs that Brown saw the B-1 prototype's first flight on December 23, 1974. During his tenure, Brown would also preside over American military involvement with crises in Cyprus, Lebanon, Cambodia, and Korea. He also worked with a staff that included his classmate Ed Rowny in the planning of the Strategic Arms Limitation Talks (SALT). The SALT talks constituted perhaps the most important issue for Brown as chairman of the Joint Chiefs of Staff. Two years before he became chairman, the United States had reached the SALT I agreement with the Soviet Union. Both countries agreed to confine deployment of antiballistic missiles and to limit the number of fixed land-based ICBMs and submarine ballistic missiles.

On May 12, 1975, two weeks after the fall of South Vietnam to the Communists, the radical Cambodian dictator Pol Pot sent a band of his Khmer Rouge guerrillas to seize the American civilian cargo ship SS *Mayaguez* on the high seas in the Gulf of Thailand. Under Brown's direction, US Marines retook the ship and stormed Koh Tang Island, where they believed the crew was being held, while jets from the carrier USS *Coral Sea* sank three Cambodian naval vessels and attacked a Cambodian air base. These actions brought release of the crew, and even though 18 US servicemen lost their lives, the operation found support in the United States and produced some measure of satisfaction after the humiliation of the fall of Saigon.

Just as Brown's time in Vietnam had been characterized by presidentially mandated restrictions on Seventh Air Force operations, so, too, did his tenure as JCS chairman coincide with President Jimmy Carter's determined effort to downsize the American military. By the time that Carter replaced Republican Gerald Ford in the White House in January 1977, it was clear that the three B-1 prototypes that Brown hoped would become the backbone of the Air Force had flown into a turbulent storm of controversy from which they would not emerge unscathed. Carter, a Democrat, had campaigned for the presidency on a platform of reducing defense spending, particularly spending on strategic weapons, and the B-1 program was a prime target. Despite George Brown's counsel that the B-52 needed to be replaced, Carter made good on his campaign promises and canceled the B-1 production program on June 30, 1977. The initial test phase, along with the completion of a fourth prototype, was allowed to continue, however. Meanwhile, although few people knew it at the time, Brown had overseen the genesis of the highly classified stealth technology that was to emerge from the veil of secrecy a decade later in the form of the Northrop B-2 stealth bomber and the extraordinary Lockheed F-117 stealth fighter bomber that delivered such dazzling performance in the 1991 Gulf War.

In 1978, George Brown ended his four years as Chairman of the Joint Chiefs of Staff, retiring from the US Air Force as a full general 37 years after that warm June day on the banks of the Hudson when he was commissioned as a second lieutenant. When he stepped down, the cancer that would claim him before the end of the year had already been diagnosed. On December 5, 1978, that cancer did what the German antiaircraft guns over Ploesti and the Messerschmitts over Berlin could not. George Scratchley Brown, the highest-ranking member of Black '41, joined the Long Gray Line.

Within two years, the Carter administration would be replaced by the more defense-conscious administration of Ronald Reagan, who was willing to address the need for a replacement for the aging B-52 fleet. On October 2, 1981, President Reagan officially asked the North American Aircraft component of the Rockwell Corporation to develop a new version of the B-1 that would retain the original's general appearance while incorporating the state-of-the-art avionics and stealth technology that had evolved during the ten years since the original B-1 design. The new aircraft would be designated B-1B. The first of a hundred flew in 1984 and became operational with the Strategic Air Command in 1986, a testament to George Brown's persistence. Three years later, both the B-2 and F-117 were also flying.

Ed Rowny's star had also ascended. He served under George Brown in the 1970s, before Brown became chairman of the Joint Chiefs of Staff, but unlike Brown, Rowny was best remembered for *dis*armament. He came out of World War II favoring retention of America's armed might, but four decades later he was a party to the negotiations that resulted in some of the deepest cuts in strategic weapons in world history.

When he left the 2d Infantry Division in Korea in 1952, Ed Rowny was assigned to the staff and faculty of the infantry school at Fort Benning for three years, and in 1955 he went to the Supreme Headquarters for Allied Powers in Europe (SHAPE) at Mons in Belgium for two years. After a stint with the National War College, Rowny was with the Army Concept Team in Vietnam from 1963 until 1966. For the next two years, he was deputy chief of staff for logistics for the US Army's European Command, after which he returned to Washington, D.C., as deputy chief of research and development for the Department of the Army.

He went back to Korea as the commanding general for I Corps until 1971. After Korea, Rowny was sent back to Europe to serve as deputy chairman of the NATO Military Committee, a natural stepping-stone to the job for which he is best remembered.

In Europe, Rowny had time to reflect on the situation that had made NATO necessary and that would make arms control such an important issue. "We should have paid more attention to the political aftermath of World War II," Rowry wrote in 1990. "Particularly, we should have paid more attention to 'the soft underbelly' of Europe. We should have moved through and beyond Berlin, and not demobilized so rapidly after the war."

The first round of negotiations for the Strategic Arms Limitation Treaty resulted in SALT I, which was signed in Moscow on May 26, 1972. Because SALT I was an "interim" agreement designed to remain in force for only five years, another round of negotiations got underway in November 1972. Its object was to forge a permanent agreement, which would ultimately be known as SALT II. Senator Henry Jackson of the state of Washington insisted on a strong military presence on the negotiating team. Because of his background with SHAPE and the National War College, plus the fact that he was chairman of the NATO Military Committee at the time, General Ed Rowny was a natural choice to represent the Joint Chiefs of Staff for the second round of SALT negotiations in 1973.

Initial stumbling blocks in the SALT II negotiations were overcome in November 1974 at the Vladivostok summit between President Ford and General Secretary Brezhnev, and for the next five years, Rowny and his colleagues met almost continuously with their Soviet counterparts in Geneva. The completed treaty was signed by Brezhnev and President Carter in Vienna on June 18, 1979. Having finished his work on SALT II, Rowny retired from the US Army as a lieutenant general—one of a handful of men in Black '41 to wear three stars—after 38 years of service. In the meantime, however, he had earned a Ph.D. in International Studies from American University (1977). In 1979 and 1980, he was a fellow at the Woodrow Wilson Center of the Smithsonian Institution.

The story of SALT II and of Ed Rowny's involvement with arms control was, however, far from over. After the Soviet Union invaded Afghanistan on Christmas Eve in 1979, the US Senate decided not to ratify SALT II, so it was back to the drawing board. While both sides agreed to abide by the provisions of the as-yet unratified SALT II agreement, Afghanistan put Soviet-American dialogue on the subject of arms control on the back burner.

During the presidential campaign in the autumn of 1980, Ronald Reagan invited Rowny and then-chief negotiator Paul Nitze to a private meeting at the Mayflower Hotel in Washington, D.C., which laid the foundation for the round of arms control negotiations that would begin when Reagan took office. The negotiations that began in November 1981 were directed at the serious problem of Intermediate Nuclear Forces (INF) in Europe as well as the Strategic Arms Reduction Talks (START). In 1981, Ed Rowny became chief US negotiator for START, with the rank of ambassador, and head of the US delegation to the Strategic Arms Limitation negotiations. He had already become one of President Reagan's chief advisors on the subject of arms control.

Negotiations continued for two years but collapsed in November 1983, not to resume until January 1985, when Secretary of State George Shultz and Soviet Foreign Minister Andrei Gromyko agreed to resume parallel INF and START negotiations. This time, the talks moved ahead unabated, with Rowny as special advisor to the president and secretary of state for arms control matters. Soviet President Gorbachev was, by this time, motivated by the knowledge that the precariously balanced Soviet economy was teetering on the edge of a calamitous implosion. It had been grossly top-heavy with military spending, and a major arms limitation was necessary to divert resources and salvage the Soviet economy from impending disaster.

On December 8, 1987, the treaty resulting from the INF talks—known as the Treaty between the USA and the USSR on the Elimination of their Intermediate-Range and Starter-Range Missiles—was signed by Presidents

Gorbachev and Reagan at the White House in Washington, D.C. Ambassador Rowny was present. Two years later, when George Bush was inaugurated, he asked Ed Rowny to stay on in his role as special advisor.

Two days before leaving office, on January 18, 1989, President Reagan awarded Ambassador Rowny the Presidential Citizens Medal. The citation read in part: "Edward L. Rowny has been one of the principal architects of American policy of peace through strength. As an arms negotiator and as a presidential advisor, he has served mightily, courageously, and nobly in the cause of peace and freedom."

After 11 years of work on the civilian side of arms control, in June 1990 Ed Rowny resigned as special advisor to the president and secretary of state for arms control matters. He became distinguished visiting professor of international relations at George Washington University and senior associate at the Center for Strategic and International Studies.

Throughout the years, wherever Ed Rowny went—from West Point to Korea, from Washington to Moscow—he was remembered most often for, of all things, his harmonica playing. As a boy in Baltimore, Rowny had been a member of the Baltimore Harmonica Band along with his good friend Larry Adler, who became the first and probably the greatest virtuoso of classical harmonica music. Even as late as 1985, Rowny and Adler performed together in a benefit concert in Baltimore. As chief US arms control negotiator, when "the ice got particularly thick," Rowny would occasionally whip out his Marine Band Harmonica and entertain the Soviet negotiators with Russian folk songs and even *Miy Kommunisti* ("We Are the Communists"), a marching song.

Pete Crow had received his third star when he became comptroller of the US Air Force in 1968. "This was a significant accomplishment, but nothing spectacular," he said later. "However, by the time I left the job, the Air Force was generally accorded high marks for financial management."

Crow had been comptroller of the Central Air Defense Force between 1955 and 1957, and he had attempted, with no success, to eliminate the requirement for bonding finance officers, arguing that the US Air Force was large enough to be self-bonding. When he became comptroller, he took the proposal to the Secretary of Defense. Secretary Melvin Laird's office logically took the position that if it was good for the US Air Force, it was good for all the services. The Bureau of the Budget, in turn, decided that if it were good for Department of Defense, it was good for all of government. So, 15 years after the Pete Crow's original suggestion, what was once the unthinkable notion of an upstart became national policy.

In July 1973, Crow was awarded a Distinguished Service Medal for duty as comptroller and was assigned as assistant vice-chief of staff, with front-row seats on the October 1973 Israeli-Egyptian War.

In September 1974, Pete Crow joined NASA as associate deputy administrator, coordinating various NASA programs with the Department of Defense and other government agencies. He helped to coordinate the historic Apollo-Soyuz project in July 1975, a high-water mark in the era of *detente*, when a three-man American Apollo spacecraft docked in orbit with the Soviet Soyuz spacecraft and its two cosmonauts. As others in NASA, he sweated out the coming of age of the space shuttle. Upon retiring from NASA, he was awarded its Distinguished Service Medal.

Civilian Life

By 1962, members of the Class of 1941 had passed the 20th anniversary of their graduation, and during those 20 years, 60 of the men had retired. Nevertheless, over half of the original 424 graduates were still on active duty in 1964. In 1967, 34 men retired, followed by 19 in 1968, and 36 in 1969. Only three retired in 1970, but 1971, the long-awaited 30-year mark, saw the retirement of 52 classmates, leaving only 20 on active duty. The last three to retire were George Scratchley Brown (1978), Ed Rowny (1979), and Charles Schilling, who had returned to West Point to teach in 1956, ultimately became head of the Engineering Department, and retired as a brigadier general in 1980. The boys in gray who had become the men in khaki were now all civilians.

For most, retirement would be a quieter time, but for others it would mean new careers that would lead off in new directions. After a career in the US Air Force and two years as commandant of cadets at the new Air Force Academy at Colorado Springs, Bill Seawell retired as a brigadier general in 1963, only to start a new career in commercial aviation that led to his serving as chairman of the board and chief executive officer of Pan American World Airways from 1972 to 1982.

Pete Crow retired to San Antonio in 1978 with his wife Tulah Dance Crow, the striking blonde with the big red horse who had been 16 when Pete fell in love with her in 1941. Crow continued to be involved with NASA and with the Air Force space activities, served on various boards, brokered real estate, and consulted. He said later that his greatest achievement had been to save a daycare center for the mentally handicapped from going bankrupt.

The day after he retired in 1967, Joe Gurfein took a job as an administrator at the University of Maryland in the Department of Mathematics. Less than a year later, when the university opened its Federal City College in the District of Columbia, Gurfein went there as a director of planning and was immediately promoted to dean of administration, a post that he held for the next three years. In July 1972, he went to George Mason University, first as the director of planning during campus construction and later as an associate professor.

But retirement was not always a pleasant experience. George Pittman pointed out that the USMA "taught us duty, honor, country, and to accept people at their word. In the pre-World War II years, there was little paperwork and few signatures were required, since there was no place for such on a battlefield. After a career in the military, many military retirees were taken by confidence men and entrepreneurs interested only in money, with no regard to how or from whence it came. The military has a share, but I believe a far smaller percentage of its officers are so involved. In civilian life, too many career military people with three decades sheltered within the discipline of military life have been faced with denials of promises, unless they had it in writing with signatures. After retirement, I entered civil aviation as the assistant manager of a small civilian airport. Twenty-one years later [in 1988] I retired again."

Before he retired in 1965, Lynn Lee spent six years as a witness to congressional committees for the Corps of Engineers budgets, two of those years presenting classified—mostly Nike air defense missile—budgets, and a year and a half promoting the civil works program for 13 Western states, or, as his wife put it, "defending the pork barrel." When he retired, Lee went into the civilian construction business, but, he said, "I felt rather naive about the civilian business world."

Paul Skowronek retired in 1971, finishing his 30 years in the Army as senior military attaché in Bulgaria, another one in his series of what he likes to call "mysterious, exotic assignments." He was immediately accepted at Oxford for a postgraduate diploma program in Slavonic Studies. Diploma from Oxford in hand, he spent three more years at the University of Colorado earning his Ph.D. in history before he was off once more for "mysterious, exotic" activities in Algeria, Iran, the Philippines, and Guatemala.

Duty, Honor, West Point

The character of reunions, like that of the men who attend them, changes over the years. In the 1940s, the topic of conversation was the war, the pride of having played a role in the century's biggest event, the relief of those who survived, and a sadness for those who did not. In the 1950s and early 1960s, talk always turned to comparison of careers and the natural boasts of how important each man's role was in safeguarding the peace and prosperity of the free world. By 1967, half of the surviving members of Black '41 had retired. Over the next five years, culminating in their 30th anniversary in 1971, nearly 150 men retired. From that point on, the number-one topic of conversation was not the war, the degradation of the US Army or Air Force, nor the specifics of careers. Rather, the topic was children—notably, those with military careers—and, of course, grandchildren.

In 1991, as the Class of 1941 reached the half-century mark, well over half of its members were still alive, most were still active, and most had shared the details of their grandchildren—and great-grandchildren—with their classmates. Talk now turned to reflection on the army and on the world as it had evolved during their lifetimes and speculation on how it would evolve in the future.

The members of Black '41 left the USMA, but the USMA has never, and will never, leave them. Some fifty years after he was called "the little man with the big name," Mike Greene ran into a fellow from the Class of 1939, whom he hadn't seen since his days at West Point. Mike said, "Hello. I'm Mike Greene."

"Michael Joseph Lenihan Greene?" was the reply.

Mike Greene's relationship with the USMA was initially limited to class reunions, football games, and association of graduates functions. Then his two daughters married West Pointers (Classes of 1968 and 1972) who were later on assignment with the faculty. As a result, Mike and his wife Eileen were frequent visitors at West Point for about seven years. From 1975 to 1976, and again from 1985 to the present, he was class president and has had frequent contact with the Academy.

In looking at the USMA today, Harry Ellis grumbles that "I am not in much agreement with the way it has changed. The young men and women there today are looking for an education. We are training managers, not leaders. I went there to be a soldier. The education was secondary in my mind."

"West Point has deteriorated because the Academy was deliberately sabotaged," George Pickett growled. "We have a bunch of undisciplined kids running around unable to drill. I watched them march, and my God, it's a disgrace! They bump along. The lines are crooked. They look like Jack and Jill. They don't 'size' the corps of cadets anymore; they just have them all bunched up. If they were 'sized' as we were, most of the women would be isolated. Consequently, since they can't size the corps, they can't march. You cannot have a parade where you have a cadet who is five feet four inches tall next to a guy who is six feet three inches."

Pete Tanous, who was at the USMA from 1956 to 1959 as deputy chief of staff for the Logistics Office, disagrees. "USMA had changed and become more oriented to an academic life, resembling an Ivy League college. This was due to the restructuring of the academic curriculum, with less emphasis on military training. This was good, but the emphasis on 'Duty, Honor, Country' was still paramount, as it should be."

As Walter Mather saw it, over time the changes at West Point have not all been for the better. "Regardless of the charges made against the prewar system, West Point had produced the commanders and staff officers who had led that mighty World War II Army, as it had produced the key men in all previous wars of this nation. West Point has been forced to bow to excessive liberal pressure in several respects, which has weakened the fabric. First was the elimination of compulsory chapel, a very grave error. Fighting men need the solace of a sense of the Almighty, regardless of what we may title Him. Fighting men are not atheists. Second was the proliferation of required courses in the curriculum, to the point that the excessive academic load is producing graduates who know a little about many subjects, not much about any one discipline. As a result, they lack some of the basics which the previous system built into graduates. Note, for example, the decreasing percentage of West Pointers selected for general officer. Third, the admission of women, while denying their availability as combat leaders, makes a mockery of the mission of the Academy. How many possible Eisenhowers or Pattons or such did not become cadets to make possible the entry of a woman who, if she remains in service, will never be a combat leader?"

Ben Spiller could not have disagreed more with Pickett and Mather. When his daughter was 16 and a junior in high school, she wrote to the director of admissions at West Point—without her father's knowledge—and asked why women were not allowed at West Point. The answer she received— and she remembers it today, at age 32, was: "We cannot allow women to occupy spaces which could be filled by fully qualified men."

Spiller later wrote the book *Indomitable: The Story of the First Women at West Point*. To help him, he enlisted the support of two women, Captain Kathy Silvia (Class of 1980) and Captain Barbara Grofic (Class of 1982) as well as a male member of the Class of 1980, Captain Keith Emberton. Published in 1989, *Indomitable* tells a great deal about how West Point has changed since the Class of 1941 was there, especially about the most controversial changes.

"We were trained to be officers and gentlemen—not a bad thing," Spiller wrote. "I cherish that training. But today, they are trained in communication, the social sciences—why people think and function as they do—and a myriad of other things pertinent and related to real life. I studied some of the same subjects as had my grandfather, Class of 1882, and my great-grandfather, Class of 1833. On the other hand, West Point's top students were always as good or better than MIT and Harvard's top students, else why would we rival them in Rhodes Scholars and other achievements?"

Whatever West Point's impact on the Class of 1941, few of the men will deny today that it gave them a sense of duty and responsibility that they just wouldn't have found anywhere else.

Tuck Brown retired from the Army in 1964 and became senior advisor to the Remote Area Conflict Information Center program in Columbus, Ohio, an Advanced Research Projects Agency program at the Battelle Laboratories. Jess Unger, who had been in Brown's Military History class at West Point throughout their First Class Year and who had served with the Sixth Army during World War II and later as associate dean at West Point between 1960 and 1964, retired in 1965. Brown had always enjoyed Unger's sense of humor, so when he saw that Unger had retired, he asked him to come to Battelle to help run the program. They worked together for the next three years and got a great deal accomplished.

Unfortunately, Jess's health began going downhill. First he transferred to the Battelle Seattle Research Center to be near some land that he owned in the Northwest. Then he transferred to the Battelle Washington office to be near Walter Reed Hospital. One day shortly before his death in March 1975, his daughter Rosa Lee was driving him to Walter Reed and asked him, "Under what precepts would you like me to raise your grandchildren?"

Jess thought a while, then said, "What's wrong with 'Duty, Honor, Country?' "

As Tuck Brown recalled later, "I was lucky to have known the gentleman."

Epilogue:
The Brothers

It was a very pleasant day, a good spring day. June is still springtime in the Hudson River Valley, but just barely. The trees have leafed out and the hillsides are green, but the nights are still cool, the days still just warm—not sticky hot.

Five decades had now passed. It was June 1991, and it was the autumn of the century. More than 300 members of the Class of 1941 were still alive, and many had come back to this valley to mark the passing of those five decades. The classmates were assembling on the Plain for a march to the Sylvanus Thayer Monument for the alumni memorial service when several men noticed the sound of an aircraft approach. Shading their eyes against the morning sun, they craned their necks skyward as a Cessna 182 crossed the Hudson at 3,500 feet.

As they watched, a man leapt from the Cessna. A parachute blossomed. Who could it be?

A murmur rippled through the crowd. Was it a prankster from the Class of 1991 poking fun at the Class of 1941? Was it a prankster from Annapolis poking fun at West Point? Could it be a terrorist? The Gulf War had, after all, ended only three months before. The military police providing security for the event bristled as the man in the gold jumpsuit descended directly onto the Plain.

All eyes were on the intruder as he pulled off his helmet and goggles. Suddenly, like a flash of sunlight upon the breast of newly fallen snow, they could see the silver hair and broad grin of Paul Skowronek, from the class they called Black '41.

"Some people will call it a typical Black '41 disregard for conventional procedure," he said a few weeks later. "I did not ask for permission, but because it was successful—even though highly unprecedented—there was no official criticism."

"Us Black '41ers are always going to pull *some* stunt," laughed Larry Greene.

The Class of 1941 had always been a class apart, an unusual mix of personalities that earned the mysterious appelation of Black '41. Its members were special, and they were unique. They were the class that graduated at the moment of their nation's transition from peace to the global war that would forever change the role of the United States in the world.

They graduated in the springtime of the century and were warned of passing of that springtime by the very man who would soon choreograph their nation's transformation. Henry Lewis Stimson, who had sent the West Point Class of 1941 forth from this Plain so long ago was gone, but not forgotten. His spirit lived on, not only in the memory of the men who had gathered at West Point, but also in the memory of the man who sat in the White House to direct a war that portended a change in the global world order unlike any since the war to which Henry Stimson sent the Class of 1941. When George Bush graduated from Andover in 1940, Henry Stimson had been the speaker. His words echoed in the president's ears as he prepared for war in 1991.

The 50 years that began with America's transition from isolated innocence to global power were born of the blood and sweat of the men who came of age in 1941. Nowhere in the United States had there been a group of young men that typified the dawn of an American half-century more than those who had graduated on June 11, 1941. They had helped to win mankind's most deadly war and had helped to keep the peace through the fearsome years of the Cold War. Now they had come back to reflect on those years, their battles, and their wars.

They came together at dawn on a June day in 1991 with that assured familiarity that is part of the special bond between brothers that those who have not had a brother cannot know.

Black '41's class president at the time of its 50th anniversary, Michael Joseph Lenihan Greene, and his older brother, Lawrence Vivans Greene, do not look like brothers. They didn't then and they don't now. Nor do they look much like their *other* brothers, the rest of the Class of '41. When they first arrived at West Point on that sweltering day in 1937, the Greenes were just two brothers amid hundreds of strangers, a little island in the midst of a turbulent sea of humanity. Four years later, they stood here on the green fields below the cliffs of the Hudson as part of a brotherhood that then numbered 424, and half a century later still numbers nearly 350.

In 1937, the United States Military Academy had taken a group of people from vastly diverse backgrounds and—for four years—had compelled them to live together in these gray buildings, forced them to grow into manhood together. It was an experience, and an education, they would never forget.

Each man left the Academy knowing that the man next to him could be depended upon to respond in a certain way that no outsider would ever fully understand. Each one of them knew that if he had to put his life on the line, he could depend on his brothers, and his brothers could depend on him.

It was a bond that for each of them would never be broken.

Appendix A

Directory of the USMA Class of 1941

Adams, Harwell Leon Cadet Company H; resigned in 1954

Adams, Howard Frank Cadet Company I; killed in action on February 26, 1943

Adams, Jonathan Edwards, Jr. Cadet Company G; retired as lieutenant colonel, US Army, in 1962

Adjemian, George Roopen Cadet Company G; retired as colonel, US Army, in 1967

Ahern, Joseph Patrick Cadet Company I; retired as colonel, US Air Force, in 1962

Aldridge, Richards Abner Cadet Company I; retired as lieutenant colonel, US Air Force, in 1961

Aliotta, Michael Frank Cadet Company H; retired as lieutenant colonel, US Army, in 1969

Anderson, Windsor Temple Cadet Company C; retired as lieutenant colonel, US Air Force, in 1961

Andrews, George Lincoln Cadet Company C; died in 1945

Andrus, Burton Curtis, Jr. Cadet Company M; retired as colonel, US Air Force, in 1969

Armstrong, Clare Hibbs, Jr. Cadet Company E; retired as lieutenant colonel, US Army, in 1969

Ascani, Fred John Cadet Company C; retired as major general, US Air Force, in 1973

Atkinson, John Earl Cadet Company K; retired as colonel, US Air Force, in 1967

Atteberry, Roy Leighton, Jr.   Cadet Company B; retired as brigadier general, US Army, in 1971

Austin, Emory Ashel, Jr.   Cadet Company E; killed in action on May 15, 1943

Avery, Hamilton King   Cadet Company D; retired as lieutenant colonel, US Air Force, in 1961

Bagshaw, Harry Kendall   Cadet Company E; resigned in 1954

Bailey, Leslie Wilmer   Cadet Company C; retired as lieutenant colonel, US Army, in 1969

Baker, Frederick John   Cadet Company G; retired as colonel, US Air Force, in 1965

Ball, Clinton Field   Cadet Company L; retired as colonel, US Air Force, in 1961

Barnett, Cargill Massenburg   Cadet Company G; died in 1942

Barney, John Coles, Jr.   Cadet Company H; retired as colonel, US Army, in 1971

Barrow, Sam Hardy   Cadet Company B; retired as colonel, US Army, in 1969

Bentley, Jack Leith   Cadet Company H; retired as colonel, US Air Force, 1961

Berger, Leon Herman   Cadet Company E; retired as colonel, US Air Force, in 1967

Besancon, Harry Charles   Cadet Company B; retired as colonel, US Army, in 1969

Betts, Curtis Francis   Cadet Company H; retired as lieutenant colonel, US Air Force, in 1962

Birdseye, Mortimer Buell, Jr.   Cadet Company L; retired as lieutenant colonel, US Army, in 1960

Blalock, Hill   Cadet Company C; resigned in 1947

Blanchard, Henry Nathan, Jr.   Cadet Company E; killed in action on June 17, 1944

Boatwright, Linton Sinclair   Cadet Company D; retired as major general, US Army, in 1972

Bodson, Henry Richard   Cadet Company F; retired as colonel, US Army, in 1969

Boggs, Edgar Clayton   Cadet Company I; killed in action on February 5, 1945

Borman, Robert Channing   Cadet Company K; retired as colonel, US Army, in 1969

Boswell, Henry, Jr.   Cadet Company A; retired as colonel, US Army, in 1970

Brier, William Wallace, Jr. Cadet Company H; retired as colonel, US Air Force, in 1969

Briggs, Leon Arthur Cadet Company L; retired as major, US Air Force, in 1961

Brinson, Robert Hendrick, Jr. Cadet Company L; retired as captain, US Army, in 1944

Brooks, John Adams, III Cadet Company H; retired as brigadier general, US Air Force, in 1969

Brown, Earl Vincent Cadet Company I; retired as colonel, US Army, in 1967

Brown, Earle Wayne, II Cadet Company E; resigned in 1947

Brown, Edwin Watson Cadet Company M; retired as colonel, US Air Force, in 1969

Brown, George Scratchley Cadet Company L; retired as general, US Air Force, in 1978

Brown, Horace Maynard, Jr. Cadet Company C; retired as colonel, US Army, in 1970

Brown, Joseph Tuck Cadet Company A; retired as colonel; US Army, in 1964

Brown, Robert Duncan, Jr. Cadet Company A; retired as colonel, US Army, in 1963

Buchanan, Earl K. Cadet Company K; retired as colonel, US Army, in 1971

Burtchaell, John William Cadet Company L; retired as colonel, US Army, in 1966

Busbee, Charles Manly, Jr. Cadet Company H; retired as colonel, US Army, in 1969

Buttery, Edwin Boynton Cadet Company F; retired as colonel, US Army, in 1969

Callaway, John Wilson Cadet Company C: retired as colonel, US Army, in 1971

Camp, John Holmes Cadet Company G: retired as lieutenant colonel, US Army, in 1969

Campana, Victor Woodrow Cadet Company F; retired as lieutenant colonel, US Army, in 1961

Campbell, Raymond Potter, Jr. Cadet Company L; retired as colonel, US Army, in 1968

Canella, Charles Joseph Cadet Company C; retired as colonel, US Army, in 1968

Cannon, Charles Arthur, Jr. Cadet Company G; retired as colonel, US Army, in 1971

Carlson, Vincent Paul Cadet Company B; retired as colonel, US Army, in 1964

Carman, Charles MacArthur, Jr.   Cadet Company I; resigned in 1948

Carney, Marshall Warren   Cadet Company M; died in 1943

Carroll, James Henry   Cadet Company K; retired as lieutenant colonel, US Army, in 1961

Cator, Bruce Campbell   Cadet Company G; retired as lieutenant colonel, US Army, in 1959

Celmer, Theodore Bernarr   Cadet Company M; retired as major, US Army, in 1961

Chapman, Curtis Wheaton, Jr.   Cadet Company L; retired as major general, US Army, in 1975

Chavez, Atanacio Torres   Cadet Company G; retired as lieutenant colonel, Philippine Army

Cheaney, Ira Boswell, Jr.   Cadet Company A; killed in action on January 30, 1942

Christensen, John Moore, Jr.   Cadet Company I; retired as colonel, US Army, in 1964

Clapp, Wadsworth Paul "Bill"   Cadet Company G; killed in action on February 22, 1945

Clark, Howard Warren   Cadet Company I; retired as colonel, US Army, in 1968

Clark, John Calvin   Cadet Company I; resigned in 1946

Clark, Robert Evarts   Cadet Company A; retired as colonel, US Army, in 1967

Cleary, Thomas James, Jr.   Cadet Company A; retired as colonel, US Army, in 1968

Clendening, Herbert Campbell   Cadet Company M; resigned in 1954

Clifford, William Eugene   Cadet Company D; retired as colonel, US Army, in 1971

Clinton, Roy J.   Cadet Company C; retired as colonel, US Army, in 1968

Coakley, Robert John   Cadet Company D; retired as colonel, US Army, in 1971

Cochran, Harrington Willson, Jr.   Cadet Company L; retired as colonel, US Army, in 1971

Cochran, Wharton Clayton   Cadet Company C; retired as colonel, US Air Force, 1962

Cofer, Floyd Sturdevan, Jr.   Cadet Company M; retired as colonel, US Air Force, in 1961

Coker, Norman Kitchner   Cadet Company M; resigned in 1946

Coker, Sears Yates   Cadet Company L; retired as colonel, US Army, in 1967

Cole, Clifford Elbert   Cadet Company D; retired as colonel, US Air Force, in 1969

Colleran, Robert James   Cadet Company L; resigned in 1947

Collins, Leroy Pierce, Jr.   Cadet Company D; retired as colonel, US Army, in 1972

Collison, Tom Depher   Cadet Company M; retired as lieutenant colonel, US Army, in 1961

Connally, Lanham Carmel   Cadet Company K; killed in action on July 4, 1945

Cooper, David   Cadet Company M; retired as colonel, US Army, in 1968

Cooper, George William   Cadet Company C; retired as lieutenant colonel, US Army, in 1955

Cooper, Robert Lawrence   Cadet Company H; discharged upon graduation

Corbin, Thomas Goldsborough   Cadet Company B; retired as major general, US Air Force, in 1974

Couch, Richard Waggener   Cadet Company A; resigned in 1954

Cox, James Issac   Cadet Company M; retired as colonel, US Air Force, in 1961

Cramer, Thomas Rees   Cadet Company C; killed in action on July 2, 1943

Crow, Duward Lowery "Pete"   Cadet Company A; retired as lieutenant general, US Air Force, in 1974

Cummings, Robert Lloyd   Cadet Company B; killed in action on November 30, 1944

Cummins, William Kneedler   Cadet Company G; retired as lieutenant colonel, US Air Force, in 1958

Curley, Thomas Winston   Cadet Company A; retired as lieutenant colonel, US Army, in 1962

Curtis, Gwynne Sutherland, Jr.   Cadet Company C; retired as colonel, US Air Force, in 1961

Dalby, Albert Samuel   Cadet Company K; retired as lieutenant colonel, US Army, in 1962

Danforth, Carroll Freemont   Cadet Company B; retired as lieutenant colonel, US Army, in 1962

Day, Paul Chester   Cadet Company B; retired as colonel, US Army, in 1971

Deane, John Breed   Cadet Company M; retired as colonel, US Army, in 1965

de Jonckheere, Eric Thomas   Cadet Company M; retired as colonel, US Air Force, in 1969

Delaney, Richard   Cadet Company F; retired as lieutenant colonel, US Army, in 1962

deSaussure, Edward Harleston, Jr.   Cadet Company C; retired as major general, US Army, in 1972

deSilva, Peer   Cadet Company A; resigned in 1953

D'Esposito, John Vincent   Cadet Company D; retired as lieutenant colonel, US Army, in 1961

Dessert, Kenneth O'Reilly   Cadet Company F; retired as colonel, US Air Force, in 1966

Detwiler, Robert Putnam   Cadet Company K; retired as colonel, US Army, in 1969

Deyo, Truman Eugene   Cadet Company I; retired as lieutenant colonel, US Army, in 1962

Dienelt, James Henderson   Cadet Company E; killed in action on June 11, 1943

Dillard, Junius Edward   Cadet Company I; resigned in 1947

Dilts, Peter Kirkbride   Cadet Company A; retired as major, US Army, in 1962

Dixon, Robert Toombs   Cadet Company G; retired as colonel, US Army, in 1969

Driscoll, Donald Lyons   Cadet Company I; killed in action in December 1, 1950

Drum, Heister Hower   Cadet Company F; retired as major, US Army, in 1947

Due, Kenneth Oswalt   Cadet Company B; retired as captain, US Army, in 1945

Duke, Paul Demetrius   Cadet Company D; killed in action on August 4, 1944

Durr, Ernest, Jr.   Cadet Company M; killed in action on April 26, 1945

Easton, John Jay   Cadet Company M; retired as colonel, US Air Force, in 1961

Eaton, Dan Holton   Cadet Company H; died in 1941

Edger, Robert Huff   Cadet Company F; retired as colonel, US Army, in 1971

Edgerton, Bruce Wilds P.   Cadet Company K; resigned in 1949

Elder, Clarence Lewis   Cadet Company L; retired as colonel, US Air Force, in 1965

Ellis, Harry Howard   Cadet Company F; retired as colonel, US Army, in 1966

Ellis, Harry Van Horn, Jr.   Cadet Company D; retired as lieutenant colonel, US Army, in 1969

Elsberry, Robert Vaughn   Cadet Company D; retired as lieutenant colonel, US Army, in 1961

Evans, Andrew Julius, Jr.   Cadet Company F; retired as major general, US Air Force, in 1974

Faulkner, Lyman Saunders   Cadet Company A; retired as colonel, US Army, in 1963

Felchlin, Howard Lawrence   Cadet Company F; retired as colonel, US Army, in 1963

Fisher, Thomas Legate, II Cadet Company L; retired as colonel, US Air Force, in 1967

Fitzpatrick, Francis Cornelius Cadet Company M; retired as colonel, US Army, in 1971

Flanders, Charles Llewellyn, Jr. Cadet Company E; retired as colonel, US Army, in 1967

Fletcher, Charles William Cadet Company E; retired as brigadier general, US Army, in 1971

Forsyth, James Paul Cadet Company I; retired as lieutenant colonel, US Army, in 1968

Foster, Horace Grattan "Race", Jr. Cadet Company M; killed in action on August 24, 1943

Foster, Hugh Franklin, Jr. Cadet Company A; retired as major general, US Army, in 1975

Fowler, James Daniel Cadet Company E; retired as colonel, US Army, in 1967

Franklin, Elkin Leland Cadet Company D; killed in action on April 20, 1944

Frawley, Herbert Welcome, Jr. Cadet Company D; died in 1942

Freese, Ralph Earl Cadet Company F; retired as lieutenant colonel, US Air Force, in 1956

Gardner, William Gardner Cadet Company F; killed in action on June 6, 1944

Garrett, Robert Willoughby Cadet Company A; retired as colonel, US Army, in 1967

Gauvreau, David Gabriel Cadet Company K; retired as colonel, US Army, in 1971

Geldermann, Edward Joseph Cadet Company I; retired as colonel, US Army, in 1961

Gerace, Felix John Cadet Company L; retired as major general, US Army, in 1970

Gerig, Frank Austin, Jr. Cadet Company E; retired as lieutenant colonel, US Army, in 1961

Gilbert, Willard Russell Cadet Company G; retired as colonel, US Air Force, in 1966

Gillis, William Graham, Jr. Cadet Company M; killed in action on October 1, 1944

Gleason, William Thomas Cadet Company D; retired as brigadier general, US Army, in 1971

Goddard, Guy Harold Cadet Company D; retired as major general, US Air Force, in 1971

Goodell, Howard Clarke Cadet Company B; retired as lieutenant colonel, US Air Force, in 1962

Gould, Gordon Thomas, Jr. Cadet Company I; retired as lieutenant general, US Air Force, in 1974

Grace, Denis Blundell Cadet Company M; retired as colonel, US Army, in 1971

Graham, James Wetherby Cadet Company H; retired as colonel, US Army, in 1967

Gray, Paul, Jr. Cadet Company M; retired as colonel, US Army, in 1967

Green, James Oscar, III Cadet Company F; retired as colonel, US Army, in 1965

Greene, Lawrence Vivans Cadet Company A; retired as brigadier general, US Army, in 1971

Greene, Michael Joseph Lenihan Cadet Company G; retired as brigadier general, US Army, in 1971

Gribble, William Charles, Jr. Cadet Company K; retired as lieutenant general, US Army, in 1976

Grygiel, Joseph Stanley Cadet Company K; retired as colonel, US Army, in 1971

Gurfein, Joseph Ingram Cadet Company B; retired as colonel, US Army, in 1967

Gurnee, William Harold, Jr. Cadet Company B; retired as colonel, US Army, in 1963

Hall, Max Woodrow Cadet Company G; retired as lieutenant colonel, US Air Force, in 1967

Hampton, Fred Milas Cadet Company I; died in 1942

Harding, Edwin Forrest, Jr. Cadet Company G; retired as colonel, US Air Force, in 1955

Harper, Matthew Gordon, Jr. Cadet Company F; resigned in 1954

Harris, Charles Knighton Cadet Company E; retired as lieutenant colonel, US Army, in 1967

Harris, John Frederick Cadet Company A; retired as colonel, US Air Force, in 1967

Harrison, Matthew Clarence Cadet Company G; retired as colonel, US Army, in 1964

Harvey, Harry Canavan Cadet Company M; retired as colonel, US Air Force, in 1963

Hatfield, Mills Carson Cadet Company B; retired as colonel, US Army, in 1964

Hauser, Auburon Paul Cadet Company D; retired as colonel, US Army, in 1969

Hauser, John Nathaniel, Jr. Cadet Company F; died in 1944

Hayduk, Alfred George Cadet Company D; retired as colonel, US Air Force, in 1967

Healy, James Gerard Cadet Company H; retired as colonel, US Army, in 1971

Heaton, Donald Haynes Cadet Company F; retired as colonel, US Air Force, in 1969

Hendrickson, Roy George Cadet Company A; retired as major, US Army, in 1966

Henschke, John Miles Cadet Company K; retired as colonel, US Air Force, in 1960

Henzl, Leo Charles Cadet Company K; resigned in 1956

Hershenow, William, John, Jr. Cadet Company I; retired as lieutenant colonel, US Air Force, in 1956

Hetherington, Ralph Robinet Cadet Company C; killed in action on December 1, 1944

Hewitt, Merritt Lambert Cadet Company M; retired as lieutenant colonel, US Army, in 1961

Hicks, George Luther, III Cadet Company G; retired as colonel, US Air Force, in 1967

Hoebeke, Arnold Jacob Cadet Company E; retired as colonel, US Army, in 1961

Hoge, William Morris, Jr. Cadet Company M; retired as lieutenant colonel, US Army, in 1967

Home, Justus MacMullen Cadet Company L; died in 1944

Horn, Robert William Cadet Company K; retired as lieutenant colonel, US Air Force, in 1962

Howze, Frank Benton Cadet Company L; killed in action on September 15, 1950

Huffman, Burnside Elijah, Jr. Cadet Company E; retired as major general, US Army, in 1975

Humber, Charles Herbert Cadet Company G; discharged upon graduation

Hume, Thomas Abbott Cadet Company G; killed in action on July 31, 1951

Hutson, Stanton Claude Cadet Company B; retired as colonel, US Army, in 1964

Irwin, Henry Durand Cadet Company L; resigned in 1947

Jarvis, Harry Lee "JoJo", Jr. Cadet Company D; killed in action on August 1, 1943

Jensen, Allen Cadet Company K; retired as colonel, US Army, in 1965

Johnson, Allan George Woodrow Cadet Company L; retired as colonel, US Army, in 1968

Johnson, Malcolm Corwin   Cadet Company I; retired as colonel, US Army, in 1971

Johnson, Robert Paul   Cadet Company C; retired as colonel, US Army, in 1967

Jones, Charles Edwin   Cadet Company K; killed in action on March 16, 1943

Jones, Morton McDonald, Jr.   Cadet Company I; retired as brigadier general, US Army, in 1970

Jones, Perry Thompson   Cadet Company F; killed in action on April 12, 1945

Kaiser, James Lawrence   Cadet Company K; retired as colonel, US Army, in 1968

Keagy, Robert Bernard   Cadet Company A; retired as colonel, US Army, in 1964

Keleher, Reynolds Robert   Cadet Company M; retired as lieutenant colonel, US Army, in 1962

Kelley, Roy Skiles   Cadet Company G; retired as brigadier general, US Army, in 1971

Kelsey, Straughan Downing   Cadet Company B; retired as colonel, US Air Force, in 1963

Kemp, Paul Richard   Cadet Company A; resigned in 1949

Kennedy, Kenneth, Wade   Cadet Company B; retired as brigadier general, US Army, in 1971

Kercheval, Benjamin Berry   Cadet Company E; retired as colonel, US Army, in 1967

King, James Henry   Cadet Company C; retired as colonel, US Army, in 1971

King, Riley Smith   Cadet Company H; resigned in 1948

Kisiel, Edwin Charles   Cadet Company D; retired as colonel, US Army, in 1970

Kline, Richard William   Cadet Company B; retired as colonel, US Air Force, in 1961

Knowles, Wendell Pollitt   Cadet Company D; retired as colonel, US Army, in 1965

Knowlton, Joseph Lippincott   Cadet Company D; retired as colonel, US Army, in 1966

Kosiorek, Stephen Thaddeus   Cadet Company G; retired as lieutenant colonel, US Army, in 1964

Kramer, Robert Sealey   Cadet Company K; retired as colonel, US Army, in 1971

Kromer, William Annesley   Cadet Company A; killed in action on December 30, 1944

Kunkel, David Ernest, Jr. Cadet Company L; retired as colonel, US Air Force, in 1961

Kuzell, Ralph Edward Cadet Company E; retired as colonel, US Army, in 1971

Laney, James Raine, Jr. Cadet Company C; retired as colonel, US Army, in 1965

Lanigan, Robert Edward Cadet Company C; retired as colonel, US Army, in 1970

LaRocca, Gerard Anthony Cadet Company H; retired as colonel, US Air Force, in 1962

Larson, Paul Rutherford Cadet Company M; killed in action on November 17, 1942

Laudani, Angelo Augustine Cadet Company I; retired as lieutenant colonel, US Army, in 1962

Lauterbach, Wallace Michael Cadet Company D; retired as colonel, US Army, in 1965

Lawson, Roger Longstreet Cadet Company E; retired as colonel, US Army, in 1971

Lawson, Thomas Rodgers Cadet Company B; retired as major, US Army, in 1946

Layfield, Moody Elmo, Jr. Cadet Company K; retired as colonel, US Army, in 1964

Ledford, Lee Bradley, Jr. Cadet Company E; retired as lieutenant colonel, US Army, in 1961

Lee, Glenn Alfred Cadet Company H; resigned in 1951

Lee, John Clifford Hodges, Jr. Cadet Company K; retired as colonel, US Army, in 1970

Lee, Lynn Cyrus Cadet Company A; retired as colonel, US Army, in 1965

Levy, Richard Mar, Jr. Cadet Company L; retired as lieutenant colonel, US Army, in 1965

Liles, Paul von Santen Cadet Company C; retired as lieutenant colonel, US Army, in 1962

Linderman, John Charles Cadet Company A; retired as lieutenant colonel, US Army, in 1947

Linnell, Frank Holroyd Cadet Company L; retired as brigadier general, US Army, in 1971

Linton, William Miles Cadet Company I; retired as colonel, US Army, in 1971

Locke, Frank Ely Cadet Company L; died in 1942

Locke, John Langford Cadet Company F; retired as major general, US Army, in 1974

Lokker, Clarence John   Cadet Company F; killed in action on November 20, 1944

Longino, Mercer Presley   Cadet Company G; retired as lieutenant colonel, US Army, in 1964

Loring, Robert Gilman   Cadet Company G; resigned in 1943

McCaffery, Benjamin, Jr.   Cadet Company B; retired as lieutenant colonel, US Army, in 1966

McClure, Jack Curtright, Jr.   Cadet Company E; died in 1964

McCool, Ralph Allen   Cadet Company K; resigned in 1948

McCulloch, Joseph Andrew, Jr.   Cadet Company F; retired as colonel, US Army, in 1971

McDaniel, William Thomas   Cadet Company K; killed in action on October 20, 1950

McElroy, James Edward   Cadet Company L; retired as lieutenant colonel, US Army, in 1960

McGrane, Edward Joseph, Jr.   Cadet Company H; retired as colonel, US Army, in 1966

McIntyre, George William   Cadet Company E; retired as colonel, US Army, in 1969

McIntyre, John Carl   Cadet Company G; retired as lieutenant colonel, US Army, in 1962

McKee, Gregg LaRoix   Cadet Company H; retired as colonel, US Army, in 1962

McKinley, James Fuller, Jr.   Cadet Company L; retired as colonel, US Army, in 1961

McMillan, Donald Leroy   Cadet Company C; retired as colonel, US Army, in 1965

McNagny, Rob Reed, Jr.   Cadet Company H; died in 1943

Magruder, Samuel Bertron   Cadet Company L; retired as colonel, US Army, in 1971

Male, Clinton Earle   Cadet Company C; retired as colonel, US Army, in 1971

Manley, John Benjamin, Jr.   Cadet Company D; resigned in 1956

Marsh, Harley Truman, Jr.   Cadet Company H; retired as colonel, US Army, in 1971

Matheisel, Rudolph Adolph, Jr.   Cadet Company E; discharged upon graduation

Mather, Walter Edward   Cadet Company G; retired as colonel, US Army, in 1969

Matheson, Charles Fuller   Cadet Company H; retired as colonel, US Air Force, in 1961

Maxwell, Thomas Ward Cadet Company B; retired as colonel, US Army, in 1969

Maynard, Charles Dorsey Cadet Company H; retired as colonel, US Army, in 1965

Mayo, Ben Isbel, Jr. Cadet Company L; retired as colonel, US Air Force, in 1963

Meador, John William Cadet Company F; retired as colonel, US Air Force, in 1963

Meyer, Arthur Lloyd Cadet Company K; retired as colonel, US Army, in 1971

Michel, John Field Cadet Company B; resigned in 1957

Michels, LeMoyne Francis Cadet Company I; resigned in 1947

Miller, Maurice Guthrie Cadet Company D; retired as colonel, US Army, in 1971

Millikin, John, Jr. Cadet Company K; retired as colonel, US Army, in 1968

Mitchell, William LeRoy, Jr. Cadet Company D; retired as colonel, US Air Force, in 1970

Molesky, Walter Francis Cadet Company G; retired as colonel, US Army, in 1967

Monson, Nelson Paul Cadet Company M; retired as major, US Army, in 1961

Moody, Alfred Judson Force Cadet Company E; died in 1967

Moore, George Bissland Cadet Company E; retired as colonel, US Army, in 1965

Moore, Walter Leon, Jr. Cadet Company F; retired as colonel, US Air Force, in 1967

Moucha, Miroslav Frank Cadet Company M; retired as colonel, US Army, in 1971

Moyer, Maynard George Cadet Company I; retired as lieutenant colonel, US Army, in 1957

Mullane, Walter Raleigh Cadet Company M; retired as lieutenant colonel, US Army, in 1960

Mullins, Charles Love "Moon" Cadet Company E; died in 1943

Murrah, Charles Robert Cadet Company I; resigned in 1946

Murray, John Francis "Jack" Cadet Company E; retired as colonel, US Army, in 1964

Muzyk, Alexander Frank Cadet Company L; retired as colonel, US Army, in 1964

Myers, Francis Joseph, Jr. Cadet Company C; retired as lieutenant colonel, US Army, in 1968

Nankivell, Harold Edward Cadet Company C; died in 1942

Neumeister, Roger Stevens Cadet Company K; resigned in 1955

Niles, Gibson   Cadet Company E; retired as colonel, US Army, in 1965

Nininger, Alexander Ramsey "Sandy", Jr.   Cadet Company L; killed in action on January 12, 1942

Norton, Harold Wesley   Cadet Company C; retired as colonel, US Air Force, in 1964

Norton, John   Cadet Company M; retired as lieutenant general, US Army, in 1975

O'Brien, Paul James   Cadet Company A; killed in action on December 1, 1943

O'Connell, Thomas Courtenay   Cadet Company F; retired as major, US Army, in 1960

O'Connor, Roderic Dhu   Cadet Company B; retired as colonel, US Air Force, in 1967

Osgood, Richard Magee   Cadet Company A; resigned in 1955

Oswalt, John Roy, Jr.   Cadet Company A; retired as colonel, US Army, in 1971

Panke, Robert Edward   Cadet Company L; retired as colonel, US Army, in 1968

Parks, Samuel Wilson   Cadet Company C; retired as lieutenant colonel, US Air Force, in 1961

Peabody, Hume Jr.   Cadet Company E; died in 1942

Peddie, Joseph Scott   Cadet Company A; retired as colonel, US Air Force, in 1961

Peirce, Charles Leonard   Cadet Company G; killed in action on September 30, 1944

Perkin, Irving Richard   Cadet Company A; retired as colonel, US Army, in 1967

Petre, William McVay   Cadet Company B; resigned in 1946

Pickett, George Bibb, Jr.   Cadet Company H; retired as major general, US Army, in 1973

Pierpont, Robert Patterson   Cadet Company B; killed in action on October 24, 1944

Pique, Paul Edgar   Cadet Company E; retired as colonel, US Army, in 1971

Pittman, George Henry, Jr.   Cadet Company D; retired as colonel, US Air Force, in 1967

Plume, Stephen Kellogg, Jr.   Cadet Company L; resigned in 1953

Poff, Ernest Franklin   Cadet Company H; retired as major, US Army, in 1962

Polk, Richard Bradford   Cadet Company F; resigned 1946

Polla, Hector John   Cadet Company M; killed in action on January 21, 1945

Poole, Edgar Thornton, Jr.   Cadet Company C; retired as lieutenant colonel, US Air Force, in 1962

Powell, Edwin Lloyd, Jr. Cadet Company I; retired as brigadier general, US Army, in 1971

Pratt, William Doyle Cadet Company I; retired as colonel, US Army, in 1969

Price, Max Cadet Company K; died in 1943

Purdy, William Augustus Cadet Company I; retired as colonel, US Army, in 1971

Ramee, Paul Wyman Cadet Company L; retired as colonel, US Army, in 1968

Ramey, Stanley Meriwether Cadet Company G; retired as colonel, US Army, in 1971

Rastetter, Richard John Cadet Company L; retired as colonel, US Army, in 1964

Reagan, Thomas Edwin Cadet Company I; killed in action on August 1, 1944

Redmon, John Gabriel Cadet Company H; retired as colonel, US Army, in 1969

Reed, Wilson Russell "Ted" Cadet Company B; retired as brigadier general, US Army, in 1971

Reilly, Robert Stanley Cadet Company D; retired as lieutenant colonel, US Army, in 1964

Rhynard, Wayne Edgar "Fox" Cadet Company G; retired as colonel, US Air Force, in 1969

Richards, John Rose Cadet Company M; retired as colonel, US Air Force, in 1963

Richardson, Herbert, Jr. Cadet Company I; died in 1962

Richardson, James Cadet Company C; retired as lieutenant colonel, US Army, in 1962

Rising, Harry Niles, Jr. Cadet Company H; retired as lieutenant colonel, US Army, in 1968

Robinson, John Leonard Cadet Company C; retired as major, US Army, in 1951

Root, Paul Crawford, Jr. Cadet Company I; retired as colonel, US Army, in 1971

Rosen, Robert Harold Cadet Company M; killed in action on September 20, 1944

Rosenbaum, Bert Stanford Cadet Company B; retired as colonel, US Air Force, in 1962

Rossell, John Ellis, Jr. Cadet Company M; retired as lieutenant colonel, US Army, in 1962

Roton, William Faye Cadet Company E; retired as colonel, US Army, in 1967

Rowny, Edward Leon   Cadet Company A; retired as lieutenant general, US Army, in 1979

Roy, James William   Cadet Company H; retired as colonel, US Army, in 1963

Salinas, Daniel   Cadet Company K; died 1962

Salisbury, Lloyd Robert   Cadet Company B; retired as colonel, US Army, in 1971

Samz, Robert Walter   Cadet Company K; retired as colonel, US Army, in 1971

Sands, John Raymond, Jr.   Cadet Company B; died in 1947

Sawyer, Willis Bruner   Cadet Company H; retired as colonel, US Air Force, in 1969

Schilling, Charles Henry   Cadet Company I; retired as brigadier general, US Army, in 1980

Schnittke, Raymond Ira   Cadet Company B; retired as colonel, US Army, in 1965

Schremp, John Edward   Cadet Company A; retired as colonel, US Army, in 1967

Schultz, Bernard   Cadet Company F; resigned in 1947

Scott, Richard Pressly   Cadet Company K; retired as brigadier general, US Army, in 1968

Seamans, Charles Sumner, III   Cadet Company C; retired as colonel, US Air Force, in 1969

Seawell, William Thomas   Cadet Company A; retired as brigadier general, US Air Force, in 1963

Seneff, George Philip, Jr.   Cadet Company K; retired as lieutenant general, US Army, in 1974

Shadday, Martin Andrew   Cadet Company F; retired as lieutenant colonel, US Army, in 1969

Sharkey, Thomas Wilson   Cadet Company I; retired as colonel, US Army, in 1961

Shelton, Thaddeus Joseph   Cadet Company B; discharged on graduation

Silk, Joseph Meryl   Cadet Company D; retired as colonel, US Air Force, in 1965

Singles, Walter, Jr.   Cadet Company G; resigned in 1947

Skowronek, Paul George   Cadet Company G; retired as colonel, US Army, in 1971

Sliney, Edgar Mathews   Cadet Company E; retired as lieutenant colonel, US Air Force, in 1961

Slocum, George Lawrence   Cadet Company F; resigned in 1949

Smith, Bradish Johnson   Cadet Company D; retired as colonel, US Army, in 1965

Smith, Cecil Leo Cadet Company E; retired as lieutenant colonel, US Army, in 1959

Snider, Albert Howell Cadet Company E; retired as colonel, US Air Force, in 1967

Spiller, Benjamin Alvord Cadet Company K; retired as colonel, US Army, in 1969

Stainback, Frank Pleasants, Jr. Cadet Company H; retired as colonel, US Air Force, in 1968

Stalnaker, George Winfield Cadet Company D; retired as colonel, US Air Force, in 1969

Stanford, Frederick Clinton Cadet Company H; retired as colonel, US Army, in 1963

Starr, William Frank Cadet Company D; retired as colonel, US Army, in 1966

Stern, Herbert Irving Cadet Company F; retired as colonei, US Army, in 1968

Stigers, James William Cadet Company G; retired as colonel, US Army, 1959

Stillson, George Hamilton, Jr. Cadet Company B; retired as colonel, US Air Force, in 1966

Strain, James William Cadet Company C; retired as colonel, US Army, in 1967

Sullivan, Maxwell Weston, Jr. Cadet Company M; killed in action on January 27, 1943

Sykes, James Rayford Cadet Company K; resigned in 1955

Taggart, David Burch Cadet Company M; killed in action on January 15, 1943

Tanous, Peter Schuyler Cadet Company F; retired as colonel, US Army, in 1966

Tansey, Patrick Henry, Jr. Cadet Company F; retired as colonel, US Army, in 1966

Tarbox, Robert Mack Cadet Company B; retired as brigadier general, US Army, in 1971

Tate, Joseph Scranton, Jr. Cadet Company K; killed in action on December 22, 1943

Theisen, George Lawrence Cadet Company H; retired as lieutenant colonel, US Army, in 1962

Thigpen, Joseph Jackson Cadet Company D; resigned in 1947

Thomas, Arnold Ray Cadet Company D; retired as lieutenant colonel, US Army, in 1961

Thomas, Charles Edwin, III Cadet Company F; died in 1942

Thompson, Alden George Cadet Company L; retired as colonel, US Air Force, in 1968

Thompson, Clyde Arnold Cadet Company K; retired as colonel, US Air Force, in 1969

Thompson, Donald Vincent Cadet Company G; died in 1942

Thompson, Jesse Duncan Cadet Company E; retired as lieutenant colonel, US Air Force, in 1964

Tidmarsh, Harold Alexander Cadet Company D; retired as colonel, US Army, in 1969

Tindall, Richard Gentry, Jr. Cadet Company K; killed in action on February 9, 1945

Tonetti, Oscar Charles Cadet Company F; retired as colonel, US Army, in 1967

Torgerson, Arnold Svere Cadet Company C; discharged on graduation

Towers, Jacob Heffner Cadet Company E; retired as major, US Army, in 1961

Travis, Richard Van Pelt Cadet Company C; retired as colonel, US Air Force, in 1969

Trimble, Harry White Cadet Company C; retired as lieutenant colonel, US Air Force, in 1968

Troup, Malcolm Graham Cadet Company E; retired as colonel, US Army, in 1966

Troy, Francis Joseph Cadet Company H; killed in action on January 25, 1945

Tuttle, Robert Merrill Cadet Company I; retired as colonel, US Air Force, in 1965

Tyler, Max Campbell Cadet Company H; retired as colonel, US Army, in 1966

Tyndall, John Gavin, II Cadet Company M; retired as lieutenant colonel, US Army, in 1947

Unger, Jess Paul Cadet Company C; retired as colonel, US Army, in 1965

Upton, Ralph Reed Cadet Company E; retired as colonel, US Army, in 1965

Van Hoy, John Webb, Jr. Cadet Company I; retired as colonel, US Army, in 1964

Vaughan, William John Dooley Cadet Company F; retired as colonel, US Army, in 1971

von Schriltz, Dick Stanley Cadet Company M; retired as lieutenant colonel, US Army, in 1969

Waitt, Robert Graham Cadet Company I; retired as lieutenant colonel, US Army, in 1950

Walker, James Philip Cadet Company A; killed in action on September 7, 1943

Walters, Edison Kermit   Cadet Company G; retired as lieutenant colonel, US Air Force, in 1954

Ward, Joseph Hester   Cadet Company B; killed in action April 5, 1945

Ward, Thomas Martin   Cadet Company A; Cadet Company A; resigned in 1946

Watson, Leroy, Hugh, Jr.   Cadet Company E; died in 1959

Weidner, Joseph John   Cadet Company D; retired as lieutenant colonel, US Air Force, in 1964

Welles, George Hollenback   Cadet Company H; retired as lieutenant colonel, US Army, in 1965

West, Ben Marshall   Cadet Company H; retired as lieutenant colonel, US Air Force, in 1962

West, DuVal   Cadet Company L; resigned in 1947

Whitaker, Ernest Jeunet   Cadet Company C; retired as colonel, US Army, in 1965

White, Alpheus Wray   Cadet Company A; retired as colonel, US Air Force, in 1965

White, Lester Strode   Cadet Company A; died in 1943

White, Theodore Knox   Cadet Company K; died in 1953

Willes, Charles Gleeson   Cadet Company G; died in 1956

Winfree, Issac Owen   Cadet Company B; retired as colonel US Air Force, in 1965

Woodruff, Roscoe Barnett, Jr.   Cadet Company D; retired as colonel, US Air Force, in 1967

Woods, David Seavey   Cadet Company A; retired as colonel, US Air Force, in 1969

Woodward, William Hunter   Cadet Company L; died in 1966

Woolwine, Walter James   Cadet Company A; retired as lieutenant general, US Army, in 1975

Yates, Elmer Parker   Cadet Company F; retired as major general, US Army, in 1971

Zarembo, Edward Benedict   Cadet Company D; retired as lieutenant colonel, US Army, in 1960

Zott, John Henry, Jr.   Cadet Company H; resigned in 1953

Appendix B

US Army Organization in World War II

(Showing Commanders and Their USMA Graduation Dates)

European Theater

Supreme Commander: Dwight David Eisenhower (Class of 1915)

6th Army Group: Jacob Loucks Devers (Class of 1909)
7th Army Group: Alexander McCarrell Patch (Class of 1915)*
12th Army Group: Omar Nelson Bradley (Class of 1915)
First US Army: Courtney Hicks Hodges (Class of 1908)**
Third US Army: George Smith Patton (Class of 1909)
Seventh US Army: Alexander McCarrell Patch (Class of 1914)*
Ninth US Army: William Hood Simpson (Class of 1909)
Fifteenth US Army: Leonard T. Gerow***

Mediterranean Theater

5th Army Group and Fifth US Army: Mark Wayne Clark (Class of 1917)
Seventh US Army: Alexander McCarrell Patch (Class of 1914)*

Pacific Theater

Supreme Commander: Douglas MacArthur (Class of 1903)

Sixth US Army: Walter Krueger***
Eighth US Army: Robert Lawerence Eichelberger (Class of 1909)
Tenth US Army: Simon Bolivar Buckner (Class of 1908);
later Joseph Warren Stillwell (Class of 1904)

---

*The Seventh US Army moved from Italy to France in 1944.
**General Hodges entered USMA with the Class of 1908 but did not graduate.
***Not a USMA graduate.

Appendix C

US Army Organization in the Korean War

(Showing Commanders and Their USMA Graduation Dates)

Supreme Commanders

(US Far East Command/United Nations Command)

1950–1951: Douglas MacArthur (Class of 1903)
1951–1952: Matthew Bunker Ridgeway (Class of 1917)
1952–1953: Mark Wayne Clark (Class of 1917)

Eighth US Army

1950: Walton Harris Walker (Class of 1912)
1951: Matthew Bunker Ridgeway (Class of 1917)
1951–1953: James Alward Van Fleet (Class of 1915)
1953–1954: Maxwell Davenport Taylor (Class of 1922)

X Corps

1950: Edward Almond*

---

*Not a USMA graduate.

Appendix D

Decorations Awarded to Members of the Class of 1941

during World War II and the Korean War

Congressional Medal of Honor	1 (Alexander R. "Sandy" Nininger)
Distinguished Service Cross	7*
Bronze Star	275
Silver Star	85
Distinguished Flying Cross	54
Air Medal	93
Purple Heart	85

---

*John Adams Brooks III; George Scratchley Brown; Ira Boswell Cheaney, Jr.; Wadsworth Paul Clapp; William Graham Gillis, Jr.; James Lawrence Kaiser; William Thomas McDaniel.

Bibliography

Arnold, General Henry Harley. *Global Mission*. New York: Harper & Brothers, 1949.

Black, Bold and Grey: Fifteenth Anniversary Yearbook of the USMA Class of 1941. Chicago: American Yearbook Company, 1956.

Blacker, Bolder and Greyer: The Silver Anniversary Yearbook of the Class of '41, USMA. 1966.

Carter, Kit C., and Robert Mueller. *The Army Air Forces in World War II*. Albert F. Simpson Historical Research Center, Air University, Maxwell Air Force Base, Alabama/United States Government Printing Officer, Washington, D.C., 1973.

Cole, Hugh M. *The US Army in World War II: The Ardennes: Battle of the Bulge*. Washington, D.C.: US Army Center of Military History, 1965.

Drew, Dennis M., and Donald M. Snow. *The Eagle's Talons: The American Experience at War*. Maxwell Air Force Base, Alabama: Air University Press, 1988.

Dugan, James, and Stewart Carroll. *Ploesti: The Great Ground-Air Battle of 1 August 1943*. New York: Random House, 1962.

Dupuy, Colonel Trevor Nevitt. *The Military History of World War II*. (18 vols). New York: Franklin Watts, 1965.

Eisenhower, General Dwight David. *Crusade in Europe*. New York: Doubleday & Company, 1948.

Eliot, George Fielding. *The Ramparts We Watch*. New York: Reynal & Hitchcock, 1938.

Farago, Ladislas. *The Last Days of Patton*. New York: McGraw Hill, 1981.

Fehrenbach, T.R. *This Kind of War: A Study in Unpreparedness.* New York: Macmillan, 1963.

Galbraith, John Kenneth, and Burton H. Klein. *The United States Strategic Bombing Survey.* Washington, D.C.: Overall Economics Division, 1945.

Kurzman, Dan. *Day of the Bomb.* New York: McGraw Hill, 1986.

LeMay, General Curtis E., and Bill Yenne. *Superfortress: The B-29 and American Air Power.* New York: McGraw Hill, 1986.

MacArthur, General Douglas. *Reminiscences.* New York: McGraw Hill, 1964.

MacDonald, Charles B. *The Army in World War II: The Last Offensive.* Washington, D.C.: US Army Center of Military History, 1973.

Marshall, Colonel Samuel Lyman Atwood. *Bastogne: The Story of the First Eight Days.* Washington, D.C.: Center of Military History, 1946.

Ridgeway, General Matthew B., and Harold Martin. *Soldier: The Memoirs of Matthew B. Ridgeway.* New York: Harper & Brothers, 1956.

Morton, Louis. *The US Army in World War II: The Fall of the Philippines.* Washington, D.C.: Center of Military History, 1953.

Skowronek, Colonel Paul George. "United States–Soviet Military Liaison in Germany Since 1976." Ph.D. diss., University of Colorado, Boulder, 1976.

Spiller, Colonel Benjamin Alvord. *Indomitable: The Story of the First Women at West Point.* Colorado Springs: Pikes Peak Publishing, 1989.

Stimson, Henry L., and McGeorge Bundy. *On Active Service in Peace and War.* New York: Harper & Brothers, 1948.

US Military Academy. *Assembly* West Point: USMA Association of Graduates, 1941–90.

US Military Academy. *1990 Register of Graduates and Former Cadets.* West Point: USMA Association of Graduates, 1990.

US Military Academy. *Howitzer.* West Point, USMA, 1941.

Yenne, Bill. *The History of the US Air Force.* New York: Simon & Schuster, 1984.

Index